Project Management Bootcamp

Project Management Bootcamp is a pragmatic guide for those who need to understand how to deliver projects successfully. The reader journeys through a project stage by stage, discovering what project managers commonly need to achieve at each step. Each step is supported by tables, charts, tips, and tools, which readers may adopt or adapt to their needs, and different ways of organising and delivering projects, including agile approaches, are considered.

Because theory can only get you so far, a key element of the book is learning from real projects drawing on the experience of project managers working across three continents. Each chapter ends with challenges to readers to reflect on their learning, which can be based on a theoretical case study or their own project. The result is a reflective framework that charts their learning and their project management journey from initiation to closure.

Project Management Bootcamp is essential reading for junior and mid-level career project managers, as well as any professionals who find themselves in charge of a project and are unsure how to get the best result. Students in business and management courses at undergraduate and postgraduate levels will also value its setting of theory into a practical context.

Peter Cross began his career in local government and became an "accidental" project manager when he was made responsible for delivering IT and business change projects. Later, he moved on to computer audit for an international accounting firm, where he had the opportunity to review and comment on IT project failures in both the public and private sectors. He qualified as a programme manager and went on to lead a wide variety of projects and programmes. He now works in the UK and Europe, delivering, advising on, and rescuing projects with construction elements as well as IT and business change.

Project Management Bootcamp

A Step-by-Step Guide

Peter Cross

LONDON AND NEW YORK

Designed cover image: © Getty Images

First published 2024
by Routledge
4 Park Square, Milton Park, Abingdon, Oxon OX14 4RN

and by Routledge
605 Third Avenue, New York, NY 10158

Routledge is an imprint of the Taylor & Francis Group, an informa business

British Library Cataloguing-in-Publication Data
A catalogue record for this book is available from the British Library

Library of Congress Cataloging-in-Publication Data
Names: Cross, Peter (Project management consultant), author.
Title: Project management bootcamp : a step-by-step guide / Peter Cross.
Description: Abingdon, Oxon ; New York, NY : Routledge, 2023. | Includes bibliographical references and index.
Subjects: LCSH: Project management.
Classification: LCC HD69.P75 C7596 2023 (print) | LCC HD69.P75 (ebook) | DDC 658.4/04—dc23/eng/20230222
LC record available at https://lccn.loc.gov/2023008064
LC ebook record available at https://lccn.loc.gov/2023008065

ISBN: 978-1-032-52123-7 (hbk)
ISBN: 978-1-032-52122-0 (pbk)
ISBN: 978-1-003-40534-4 (ebk)

DOI: 10.4324/9781003405344

Typeset in Bembo
by Apex CoVantage, LLC

Contents

About the Author

Project management was never my first choice of career – indeed, when I started out, I doubt that the term was widely used at all. I began in local government and became an "accidental" project manager when I was made responsible for delivering a few IT and business change projects. Computers were still regarded with suspicion by many in those days, so convincing a series of sceptical stakeholders that it really was necessary to enter the personal details of their clients into a database or that we really could control and print tickets for a concert hall holding 2,500 people was quite challenging!

Later, I moved on to computer audit for an international accounting firm, where I had the opportunity to review and comment on a fair number of IT project failures in both the public and private sectors. It was around then that people started saying, "You seem to know about this, you fix it." That's how I became known for rescuing failing projects. I qualified as a programme manager and went on to lead a wide variety of projects and programmes.

More recently, my work in both the UK and Europe has included delivering, advising on, and rescuing projects with construction elements as well as IT systems and business change.

Acknowledgements

Just as no one individual can deliver a project, this book is the result of learning, support, and input from many people over the course of more than two decades of managing projects. Amongst them, and in no particular order, there are multiple people called John, at least a couple of Mikes and Colins, several Ians, David, Robin, Gerry, Graham and Graeme, Cameron, Mark, Anne, Bob, Hugh, Jo, Klaus, Kathryn, Andrew, Paul, Jill, Hugh and Mark. Only I know how much I am indebted to them all.

I need to add special mention to my wife Jo for her support and her corrections to my English, to my daughter Charlotte for her publishing know-how and her editing of the text (I knew it was worth teaching her to read), and to Andrew Harrison, Helena Parkinson and the team at Routledge for transforming my thoughts on project management into this book.

To these and many others belongs all of the credit whilst I, of course, own all of the mistakes.

Introduction

What Is a Project Manager?

Have you ever sat through the credits of a major film and thought about how all those people, from the lead actors to the key grip, best boy, transport co-ordinator, and production accountant, were brought together and organised so they could create the two hours of story you just watched? Or have you ever waited at a set of temporary traffic lights and watched a group of workers doing something inexplicable with a digger and wondered what exactly was going on? Or been to a music festival and reflected on the organisation needed to get multiple bands operating on multiple temporary stages, provide food stalls and toilets[1] for the people attending, and clean up the site afterwards? All these are very visible, public manifestations of an activity by an individual who has brought together a temporary team of people with disparate skills to deliver something very specific. The activities of such individuals sit behind many human endeavours, and although they may carry many different job titles, mostly we can call them project managers.

Who Are *You*?

Much like in the famous saying about greatness, you may be born a project manager, achieve project manager status, or have project management thrust upon you. Certainly, if you are in the last group and wondering what to do next, you need to read this book. If you are in one of the others, this book may provide some tips, tricks, hacks, and devious thoughts that you find helpful. You may also be a student fresh to project management as a subject or pondering on how it may all work in real life after having completed a theory course.

Of course, you may also be responsible for commissioning projects and wanting to know more about the challenges faced by the project leadership in delivering your business vision, or maybe you have been drafted in to work on a project and want a better picture of what is going on. In all these cases, this book will give you a wider perspective, some issues to think about, and perhaps a few handy hints.

How Could This Book Help You?

Projects are everywhere, and whilst some (usually the big ones) are managed by people who are career project managers and have the pieces of paper to prove it, many are not. These projects are managed by all types of professionals, including people just drafted in to pick up a project and to make it happen. In others words, people like you and me. If you are new to having responsibility for projects, then this book and its hands-on, step-by-step approach to pragmatic project

DOI: 10.4324/9781003405344-1

management might just help you avoid the more obvious pitfalls and pratfalls that await the unwary.[2]

Some of the things we cover could also help you recognise some of the "brown sticky stuff" which, as project managers, we inevitably find flying our way from time to time, and then learn to duck at the most appropriate moment. Alternatively, this book may convince you that project management is really just not for you – that it is an area to be avoided at all costs and in the future will lead to you raising both hands and running screaming from the room if anyone ever mentions it to you. That's fine! Project management can be difficult, frustrating, and challenging, and it is not for everyone.

Need to Improve Project Management

Although there are projects everywhere, they are not generally perceived as totally successful. There are major gaps between the expectations for projects and what we as project managers actually deliver. The popular press abounds with stories of projects that fail to deliver in some way: they might be over time or way over budget, or their solutions just don't work. There are regularly inquiries and even court cases when projects collapse.

Professional bodies regularly conduct research on multiple dimensions that demonstrate that there are still gaps between the aspiration for projects and what they actually deliver in terms of time, cost, and quality of outcomes (Project Management Institute, 2020). Consequently, despite improvements over the years in our understanding of projects and their successful delivery, there is a steep hill for project managers to climb. The objectives of this book are to help you:

- Make that climb by breaking activities down into logical, achievable steps and explaining the thinking behind them.
- Survive your experiences of managing a project (and hopefully deliver on time, to cost, and to quality) by passing on practical experience from real-life projects.

Although this book is built around business experiences, projects are found in our personal lives as well. Moving house is a project, as is getting a new job, redecorating the kitchen, or going on holiday. Consequently, many of the thought processes, skills, and techniques of project management can be applied to projects in our personal lives in the same way.[3]

Not Another Methodology!

You will be pleased to hear that the book is not another project management methodology. There are quite enough excellent ones around already, and in any case, it seems to be a rule that such tomes run to at least six hundred pages. This book is also not recipe which, if followed absolutely, will manage your project for you. Instead, this book suggests ideas, topics to think about, and approaches to solving problems that you might consider when working out how best to deliver any particular project.

This book is agnostic about particular methodologies, and the practical advice given and the different steps suggested can be applied whether you have chosen to follow a specific methodology or not. We can think of methodologies, often with their large and complex textbooks and guides, as a highway code and any exams that we may have to take in them as passing a driving theory test. Instead, this book is about what happens when we actually sit alone in the car and start driving.[4]

The difference between managing a project and driving a car is that we don't need a licence to start up a project, and the similarity is that if we don't do it right, we can crash quickly, publicly, and sometimes very spectacularly. So, regard this book as a personal advisor metaphorically sitting in the car beside you and offering support, advice, tips, and warnings as we go on the journey of delivering a project. The decisions about what to do in your particular circumstances, or the judgements that you may have to make, are, of course, always your own.

In recent years, a philosophy has arisen in project management which you will see referred to as "agile," and often the older, more sequentially based project management is referred to as "waterfall." Both terms arose from particular types of IT development projects, but there are ideas in each approach which are applicable to every project. This book, however, is neutral about whether you plan your project as a series of bursts of agile activity or in more traditional sequential project phases. As a basis, this book starts from the traditional, sequential approach, but it is worth saying from the outset that however the project is labelled, all project management demands agility and flexibility to respond to changing information and circumstances in a controlled way. We also discuss many of the elements of the project manager's role which apply irrespective of the philosophy being followed – for example, running an "agile" project doesn't mean that you can dispense with managing risk, or that you can ignore team communications in a sequential project.

Underlying the approach of this book is the concept that project management is not really about techniques, or software tools, or ticking the boxes and jumping through the hoops required by a methodology or philosophy, useful though all of these can be in the right context. Instead, it is based on the pragmatic understanding that it is our behaviours as project managers – our thought processes, our creativity, and our ability to influence, organise, and manage others – that are the central factors in delivering projects successfully. I do not wish to worry you at this point, but the UK's chartered professional body, the Association for Project Management, breaks these factors down into no less than twenty-seven competencies that the project manager should have (Association for Project Management, 2015).

What you are going to need is a guidebook.

A Guidebook for the Journey

This book goes through a project's journey stage by stage, covering what projects commonly need to achieve at each stage and showing how others have responded in real life to the events, problems, and issues they have encountered in delivering their projects. Sprinkled throughout the book are practical examples of some simple tools that you may find useful for recording and communicating your thinking. As this is not a methodology, whether you use these tools, invent your own, or find a brilliant idea from a methodology or elsewhere is a completely personal choice.

One word of caution. Whilst the analogy of the project as a journey is useful, many of our activities as project managers are not quite as sequential as they may appear. For example, the fact that we cover preparing a business case in Chapter 4 does not mean that we haven't been thinking about many of the areas it covers from the start or that once we have an agreed-upon business case that we can move on and never come back to it. Equally, we talk about creating and leading a project team in Chapter 6, but we have to start thinking about our team from the outset – and maybe almost immediately recruiting some team members to share our load in creating the project. So, although there are many sequential areas of preparation, in reality we often start many of them at once, and many of our early steps will lead directly to others at different stages of the project.

This book is organised so that every chapter covers a key part of the project manager's journey, and at the end of each one, there is an opportunity to reflect on how you would have tackled managing that part of a project manager's role. Again, you may choose to experiment with some of the tools outlined in the text or use those from your organisation's methodology of choice. These reflections (you can think of them as homework, if you like) can be based either on the theoretical case study or a project of your choosing and will build into a skeleton project file that charts your thinking about your example project from start to finish.

Please regard this book as a series of dialogues or personal coaching sessions. As in any good coaching session, feel free to argue with what is said when you don't agree. You can then put together your thoughts and translate them into strategies to follow for your project that may meet your unique circumstances far better than anything generic contained in any book. This sort of a dialogue underlines the first rule of project management: always be humble. As project managers, we don't know everything, and sometimes we can be just plain wrong. We always have to rely on others for their knowledge, skills, support, and, where necessary, correction. The second rule of project management is always be proud – of the achievements of your team for successfully delivering what is never easy and often seems to vary from the very awkward through the extremely difficult to the downright impossible.

None of the discussions in the following chapters are definitive. Instead, they represent starting points for you and your journey as a thinking project manager. Real-life examples and lessons learned at the coal face of project management draw on the personal experiences of project managers as well as published information and reports on projects. Sadly, published information on projects tends to be in the category of, What the hell went wrong with this project? Someone needs to take the blame. Reports tend to be a bit thin on the ground that say things such as, "Hey, this all went great, and the project does what it says on the tin, in future you should do things this way."

Wherever possible, this book includes positive lessons as well as negative ones. Lessons, however, do not always transfer well, and what worked well in one environment or culture may be less successful or require adaption in another one. If you are in doubt about any point or want to see how others are tackling a particular issue, the internet provides a rich view of information resources, including academic research published in open-access (i.e., free to read) journals and postings by different institutes and companies. A word of warning, though: you may find that some of the materials finish by recommending that the only real solution is to buy the author's software or services.

Virtual Excursions and Real-Life Projects

Projects are basically social organisations of people from disparate backgrounds, and as we will see, both the projects and the people come in many different shapes and sizes. A construction project, for example, will require some differences in approach from a business change one. We nevertheless find that there is usually more commonality than difference in the challenges faced in managing many different types of projects. To illustrate this similarity, throughout the book are a number of virtual field trips to consider the real problems faced by projects. With the exception of the first one,[5] all the field trip scenarios are based on real-life project situations that have been faced by practising project managers. Each visit or other example in the text has been chosen to illustrate universal points about project management. This book also includes some reflections and examples from project managers working in different countries and on different continents. Where the information is unpublished, I have striven to ensure that the projects and the players within them cannot be identified;[6] where information has been published, the examples are identified.

As you might expect if you are working in cultural or economic environments that are different from the example, it might be necessary to think a bit about how to transfer their lessons to your own situation.

If you are intending to make project management a career or deliver a multimillion-dollar megaproject, please regard this book is an introduction to how projects work and a starting point for your thinking as a project manager. Afterwards, you will probably want to move on to the heavyweight guidance provided by the bodies responsible for the widely accepted project management methodologies such as those described in the PRince family of guides, the Association for Project Management publications, and the Project Management Institute PMBOK Guide.[7]

So, whatever your background, your career objectives, and your current situation, I hope you find something in this book that is informative, something that is thought provoking, and something that just sometimes raises a small smile.

Disclaimer

I feel it only fair to point out, before you read the next chapters, that my direct knowledge of sewage treatment is limited to one conversation decades ago. This conversation was with an external auditor about the valuation of the macerators owned by one UK water company. Macerators, it turns out, were pieces of plant that ground up raw sewage before it was routinely pumped directly into the sea via long pipes – a practice that I hope has largely ended, but I still won't eat shellfish or swim in UK rivers or the sea. So, my apologies to any experts in disposing of our sewage, but as an imaginative case study, it does seem to me to make a great model for thinking about projects that have many facets – in this case, research, manufacturing, construction, marketing, and business change. Alternatively, perhaps it's just my juvenile sense of humour biasing my judgement. Either way, you can select this as your continuing case study or another project of your own choosing.

Right, let's get on because we have much to do and a lot of material to get through, and it's time to make a start on your project. As one of my earliest mentors used to say,

Five minutes lost at the beginning cannot be made up at the end.

Have fun, and good luck.

Notes

1 Are there ever enough toilets?

2 As this author once was.
3 Although, on a personal project, you will not normally be needing as much documentation.
4 If you are not a car driver, my apologies, and please see Chapter 7 on use of language that is not appropriate to the audience.
5 The reasons for this will become apparent.
6 If you can work out what any particular project was, please respect confidentiality and keep it to yourself. Whatever guess you come up with, I would in any case deny it.
7 Be warned that although the books written to guide you through the detail of these methodologies are weighty, dense, and comprehensive, they usually contain less pathetic attempts at unsuitable humour than this one, have very few sneaky asides, and, as far as I can see, have nothing to say about sewage or ancient monuments. I can also almost guarantee that they will all fail to mention anyone getting stuck in a public toilet or the importance of biscuits to successful project management!

Bibliography

Association for Project Management, 2015. *Competency Framework*, Princes Risborough: Association for Project Management.

Project Management Institute, 2020. *Pulse of the Profession*, Newtown Square, PA: Project Management Institute.

1 Projects

Grasping the Basics

Aim of this chapter: To explore the issues around the differences between projects and other areas of management.

Learning outcome: To be able to recognise when project management techniques will be helpful and to be able to confidently describe the role of the project manager.

Throughout the book we will be taking a few virtual and imaginative field trips, but as project managers, we have to be creative and imaginative people, so this is not going to be a problem for us. We will also think about a few problem situations as we progress through a theoretical project, but more on that later.

But What Exactly Is a Project?

Today, projects are everywhere. They are used to design and deliver everything from a new supersonic fighter to theme parks, motorways, computer systems, office moves, business reorganisation, and new consumer products. Projects are the way humanity has organised itself for millennia in order to deliver something other than day-to-day repeated activities. Creating Stonehenge, constructing the Hoover Dam, designing the Spitfire, or even replacing an old garden shed are examples of projects. Projects turn vision into reality, and the people who make this happen are project managers.

Projects always resemble the plot of a good story. They have a beginning, a middle, and an end, and they usually involve the resolution of conflict between the characters and the overcoming of apparently insuperable challenges. What projects deliver, however, can live long after the individuals involved and the troubles in delivering the project are forgotten. If we delve beyond the ruins, we would find that behind the Colosseum, the Great Wall of China, Stonehenge, the Colossus of Rhodes, and the Hanging Gardens of Babylon were harassed project managers trying to balance time, cost, and quality. Indeed, from our point of view as project managers, understanding the "How" of delivering such "wondrous things" in ancient times should be as exciting as seeing the remains of the "What" which is left behind now.

DOI: 10.4324/9781003405344-2

But now for our first virtual field trip. There are many of these trips as we go through the book, and each one represents a visit to a real project and a chance to consider the problems faced by its project manager.

We find ourselves in a very large space, soon to be a major construction site. The area is quite flat, barren, and very dusty. It is still early in the day, already quite warm, and a cloudless sky promises that the day will soon be heating up. Sorry, I should have warned you about needing sunglasses, a hat, and sunblock.

Over to our left, we can see that many pegs have been driven into the ground to start laying out the site. On one side of the site and not far from us stands a tall, burly, weather-beaten man who is addressing a small group of similarly built people.

"OK. This is it. This is the big one. It's the most major infrastructure project for years, and we will be delivering for the top man himself, personally. This is what he wants, and so there will be extra effort needed all round to make sure that we deliver this monster as speedily as possible.

"Make no mistake, people will be coming from all over to see it for years to come, and we are the ones who are going to make it happen. I'm looking for the highest quality, really showcasing what we can do.

"Budget will always be an issue. We are looking to deliver this within estimate, so always please check with me first if it looks like you might go over, but for him, quality is the most important thing, so we should get any necessary changes through without too many difficulties.

"Understand now that this is going to be big and complex to deliver. I can tell you that the quantity surveyors have done initial estimates, and we are looking at having to bring in over six million tonnes of material, some of which have to be hauled from over eight hundred kilometres away. Don't I love architects and their fancy designs!

"Anyhow, the easiest way to shift this lot is going to be by boat, so in addition to the beast itself, we will be building docks, both here and close to the quarries. We are already working on organising the shipping, but some boats will have to be built especially for us.

"In addition, the workforce will be living on site, so we are also planning to build housing and facilities for them and their families.

"We have a ten-year contract, expandable by up to another ten years if we hit snags, which we undoubtedly will as no one has ever built one this size before.

"This is going to be remarkable; you'll be showing this to your kids and grandkids and saying 'This was us. We did this.'

"Now, does anyone have any questions?

"OK, you at the back. . . . You ask what shape it is. Good question. Well, there was talk about being radical and creating a bold new architectural statement for the modern age, but no, I think we're sticking with the traditional pyramid this time."

Did the first overseer on the Great Pyramid at Giza give such a briefing over four thousand years ago? Possibly not, and he certainly did not give a presentation with the detailed Gantt charts and risk registers and all the other documents and output that support the modern project manager. Someone, however, had to translate the vision for building the tallest structure ever built on the planet into a series of activities, work out the resources required for each activity, put together a credible schedule, and report to Pharaoh on how work was progressing.

Introducing Our Virtual Case Study

Time now for another field trip. On some of our future visits, we are going to be observers of real projects, sometimes we will be questioning other project managers, but this time we are visiting our theoretical case study for the first time and putting *you* in the hot seat. Take notes because there will be questions (and many problems) later.

Imagine you work for a major business services firm that specialises in providing private and public sector administration services to a wide range of clients in different countries. Sitting in a windowless part of the third floor, you have been quietly running a part of the business administering private pensions for a number of clients for several years. Now, with no notice, you have been called up to your boss's corner office in the executive suite on the 52nd floor.

It is quiet and calm up here, almost serene, in contrast to the noises and bustle of your normal life down on the third floor. Maybe it is the thick carpet and the side tables in the corridor with fresh flower arrangements. You barely notice these because, although you have never been up here before, you are fearing the worst. The trip up in the express executive lift was a nervous one, as you were trying to work out what you had done and how best to defend yourself. Coming along the corridor, you notice the signs on the very nicely panelled doors reading director of this or chief officer of that, and finally you are standing outside the last one. Your hands are starting to feel clammy. The boss's personal assistant waves you straight into the inner sanctum. Now you enter a huge and very tastefully but sparsely furnished room; the south wall is mainly glass looking out across the city to the hills beyond. You are not invited to sit down. Instead, you are left standing with the low winter sun shining from behind the boss's head directly into your face. You have to squint and you think this resembles the set-up for a police interrogation in a low-budget movie. You are sure the positioning was deliberate in order to put you at a disadvantage.

You try and control your breathing, which is now very fast and shallow. Your mind is racing, you are going to be told that you are being let go and should hand over the pass that you are not going to need any more as security has been called to escort you immediately from the building. It has happened to colleagues in the past. How are you going to tell your partner? You'll have to drop the idea of getting a new car. No more company healthcare plan. You can cancel the summer holiday; you'll lose the deposit, of course . . .

Instead, the one-sided conversation goes as follows:

Boss: "Ah, [insert your name here], thanks for coming in. I wanted to share some exciting news. We have decided to develop a new service, the construction, installation, maintenance, and operation of sewage treatment plants. The research and development team have come up with a fancy new way of treating sewage with engineered bacteria, which produces far more energy than anyone has ever thought possible. It also produces water clean enough to drink and a super fertiliser. It seems that, if they're right, we pump sewage in and pump money out. I don't know anything about it – far too technical for me, of course, but we're going to become first the national, and then global, leaders in this area.

"Now it's all very well having this unique technology, but establishing our world-leading presence in bio-energy from sewage is going to be a job like no other. It has complete support from my fellow board directors, including the chair, who has also discussed it with some of our principal backers. Now, I was really pleased when your name was suggested to me as the person who could make it happen. You are, of course, indispensable to me in your current job, so this is going to be an additional task. I'm sure that you can find the time, but please remember that you can call on me for support whenever you need."

Now, you are a novice at this sort of conversation, and in your state of nervous anxiety, followed by a massive surge of relief, what you heard was, "My name was suggested from the top. I have been chosen, even marked, for greatness," but you missed the real point. You have been landed in it, literally up to your neck, which is also metaphorically, and simultaneously, on the block.

So, relieved that you still have a job and flattered to be chosen, you reply, "Ah, that's great news, umm, have you got any, ah, um, background papers that I can read?"

In the lift going back down, you start to reflect on what just happened. You implicitly agreed to take on a project, but you have been given no time to do it, you have no information on what is required, you have only a sketchy idea of the business sponsor's vision, and most importantly, you have not a single clue as to what success will look like.

Congratulations. You are now a project manager, fully formed and ready to go, with the future financial success of your company riding on your back alone. Welcome to our wonderful world of project management.

Mini Glossary, or What Do We Mean by . . . ?

Given that projects are everywhere, we should have an instinctive grasp of what a project actually is, but start reading the literature and even a seasoned professional can come away confused. It seems that everyone has a slightly different meaning for any of the words involved in managing projects. As far as possible, this book avoids using those sorts of terms and acronyms that seem designed to build barriers to sharing knowledge and create a secret society for those who understand them. This is especially true of those contrived acronyms designed to make spurious words, such as COCOA – the Campaign to Outlaw Contrived and Outrageous Acronyms. In the world of project management, there are many terms that are widely used – or abused. We deal with most of these as they arise, but we do need clarity, from the outset, about what project managers mean by a few of the terms to which we will return:

Project. A series of related tasks designed to solve a problem or achieve an objective, which is delivered by a temporary alliance of individuals. Recognising what is and what is not included in a project and setting appropriate boundaries are important aspects of the project manager's role.

Programme. A series of interdependent projects, but beware, because in construction especially, the term *programme* can also mean a plan, and sometimes a large project (more properly now called a megaproject) can be referred to as a programme.

Portfolio. A series of parallel projects existing at the same time. We can also include sub-projects as frequently these are major undertakings in their own right. For example, building a railway station is obviously a project. It does not become less so because it is part of a bigger project to build a new railway line.

Project plan. A timed series of activities which, when followed, will deliver the project's objectives. But a plan is not just a series of one-line entries for each activity that make up a timeline, although in some areas it is regarded as such. For our purposes, a plan needs to include the timeline showing each activity together with a detailed description of the activity and its inputs, outputs, costs, and risks. We, as project managers, are the ultimate owners of the plan, but it is usually nego-tiated with other stakeholders and approved by the project's commissioners. We will find that as project managers, we will not only be reporting our progress against the plan, including deliveries against time and costs against deliverables, but also reporting updated projections of time and cost to complete. As project managers we do have to be clear about this definition and, for example, may have to explain to our construction colleagues that what they refer to as plans we simply call drawings.

Project manager. This is us, the person responsible for taking the vision of what the project will deliver and making it happen. Many of the skills and attributes of the project manager apply also to programme and portfolio management, so these terms are included in this simple definition. It is usually a temporary role. There are career project managers who move from one project to another, or someone within in an organisation may be seconded into the role for one project only. Often the worst-case scenario, as with our case study, is when becoming a project manager is added to our existing duties. As project managers we may feel like we carry total responsibility for the project, but in reality, we always report upwards to a responsible person in the client's business organisation – the project's business sponsor – and to the project board. Experience shows that responsibility does not stop there. There are many examples of CEOs finding themselves having to take responsibility for a failed project and it being the cause of their departure from the organisa-tion. A project failure may even resurface to haunt a former CEO in their later career – I will refer to the UK Post Office's Horizon computer system fiasco later.

Client. Throughout this book, the term *client*, or *host organisation*, is used to identify the company or organisation responsible for commissioning a project. For the sake of clarity and consistency, the term *client* is used even where the project is delivered wholly internally to an organisation.

Business sponsor. The person responsible within the client management team for the successful delivery of the project. Their role often includes taking the credit when a project is successful or, if the project fails, passing the blame on before everything explodes. In the PRince2 methodology, as used by the UK government and in many other organisations around the world, this role has the snappy title of senior responsible owner.

Supplier. Any organisation responsible for supplying goods or services to the client for use by the project. Depending upon context, industry, and country, a supplier may also be referred to by terms such as *bidder, tenderer, seller, vendor, consultant, service provider, contractor,* or *sub-contractor.* This last is a supplier to a contractor; for example, in the construction industry, this would include sup-pliers of materials or hired heavy equipment or labour only to the company with whom the client contracted, who is also often referred to as the prime contractor.

Stakeholder. A stakeholder is any individual or organisation with an interest in the outcome of our project. They can include our project team, the business sponsor, suppliers, other departments in the client organisation, its customers, funding organisations, and the wider community.

Risks and issues. Again, there is a massive amount of theory, discussion, and definition that may or may not help clarify the meaning of risks and issues and what to do about them. At its core, a risk is any event that might happen which will impact on the delivery or outcome of the project, and an issue is a risk that has already happened. You could say that an issue is a risk with a prob-ability of 100%.

Both risks and issues can have a positive effect on a project – although mostly we are concerned with managing those which will have negative ones. We could use the terms potential and actual problems instead because, psychologically, we have a built-in drive to deal with and solve problems, but risks can be taken or run and issues ignored.

Our role as project managers is one of:

- Understanding when, where, and how risks and issues are likely to arise.
- Attempting to reduce the likelihood of their occurrence.
- Planning how to mitigate their untoward effects or maximise the beneficial ones.
- If they do occur, managing the necessary alterations required to the project, its plans, and its outputs.

Like in any good design, our aim in designing a project should always be not to have to manage risks but to have identified them and designed them out at the outset. As a simple analogy of designing a risk out of a project to build an aircraft, think of a one-way valve that is needed in the fuel feed line between the fuel tank and the engine. There is a real risk that the valve can be fitted the wrong way, in which case the fuel does not get through and the engine stops. We can manage the risk of installing the valve the wrong way with extra inspections during initial installation and maintenance during the life of the aircraft (mitigation of the risk). By thinking about this risk early enough in our project, however, we can design it out completely by giving the inlet and outlet sides of the valve different sized connections. This design change would make it physically impossible to fit the valve the wrong way, have minimal cost implications, and save the cost of all those extra inspections during construction and the life of the aircraft (risk avoidance).

Solution. The diversity of results from different projects is immense, from pyramids to jet fighters and from new products to replacement customer management systems – and everything imaginable in between. Consequently, throughout this book these project results are given the umbrella term *solution*, simply on the grounds that we can describe the objective of every project in terms of solving a problem, such as "We need to get this new product to market as quickly as possible to beat the opposition" or "We are falling behind in our combat abilities and this new jet fighter has to be flying within 180 days." In fact, it is reported that Lockheed, utilising previous British research on jet fighters, delivered the airframe for the prototype of the US's first jet fighter, the P-80 Shooting Star, in only 143 days.

Other terms are often also used such as outcomes and outputs.

Deliverables. Projects generally have two types of deliverables:

- Project deliverables are generated by the project's processes and support the delivery and management of the project. They are common to all managed projects and include documentation such as project plans, business case, logs of risks, and reports. This is examined in detail in later chapters.
- Product deliverables are items produced by the project which contribute directly to the required solution. These deliverables are highly project specific but could include a set of drawings, software programmes, a jet engine, a brick wall, or even the sewage treatment plant from our case study.

Benefits. *Benefit* is a term often used quite loosely in justifying the time, effort, and expenditure on a project. In the private sector, where organisational goals are usually financial or performance related, it is possible to take a strict interpretation of benefits. Factors such as return on capital employed or measurable performance improvements, such as productivity increases or a reduction in the number of days lost through employee injuries, can be used. On one IT project, the financial justification rested solely on cost savings from reducing the large postal costs incurred when sending case papers to remotely based staff. The public sector – and infrastructure projects, in particular – tends to include calculations on the social benefit to the wider community, often by inferring benefit from indirect measures. One example of this type of social benefit being included in a business case was a desired switch from car journeys to bus passenger journeys, which was used to support the justification of street signs displaying projected bus arrival times.

Project management office. This is often called a PMO. Project management has spawned a completely new area of expertise and specialism, the project management office. Sometimes also called the enterprise project management office (EPMO), this concept seems to have developed a life of its own. It can either be the administrative centre of a project or, in larger organisations, can be used to signify a team with a role of overseeing and co-ordinating multiple projects.

Project baseline. The project's baseline is usually defined as the original scope, cost, and planned delivery schedule. It is critical to our successful management of the project's delivery that the baseline is fully defined and understood before the actual project delivery starts. The reason for this is that one of the biggest dangers that projects face is called "scope creep," where unrecognised or uncontrolled changes to what we are delivering undermine the whole project. We manage this through formal "change control" processes.

Line Management. Line management is generally referred to as being the management of staff directly responsible for delivering an organisations goods or services. In general, it is the lowest level of a hierarchy of management.

How Does Managing Service Delivery Differ from Project Management?

As the new sewage project manager, on the way back to your office, you start to reflect on what you are going to have to do, and your first thought is, "OK, I'm used to being responsible for the management of normal day to day business operations, leading my team and all that sort of thing, and this project management stuff can't be that different. Can it?"

This is our first thought exercise, and there will be others later. When you encounter one, you may find that it is worth pausing for a few moments before moving on, allowing you time to reflect on the issues posed.

In this case, how do you feel that project and line management differ? It might even be useful to make a few brief notes.

Earlier, we defined what a project is, but what about defining management?

Management. The setting of strategy and goals within an organisation; the allocation of resources, motivation, and co-ordination of staff; and the overseeing of the delivery tasks designed to achieve those goals.

Managing a project falls easily within this definition, and so our right starting answer to the question could be, "Yes, project management is different, but it shares many similar features and attributes with line management."

As we journey through the following chapters, many of the differences with line management will emerge. We are also going to consider managing projects in the context of a client or host organisation, although we should also recognise that some projects – such as self-building your own dream home – may mean that you are simultaneously project manager, business sponsor, and client. But first we need to focus on the biggest similarity of all between line and project management: people. Like all managers, project managers deliver primarily through other people, but here is the first major difference. Most of these people are not under our direct line of control, and many will probably not even work for the same organisation.

It is essential here to understand the biggest difference between managing a line of service and managing a project. Line management is basically about managing continuity, often meaning that line managers have to endlessly repeat the same actions and processes. The individual issues they face might change, but they sit within this repeating management framework.

In an imaginary role as manager of an outsourced service to the insurance industry, and maybe in our current real-life role, we can expect to operate in weekly, monthly, quarterly, and annual cycles as a line manager. For example, it's July and time for staff appraisals, or it's October and time to start business planning, or it's January and we have to get on with budget setting. Midmonth we might be setting month-end projections, and at the end of the month we might be justifying why our midmonth projections were wrong or why our figures are so different from last year.

In total contrast to line management's continuities and certainties, project management is about managing discontinuity and uncertainty. We are going to deliver changes, such as to systems, processes, the built environment, or products. When we are appointed at the start of the project, we do not inherit a department with staff and a set of processes developed over the years to deliver repetitive tasks such as paying bills and staff wages, building cars, packing goods in a warehouse, or administering insurance policies. Instead, we have to determine goals and targets, build, and motivate our team, create appropriate processes, and ensure that we have both a realistic design and the means to deliver it. Even if we are given the job of building a tenth theme park with identical rides to the other nine, we still have to deal with new stakeholders, changed budgets, different ground conditions, new contractors, and so on. No two projects will ever be the same. Obviously, our predecessors as project managers of theme parks one to nine will have overcome many of the challenges that we will face, so we can learn from the past, but our project will be unique, and we cannot just copy what they did before.

We can prepare, research, analyse, discuss, and obtain agreement to deliver the perfect project plan, but in the end, our role is about managing delivery in the real world, which is not perfect. In the real world, our information is often imprecise, inaccurate, late, or not available. Our relationships are temporary and often both contractual as well as personal, and events completely outside our control will impact our project. We are responsible for managing imperfection to deliver the best result possible.

Welcome to uncertainty and the world of the project manager.

Information Seeking Framework

Now, if, when we were in front of the boss and being given the enviable role of project manager global sewage, we had been experienced, smart, and on the ball, we would immediately have recalled the first few lines of a poem by Rudyard Kipling. If you take nothing else from this book, commit these lifesaving lines to memory. Put them on the home screen of your phone or tattoo them on your wrist:

> I keep six honest serving-men[1]
> (They taught me all I knew);
> Their names are What and Why and When
> And How and Where and Who (Kipling, 1902)

From our point of view as project managers in a complex world, we have to add a seventh serving person who is essential to us but was sadly missed by the Victorian poet. This is the double-barrelled and ever-present seventh serving person, How Much.

With these seven working for us, whenever we get caught out by our boss in this way, we will have the basic framework to start to grill them about what they really want and will be able to find out how committed they actually are to it. At the very least, we will get some information on which to base our choice of whether to accept the accolade of being noticed or decline the obviously poisoned chalice of being the conscripted project manager – in addition to everything else we were doing.

Aren't There Methodologies to Help Manage a Project?

Methodologies compete with each other and often come with manuals of many hundred pages, training courses, and exams. If you read through any of them, it can seem as if project management has become a completely new and bureaucratic discipline with its own arcane terms (which can differ between the different certifying bodies) and different procedures and methods.

Very often project management has come to mean following the administrative set of tools and checklists embedded in one or the other of these methodologies, which can seem bureaucratic. Even just using some of the simple tools suggested in this book can seem equally bureaucratic. But each tool, from whatever source, should have the dual purpose of forcing you to think carefully about the information you really require and providing an example of one way of storing it.

Project management methodologies, however, represent the distilled wisdom and experience of the many people who have delivered a wide variety of projects of differing sizes and types and have collaborated to produce their definitive views on how we should all manage a project. These methodologies also get refined and added to over time as new ideas and innovations in project delivery arise. As a result, the very fact of setting out processes and documentation that can cover many different types of projects can make the methodologies seem very dry, technical, and driven by bureaucracy. Consequently, if used without careful attention to the context and complexities of the project, any chosen methodology can become an overly bureaucratic straitjacket. The last thing we want is to spend so much time servicing the processes prescribed by the methodology that there is little time left to actually manage the project. One client senior director saw using any project management methodology as an unacceptable and bureaucratic overhead. So, the project manager had a choice on these projects: either just "wing it," as the senior director wanted, or adopt a guerrilla approach and push the methodology underground. It was still there, but the project manager just did not tell the senior management team.

Guidance from a methodology means we do not have to work out from scratch all the basic functions and tools that are required to deliver a project. In a straightforward project, developing these is not that onerous, but in a complex project, or one subject to detailed scrutiny, an established and recognised methodology provides standards guiding our work and against which the individual components of the project can be evaluated. And we can always defend ourselves by saying, "I know that problems arose, but we followed XYZ methodology as best practice."

There is a benefit to us following a methodology focused on the How of running a major project in that it frees our thinking from concentrating on developing detailed processes. In turn, this enables us focus on the What of actual delivery of the project's outcomes; the Who of managing the people who will be delivering them; the Who, How, Why, What, and When of communicating with the stakeholders interested in the project; and, as ever, the How Much this is all costing anyway. As a personal benefit, we can also study and pass the exams in the methodology, which will then qualify us to describe ourselves as a project manager and get a nice shiny new certificate in the bargain.

However, what will enable us to get the project completed on time, to budget, and to the required quality is not just following any methodology but our abilities as a manager – our abilities to think through the actions we need to take, set strategy and goals, allocate resources, motivate and co-ordinate staff, and oversee the delivery of tasks.

The Life of a Project

The flow of a project from beginning to end splits into a series of stages, each of which, in turn, breaks down into a series of project management activities. These project management activities are common across projects and are independent of what the project is setting out to achieve. There are different naming standards and interpretations of these different stages, but in order to keep it simple, this book uses the following convention:

- Inception.
- Preparation.
- Implementation.
- Closure.

These four stages taken together are often referred to as the project life cycle, and the following chapters roughly follow this life cycle throughout as well as pick up on a variety of skills, techniques, and areas that run across multiple stages of the project.

Table 1.1 sets out the sorts of activities to expect in each main life cycle stage and, critically, details the end point of the stage.

Table 1.1 Project Life Cycle Activities by Main Stage

Stage	Types of Project Activity	Stage End Point
Inception	Gaining knowledge and understanding. Obtaining approval to create the project.	Project is authorised to proceed to full preparation.
Preparation	Detailed planning and estimation, building a business case, and identifying the project team. Obtaining approval to proceed. Preparing detailed project plan, high-level risk management strategy, and other delivery strategies.	Project is authorised to commence delivery.
Implementation	Creating the project deliverables, which build into the required solution. In managing the various elements of the implementation, there will be many different combinations of activities and approvals depending on the type of project, the attitude of the client, and the chosen approach to project management. In all projects the project manager will be responsible for monitoring and reporting progress, controlling costs and risks, and managing changes to the project and its deliverables.	Acceptance of the delivered solution as completed.
Closure	Closing the project, including any necessary handover to operational management, reporting lessons learned, and disbanding the team. Summarising the lessons learned. Completing project documentation. Closing suppliers' delivery contracts and releasing the project team's final members.	Project manager leaves the building, not forgetting to turn out the lights.

Just an early word of warning on project stages. Later we examine some alternative approaches to project delivery which allow activities to be planned differently, but even if we choose to go down one of these routes, we still always have to make sure that our project will complete tasks in accordance with any necessary logical sequence dictated by the nature of the solution – i.e., we need to complete the foundations of a building before we put the roof on.

What Is the Role of the Project Manager?

Safety Moment – We Don't Come to Work to Kill or Injure Anybody

In a later chapter, we consider project risk management, which will always include any safety-related risks, but before we start to consider the detailed duties of the project manager, we should pause and consider why safety is a prime area of our responsibility. A content warning here, as we are going to discuss a real, serious, and very tragic incident in the upcoming paragraphs.

When arranging a project meeting, often one of the most challenging things we have to do is to find a convenient location that can house all the people who need to attend and is available when we need it. This may have been the case when a group of staff from the US-based Jacobs Engineering Group attended an off-site project meeting on 24 March 2005. The meeting was held in a staff office trailer about 35 metres from the isomerisation unit at the BP refinery in Texas City.

The staff involved were working on a separate unit which had been closed during a turnaround period, and the staff office trailer was just a convenient, available place for their meeting.

Due to a series of problems following a restart procedure of the isomerisation unit, nearly two hundred thousand litres of a volatile flammable fluid were discharged into the atmosphere. It ignited. At approximately 1:20 p.m. local time, a series of explosions rocked the plant. Photographs show that the staff office trailer was effectively destroyed by the blast. One hundred and eighty people were reported as injured, and fifteen people were killed, twelve of whom had been in the trailer.

Jacobs lost eleven members of their staff that day, and the CEO had to issue a statement about the loss of life. This is not a statement anyone would ever wish to have to make, or the phone calls to the families of those who died. These people were friends and colleagues who were just attending a meeting in a handy trailer. As an aside, Jacobs's stock price took an immediate hit of over 9% after the tragedy.

Sadly, safety is an area of the project manager's responsibility that often gets only light – or, worse, even no – coverage in many project management methodologies. This omission may be because ensuring safety is seen largely as a general management responsibility and not one specific to us as project managers. From our point of view, however, safety should be an area of which we are mindful from the very outset of our work. Unlike line managers, who manage mainly in certain and repetitive situations, as project managers, we are always managing novelty, change, and uncertainty, including:

- Creating new teams, often from diverse organisations, locations, and cultures.
- Potentially requiring our team members to work in unfamiliar and sometimes inherently dangerous situations.
- Possibly creating new products or services, which requires us to consider the safe operation of the outcomes of our project.

Where our project involves delivering physical work in inherently dangerous environments, such as construction, manufacturing, extraction, agriculture, and transport, the need to have clear safety policies and the processes to enforce and monitor them is self-evident. However, as the extreme Jacobs experience showed, even office-based project work carries risks, though usually these risks are only around internal office safety and travel. To complete the story, the safety culture that Jacobs built following their experience and rolled out across their global operations sought to evaluate and mitigate all risks associated with any staff activity, including office or site-based work and travel arrangements.

In summary, we should never create situations where the life, health, or well-being of others is endangered in delivering our projects, and it should be central to our role to ensure that those whose work we are organising are able to do so in a safe and protected way. We should reflect this safety-first approach throughout our project specification and management processes.

The Project Manager's Responsibilities

As a novice project manager who gained responsibility, in our case study, for delivering a new sewage treatment service globally, now we wisely consider what an experienced project manager would do next. The right answer is always to take time to think and shape or frame the project in our own head. We need to first be able to explain the project to ourselves, and then, like Pharaoh's overseer, to be able to explain it succinctly to other people. The other thing to consider, if we have a choice in the matter, is whether following a project management methodology will give us a structure around which we can deliver the project.

But what does a project manager actually have to do? Like all of our project-related definitions, that of the role of the project manager can vary greatly. Given the range of work delivered through projects in different industries, situations, and countries, this variation should not be a surprise. In considering our own role, there may be many factors that influence our activities, including the stage of the project when we are engaged. Often the project manager is not engaged at the very start. The host organisation might handle everything up to the delivery stage and then bring in a project manager. This means that we may be picking up on elements that have already been delivered and agreed. Sometimes we are expected to be the subject matter expert as well as the project manager. Alternatively, we may be teamed with a business or functional expert, which can be useful. Table 1.2 provides a list of the type of functions a project manager usually delivers, and in order to answer the question about differences with other managers, it also contrasts them with the typical functions of a line manager.

Table 1.2 Contrast between Line and Project Management Roles

Line Management Role	Project Management Role
Setting of strategy.	Framing and shaping the project.
	Owning the business case for the project and ensuring that it is updated and variations reported during the life of the project.
Setting goals.	Establishing the parameters for the project and preparing and agreeing on detailed project delivery plans.
Allocation of resources.	Identifying the project team required and negotiating the allocation of staff from internal and external sources as appropriate.
Motivation and co-ordination of staff.	Creating, co-ordinating, and motivating a cohesive project team drawn from a variety of different sources.
Overseeing of the delivery tasks designed to achieve those goals.	Monitoring progress against the delivery plans.
	Establishing criteria and processes to ensure that intermediate and final deliverables meet required quality levels.
	Financial management of the delivery of the project within agreed-upon budgets.
	Procurement and contractual management of suppliers.
	Managing changes to the project.
	Identification and quantification of risks and issues faced by the project.
	Management of risk and issue mitigation.
	Specification and management of changes to the project, including any necessary updating of plans and forecasts.
Cyclical performance reporting.	Monitoring progress and reporting clearly and truthfully to business sponsors and stakeholders throughout the project life cycle, including closure.
	Financial reporting and forecasting.
	Managing communications with stakeholders.

Just listing the main tasks of the project manager is to slightly miss the main point. Yes, the project manager is responsible for all these tasks, but often, even usually, we will be calling on others to complete them under guidance and supervision. Our role as project manager is like that of the conductor of an orchestra or the director of a film. It is the project manager's vision, motivation, and enthusiasm that provide the impetus for the project and our support and leadership that enable the team to deliver it. The factor that makes the major difference to a project's smooth running and ultimate success is the project manager's ability to communicate with, motivate, and facilitate other people. Managing our sewage treatment case study will require liaison across the company and engaging with research scientists, financial experts, marketing staff, operations managers, the public relations team, and so on.

Principal elements of the project manager's role are shown in Figure 1.1.

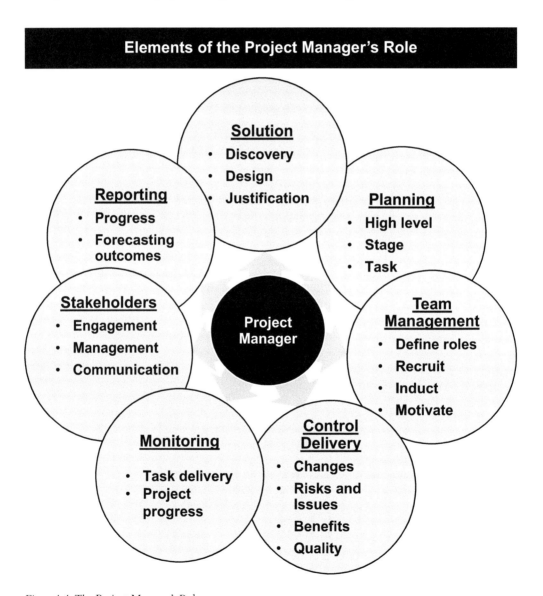

Figure 1.1 The Project Manager's Role

The balance of the project manager's role also changes over time within the project. Typically, our main work early on might be around developing the business case before moving on to initiating the project itself – building the team, agreeing on plans and budgets, and so on. The next stop might be procurement – never underestimate the effort required to procure major supplies and services. Before we know it, we are into contract management and payments. This change of focus of the project manager's role as the project progresses is shown in Figure 1.2.

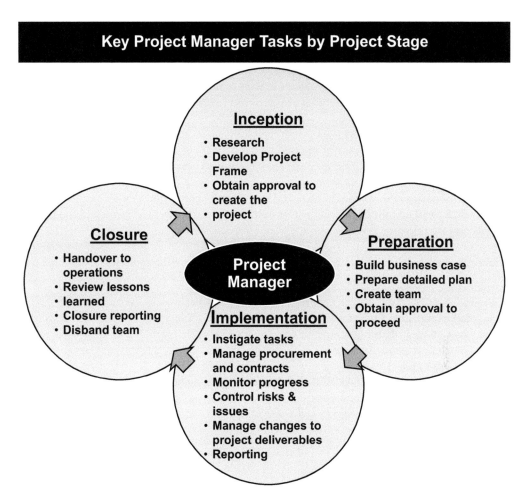

Key Project Manager Tasks by Project Stage

Inception
• Research
• Develop Project Frame
• Obtain approval to create the
• project

Preparation
• Build business case
• Prepare detailed plan
• Create team
• Obtain approval to proceed

Project Manager

Closure
• Handover to operations
• Review lessons
• learned
• Closure reporting
• Disband team

Implementation
• Instigate tasks
• Manage procurement and contracts
• Monitor progress
• Control risks & issues
• Manage changes to project deliverables
• Reporting

Figure 1.2 Project Manager Tasks by Stage

All the time we are also monitoring, reporting, preparing projections, allocating resources, chairing project team meetings, attending boards, avoiding the auditors, evaluating change requests, possibly (depending on the project) preparing responses to adverse press comments or ministerial questions, and keeping our funding organisations happy. Many of these aspects of the project manager's role are very different from that of most line managers,

Communication. The helicopter view is that the project manager is the face of the project, at least inwardly to the business and to the project itself. An ability to speak openly and frequently with all stakeholders is critical. Often, though, the business sponsor is the outward face of the project and will report to the business and other key stakeholders.

The advantage to the project manager is that although we may not get the recognition, if the project does fail, we may slip quietly away and, as if by magic, the responsibility for the failure will usually progress upwards through the client organisation and stop elsewhere.

One example of the business sponsor being the outward face of a project was when a new government agency in a European country called an external project manager in to manage a project to reform a local industry. The project had multiple related deliverables, it was highly visible, and its success was a key ministerial goal. Each week the project manager discussed the progress reports on Thursday evening with the business sponsor, who in turn discussed them on Friday morning with the senior minister. The senior minister (later prime minister) had no idea that his pet project was not being managed internally.

Framing the project. Immediately after we are appointed – or, in the case study, as soon as we have left the boss's office with our new and very inadequate sewage treatment brief – it is time for us to engage and send out those seven honest serving persons. We need to use them to gather information, start to resolve uncertainties, and enable us to put boundaries around the project. This is called project framing, and it is the first step towards understanding the project and preparing the basis for thinking about our project plan and business case.

Planning. There may be a feasibility stage to justify commencing the project, but often the first public documents we have to prepare are the:

- Project plan, which will deliver the project's vision and meet the objectives within cost, quality, and time constraints. The plan is more than a list of tasks (What) and a timeline (When). It is the explanation of (How) the team that you will create (Who) will deliver. A good plan is a living document throughout the life of the project, which permits progress to be measured and expectations to be set and adjusted.
- The business case (How Much, Why, and Where) is also a living document that is periodically updated using the greater knowledge and accuracy of information obtained during the delivery process to affirm that the project is still viable.

Facilitation and leadership. An essential part of any project manager's role is to assemble and lead the project team (Who). We will need to work out the size of the team and the skills required to complete each of the tasks in our project plan. Our team will change over time. For example, when building a traditional house, although there are overlaps, the quantity surveyor is replaced by the groundwork contractor, who in turn gets replaced by the bricklayer, roofer, carpenter, plumber, electrician, plasterer, and then the painter and decorator. When we have our initial team identified and in place, we assign tasks to each of them, with plan-driven deadlines. Like Pharaoh's overseer, we also have to ensure that the infrastructure is in place to provide the team with the necessary supplies and resources.

Change management. One of the biggest issues that all of us face as project managers is change. This can have a minor or a major impact and can even lead to projects being no longer viable. Change can be generated within the project, perhaps arising from a greater knowledge of solution requirements or an understanding of constraints, both of which we will gain as the project progresses. On one project for installing a local automated fares system, the electrical network was designed with electrical control boxes located indoors. During the detailed design task, it emerged that these boxes should only be mounted outside. This design change necessitated a whole range of wiring changes, weatherproof boxes, and additional installation works in a large number of stations. Change can also be imposed externally on a project. This might arise from legislative change, changes in technology, or changes in the client organisation's market. In a project requiring on-street electronic signage, the government imposed a rule requiring signs to be bilingual, which in turn necessitated software changes and translating over five hundred street names.

What Does a Project Manager Do Day to Day?

As we have seen, every project that we undertake will be unique and every project management role interpreted slightly differently. There will be cycles of things that commonly have to be completed: reports to project boards, analysis of risks and issues, monitoring finances, checking progress reports by our various teams, and engaging with stakeholders. Often the project manager is the only person "on the ground" with a total overview of all elements of the project. A large part of the day can be spent communicating the impact of this to different team members and resolving conflicts of understanding between them. Any plans for the day can disappear with the morning mist when the unexpected happens:

- The client manager falls out with supplier's project leader and feathers need to be smoothed urgently to keep delivery on track. In reality, as the project manager arrived, they were standing at opposite sides of a doorway, backs towards each other, scowling, with arms crossed.
- A key member of the team takes leave as the result of the death of a second parent in six weeks, and we have to get their work covered.
- An important solution component which tested fine last week has now failed, so delivery plans have to be rejigged.
- The client or business sponsor explains that although they nodded when we said A, they thought we meant B, C, or even Z, and as a result, they have done or said something that we now have to delicately unpick whilst keeping everyone on board and not grumbling (too much).

And it is still only 9:30 am and we haven't reached our desk yet, let alone sat down! Just grab a coffee and breathe deeply for a few minutes.

If you have ever seen act involving plate spinning or a blindfolded performer simultaneously juggling objects of different shapes and sizes – including axes, knives, and flaming torches – whilst balancing on a ball, you can begin to get the general idea.

In summary, as the project manager, we are the expert on the project and the troubleshooter of first and last resort. Critical to our success is providing the team with the support and leadership that are crucial to building their confidence and motivating them to deliver.

This means that two days are never alike. Our role is best summed up as doing what it takes to keep the project moving forwards. We are also the umbrella that shelters the project team members from internal distractions and the shield that protects them from grenades being thrown at the project from outside.

Project Manager's Log

The sheer momentum of a project, especially when we have engaged and successfully motivated others to deliver for us, can defeat the best intentions of any project manager. We can suddenly find ourselves:

- Focusing on the urgent instead of the important.
- Being swamped by minor issues which, if given time for a moment's consideration, we should delegate.
- Unable to remember clearly what happened last Tuesday (or Wednesday or Thursday), which can be hugely important later on when we are required to compile reports and time-lines of events. The thinking could be, well, it wasn't Monday, because someone brought in cakes, or Friday, because that was the project board meeting when the business sponsor spilt coffee over the only print of the project plan. But as to which other day it might have been . . .

Fortunately, there is a very simple tool to help us: a daily log of our activities. Whether you prefer a notebook and pencil, an app on your phone, or a programme on your laptop is a personal choice. Taking a few minutes at the end of each day to complete a project manager's log creates a space in which to reflect on the day's events and think through our agenda for the next day. This reflection helps us focus our activities and not lose sight of the important because of the urgent. Our project manager's log becomes an important tool in capturing information for us to share with the business sponsor at our review meetings. It is also a good place as we progress through the project to start capturing our first thoughts on the lessons we are learning and the risks or issues we are facing and will have to formally record later.

How Will We Know If We Are Any Good as a Project Manager?

Historically, it is true that many projects have not been well managed. Consider the 2013 critical report of Lord Browne (then the UK government's lead non-executive and formerly head of the energy company BP), who had reviewed the UK government's progress on improving project management. He reported:

> The government owns and runs 185 major projects which in total cost £414 billion. These major projects include the building of roads, railways, defence equipment and information technology systems, and they cost between £20 million and £20 billion each. Despite the level of investment, the management of these projects has been worryingly poor. When the Public Accounts Committee (PAC) reported on major projects in September 2012, it found that only one third were delivered on time and budget; that is in contrast to the highest performing private sector organisations. . . . There is still insufficient attention given prior to the initiation of projects to identifying options and risks; consistent failure to put in place project leaders with the right skills, experience and incentives; and inadequate scrutiny of the most complex and expensive projects at the centre of government.
>
> (Browne, 2013)

A good project manager is not only one who delivers a good project – i.e., one that delivers to time, cost, and quality – but also one who manages the challenges faced by their project. Now is probably the right time to introduce the bane of our life as project managers: everything seems to be trying to blow us off course and stop us delivering our project. Apart from events, which I deal with in detail in the chapter on risks and issues, every day we face disruptive forces that are often miscalled the devil's triangle. There are, in fact, four (not three) major forces which pull outwards on the project and try to deflect us from our agreed-upon objective. These forces are interlinked and often pull in direct opposition to each other.

These four disruptive forces are:

- **Time**. Contrary to popular belief, in project management, work does not normally expand to fill the time available. We try to estimate and then to manage time and effort as tightly as possible, but events frequently seem to conspire to push us to deliver a task or perhaps a whole project later than we agreed. This lateness is called slippage. It is also possible (unusual but not unheard of) for delivery periods to be shortened because of expert management by the project manager.
- **Cost**. We try hard to estimate the exact costs of the project as part of getting approval for it. Even if we were correct at the beginning of the project about the funding that we needed, the world moves on, and costs move almost inevitably upwards. We call this overspend.
- **Quality**. We may face demands to compromise on the quality of the project deliverables we originally specified either because they cannot be supplied or because meeting a higher-quality specification would take longer than the project can allow or the costs would be unacceptable. It may, of course, also be that the project's original requirements were "gold plated" and can be reduced without impacting the operability of the solution.
- **Scope**. Once we have agreed on the overall shape and content of our project, one of the greatest difficulties we have to manage is the pressure for additional items to be added to our list of deliverables or for deliverables to be changed subtly to make them "better" in some way. This is the previously mentioned scope creep, and these changes can derail the best planned and managed projects. If anyone suggests, "Wouldn't it be a good idea if we just added this extra thing in to the project, or made that thing bigger and shinier whilst we doing it anyway? It wouldn't cost much more or take much more time?" our default thought should always be no and the answer more diplomatically along the lines of, "Well, that's a great idea, thanks for bringing it to me. I suggest that we put it in our list of wants for the first upgrade."

The way these forces can interact with the project is shown in Figure 1.3.

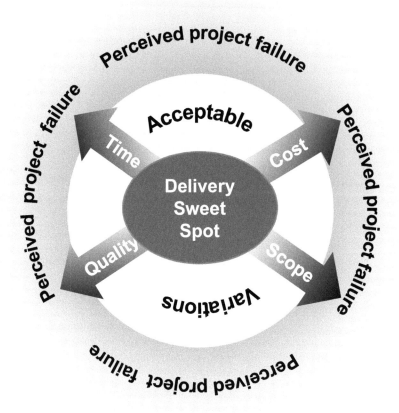

Figure 1.3 Opposing Forces Impacting a Project

Later we will explore the problems around setting project timescales and budgets, but for now, accept that they are both matters of opinion and not fact. Probably using performance against these opinions as the yardstick by which to define ourselves as good project managers is overly simplistic and may be completely misleading. Consider this simple analogy between the project manager determining the delivery plan for a project and the managers charged with running a railway on time (Table 1.3).

Table 1.3 Train Scheduling Dilemma

Train Scheduling Dilemma		
Performance Target	**100% Trains on Time**	**Minimised Journey Time**
Situation	Schedulers timetable all the trains to have artificially lengthened journey times.	Trains are scheduled to run at near maximum permitted line speed.
Effects	Trains are run more slowly so that, if there is a delay, the driver has the ability to speed up and recover lost time and may still complete the journey as scheduled.	All journeys complete in the shortest possible time. In the event of a delay, the journey will still complete in the shortest possible time. In the worst-case scenario, the journey will still complete at the same time as one scheduled to run more slowly.
	Passengers receive a poorer service because scheduled journey times are always longer than they need to be.	Most journeys will be much quicker; therefore, passengers receive a better service on average because their average journey times are shorter.
Result	More trains arrive on schedule and thereby are closer to a 100% on-time arrival target.	A greater number of trains are marked as being late because the driver had little opportunity to speed up to recover from any delay.
	The train company and its the managers are applauded for running trains on time.	The train company and its managers are criticised for the level of delays.

The lesson is equally applicable to our projects. The greater the budget and the longer the time allowed, the easier it is to deliver "successfully," even if the outcome is delivered later and at greater cost than is actually necessary.

Subsequent to Lord Browne's 2013 report, the UK Infrastructure and Projects Authority published a list of eight self-explanatory principles on which all projects depend for success. By implication, these factors provide a strong definition of what "good" as a project manager looks like:

Setting Up for Success

1. Focus on outcomes
2. Plan realistically
3. Tell it like it is
4. Prioritise people and behaviours

Project Execution

5. Control scope
6. Manage complexity and risk
7. Be an intelligent client

Overall

8. Learn from experience

(UK Infrastructure and Projects Authority, 2020)

In consequence, perhaps we should judge our performance as a project manager not just by making the project run to time and budget, although that obviously does help, but by establishing the project's real criteria for successful outcomes and then judging our behaviours and thinking when managing the competing demands on the project of time, cost, quality, and scope.

Project Manager as a Change Deliverer

A comprehensive view of successful management of change and transformation is beyond the scope of this book, and not all projects will be delivering change within an organisation, but many will have the implementation of some kind of change as an objective. If we are to deliver such an objective successfully, we may need to shape our thinking from the very outset around how such change may be achieved and what that will mean for creating and structuring our project.

One of the widely accepted views for how to achieve change successfully has been put forwards by John Kotter (Kotter, 2012). Amongst his ideas is a useful concept that we should add to our thoughts about stakeholders when we are delivering change. Kotter talks about building a volunteer army to help us deliver that change. On one project for a new community health IT system, an army of volunteers from amongst the disparate groups that would be the beneficiaries of the system was built around a hub-and-spoke model. A delegate was drawn from each team of professionals, who in turn had to recruit a team of at least ten volunteers to engage in discussions about requirements, risks, and preferred solutions and eventually participate in testing. The result was that the system, when delivered, had over one hundred engaged and knowledgeable supporters, which facilitated the user rollout.

Our Approach to Managing Projects

So far, we have covered some of the basic concepts around projects and project management. In the following chapters, we are going to consider how we can best operate as project managers in the different areas of the project. Underlying the suggested ways we need to approach each area is a set of general actions and behaviours that will allow us to successfully discharge our very special role as project managers. These behaviours reflect that we are:

- Managing novel and unique sets of circumstances, problems, and combination of individuals.
- Operating outside the "business as usual" line management of the client with very limited real power.

Figure 1.4 shows how these behaviours sit together in a simple virtuous circle.

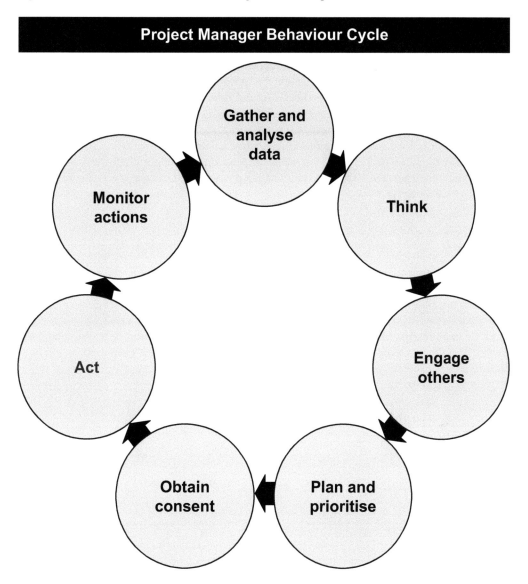

Figure 1.4 Different Aspects of Project Manager Behaviour

As the life of the project moves forwards, we find ourselves having to engage in all these different behaviours simultaneously. We may be planning the details of one series of tasks and monitoring the progress and impacts of several others, all whilst researching other tasks or engaging with a whole range of different people.

Project Route Map: Where Are We Going Next?

As this book increasingly demonstrates, all projects are different and are delivered in different ways and in different organisational environments. Each project is therefore unique and will require different balances of activities by the project manager, but there are many common themes. Beginning in Chapter 2, we are going to start looking in detail at the typical work required by the project manager throughout the life of a project, and Figure 1.5 gives a helicopter view of how we will explore this work and how it interconnects. Each of the boxes represents a series of activities covered in one of the following chapters and indicates some of the major documents we could expect to produce during that activity. Figure 1.5 is a useful overview, but it is not a map of a methodology, where we undertake one task, tick the box, and move on to the next one. The previous warning about undertaking multiple tasks at different levels at the same time still applies. If only our life as a project manager was really as straightforward as in Figure 1.5.

Using the End-of-Chapter Reflections

In the reflections section included at the end of each chapter, please draw on experience or use the theoretical project. There are no correct answers, and there are no crib sheets at the back of the book – the reflections are here solely to help you focus your thinking on the topics raised in each chapter.

Creating notes in response to each question and sticking with a single case study throughout the book will build up a record of how you choose to think through a project from its inception to its closure, and will contain some examples of how you have used the suggested basic tools, or others of your choice to structure and record that thinking.

It is not necessary to use the energy from sewage treatment case study project for your reflections, although, if chosen, it is useful to consider the widest aspects of the project, including construction, IT systems, business change, research, procurement, and manufacturing.

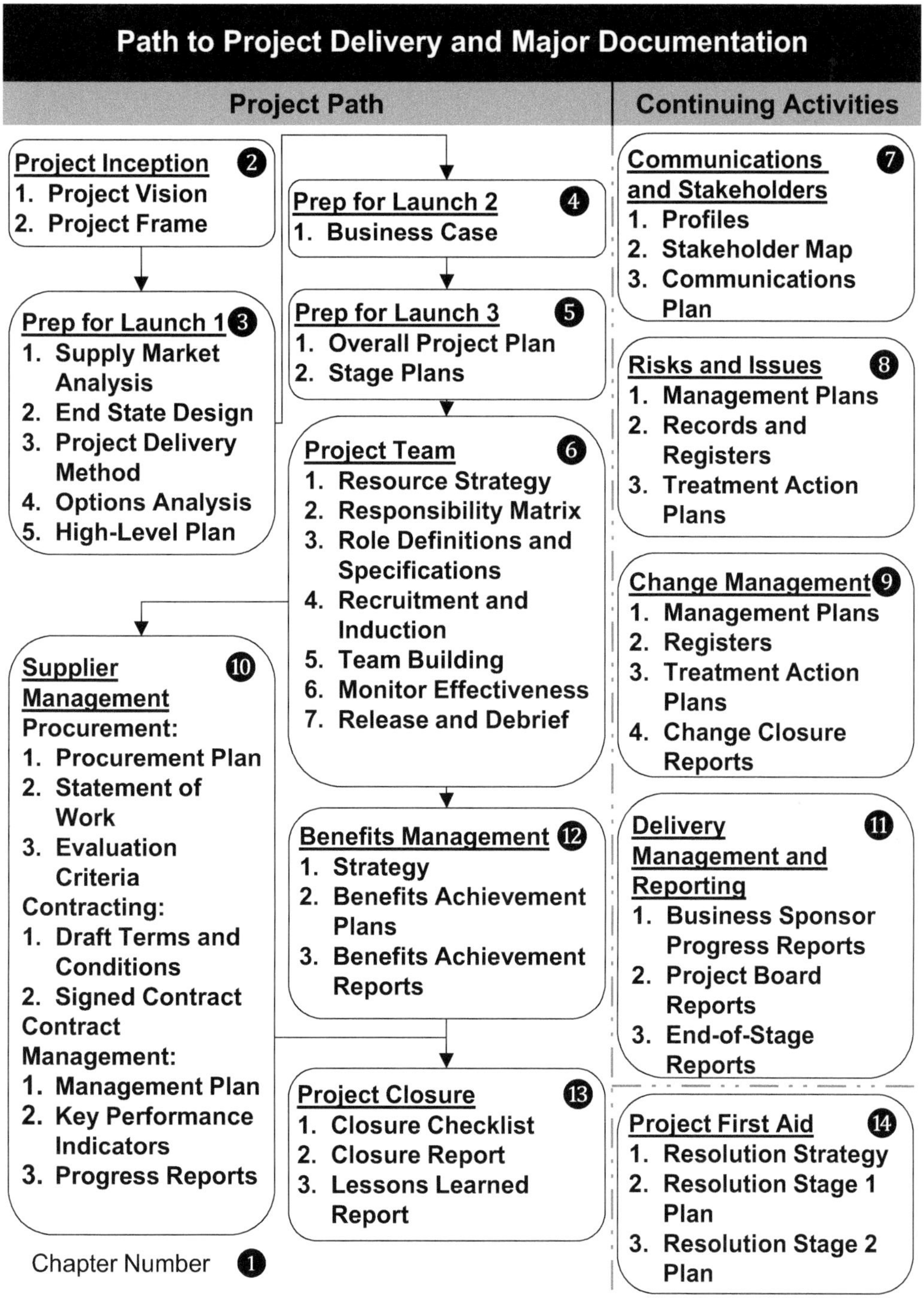

Path to Project Delivery and Major Documentation

Project Path

Project Inception ❷
1. Project Vision
2. Project Frame

Prep for Launch 1 ❸
1. Supply Market Analysis
2. End State Design
3. Project Delivery Method
4. Options Analysis
5. High-Level Plan

Prep for Launch 2 ❹
1. Business Case

Prep for Launch 3 ❺
1. Overall Project Plan
2. Stage Plans

Project Team ❻
1. Resource Strategy
2. Responsibility Matrix
3. Role Definitions and Specifications
4. Recruitment and Induction
5. Team Building
6. Monitor Effectiveness
7. Release and Debrief

Supplier Management ❿
Procurement:
1. Procurement Plan
2. Statement of Work
3. Evaluation Criteria
Contracting:
1. Draft Terms and Conditions
2. Signed Contract
Contract Management:
1. Management Plan
2. Key Performance Indicators
3. Progress Reports

Benefits Management ⓬
1. Strategy
2. Benefits Achievement Plans
3. Benefits Achievement Reports

Project Closure ⓭
1. Closure Checklist
2. Closure Report
3. Lessons Learned Report

Chapter Number ❶

Continuing Activities

Communications and Stakeholders ❼
1. Profiles
2. Stakeholder Map
3. Communications Plan

Risks and Issues ❽
1. Management Plans
2. Records and Registers
3. Treatment Action Plans

Change Management ❾
1. Management Plans
2. Registers
3. Treatment Action Plans
4. Change Closure Reports

Delivery Management and Reporting ⓫
1. Business Sponsor Progress Reports
2. Project Board Reports
3. End-of-Stage Reports

Project First Aid ⓮
1. Resolution Strategy
2. Resolution Stage 1 Plan
3. Resolution Stage 2 Plan

Figure 1.5 Path to Project Delivery

Reflections

- Does the work that you are doing meet the definition of a project, and can it be broken into the four areas of inception, preparation, implementation, and closure?
- If you are using the case study or other project, which areas of the project manager's role do you consider are your major strengths, and where do you feel that you might need to enhance your skills?
- Alternatively, if you are currently engaged in managing a project, think about how your role matches to the tasks outlined, and list your responsibilities, highlighting any areas that require you to pay more attention.
- How well does your chosen project reflect a focus on the eight success principles of the UK Infrastructure and Projects Authority?

Note

1 Apologies for the lack of inclusivity, but please give the poet some leeway for having been born in the 19th century and for spotting a good rhyme when it was needed. Incidentally, he is the first of the Nobel Prize winners we will encounter in this book.

Bibliography

Browne, L., 2013. *Getting a Grip: How to Improve Major Project Execution and Control in Government*, London: UK Government.

Kipling, R., 1902. *Just So Stories*, London: Macmillan.

Kotter, J., 2012. *Leading Business Change*, Brighton, MA: Harvard Business Review Press.

UK Infrastructure and Projects Authority, 2020. *Principles for Project Success* [Online]. Available at: https://www.gov.uk/government/publications/principles-for-project-success [Accessed 05 December 2022].

2 Project Inception

Framing Our Project

Aim of this chapter: To introduce the following:

- The importance of document control in project management.
- Project framing, including boundary definition, dealing with uncertainty, and critical success factors.
- The importance of the informal pathways within an organisation.
- Estimation, including how to reduce uncertainty in the estimation process.

Learning outcome: To be able to prepare a project frame and an initial project justification, using reasonable and defensible estimation processes.

This chapter covers the first stage of our project, which we will term inception; it is also called project initiation. During this stage we undertake initial research and build the first overview of the project and its deliverables. The importance of this early stage cannot be underestimated. It can be time consuming, and, to external observers, it may look like nothing much is happening. Using the example of a construction project for a particular building, the inception stage would be undertaking the design and then preparing the groundworks and foundations. Nothing much is visible above ground for a protracted period, but a great deal of planning and effort are involved, and failure to get it right can cause major problems later.

As Nick Smallwood, the CEO of the UK Infrastructure and Projects Authority, wrote in guidance for successfully delivering projects: "The success or failure of a project is often determined in its early stages. Whilst successful project initiation can take more time at the start, this will be repaid many times over later on in delivery – so you must get it right from the start" (UK Infrastructure and Projects Authority, 2020).

Strict Discipline: Document Control

Before we can get down into the weeds of creating a successful project, we need to think carefully about the least glamourous area of project management. How are we going to control the often very large numbers of documents and document versions that we are about to start producing? Although our project is going to be shaped by the creativity, ingenuity, and flair of our project team, it will only move forwards by creating, sharing, authorising, and updating documents. The result is that as we progress through the project, we will build a veritable mountain of discussion documents, emails, correspondence, progress reports, communications with stakeholders,

DOI: 10.4324/9781003405344-3

quotations and tenders, contracts, schedules, plans, designs and drawings, specifications, and papers for approval by the project board or business sponsor. The list of documents produced by most projects is almost endless.

Consequently, enforcing strict discipline over documents is essential to the smooth running of our projects. It is always embarrassing to find that all the participants in a meeting have different versions of a document and then to waste most of the time in the meeting contesting which version is the authorised one that should be under discussion. If the disputed version of the document in question is a draft contract, it can result in protracted legal arguments. If it is a design document, the consequences of using a superseded version could have major implications for our project. To avoid such situations, we need to ensure that there is only one (current) version of the truth and that we, as the project manager, control it.

All sorts of software systems can help, but document control can also be done manually. In either a digital or manual system, there are some very simple rules that we must ensure are followed:

- All project documentation, including emails, must be filed at a location accessible centrally by authorised individuals.[1]
- The status of any particular copy of a document should be readily apparent.
- The files should be organised in a logical manner.

Centrally Accessible Filing System

Document filing standards and systems may be mandated as part of an organisation-wide system. Equally, we may be able to choose a system designed specifically for projects, often including document sharing with third parties, or we may decide to use one that relies on Windows file directories or even manual files. Our choice will depend on the organisation's policies and our situation. For example, a central manual system is difficult to operate if our project has dispersed teams or remote workers. Whatever process or technology we choose, in relation to different classes of documents, we will need to decide:

- Who will have the right to submit documents for inclusion?
- How will documents be checked, verified, authorised, and reviewed?
- Who will be able to read, update, amend, or delete documents?
- How will documents be filed and archived?
- Who will carry day-to-day responsibility for ensuring that all documents are filed correctly?

Document Standards, Status, and Version Control

Each document needs to readily identify its status, and a useful protocol is to include a standard document control page as part of each document. For every version of the document, this page details the names and dates of the original authorship and those responsible for checking, reviewing, and authorising this document version. By retaining the history of previous versions, this page also acts as a document history. This control page can also be used to include additional information such as distribution lists.

Client organisations may have standards for the presentation of documents of different classes that must be followed, and often these can be formulated as standard templates. If such standards and templates do not exist, it is worthwhile to create them from scratch for the project to ensure that all versions of all documents are standardised and immediately identifiable as part of the project.

There are also many schemes that we can use for naming the document versions, and these can become complex. One very simple convention starts from an initial draft document which is numbered as version 0.1, and subsequent drafts are numbered 0.2 onwards. When the document is approved – for example, when a business case is first approved by the project board – it acquires a new main version number of 1.0. If, as the project progresses, we need to update the document, our process repeats, with the first draft update to the authorised version being numbered version 1.1, subsequent drafts numbered 1.2, and so on. The next version to be approved by the project board would then become 2.0. You get the idea.

Logical File Organisation

Once again, there are many different ways to organise a filing system hierarchy, and we may not have a choice as the client organisation may mandate a structure that we have to use. If we are able to create our own hierarchy, we can directly mirror the logical structure of the project's tasks in our filing structure.[2] This approach has the benefit of reinforcing the way that the project works to the whole project team. One project went so far as to include both the reference to the task and the version number in the electronic file name as follows: 2.2.3.2 Casework Replacement Project Business Case V1.0. Project staff routinely referred to the document by its full file name, making it very easy to find in the central file directory.

Project Management Step by Step

The main activities that are usually required during project inception are set out step by step in Figure 2.1. Similar figures showing the steps required for different project management responsibilities are set out in the following chapters. Each figure has been developed to fit the needs of a moderately complex project in a mature and complex organisation. This might be a medium- or large-sized company, a government department, or another public sector body. Remember, in real-world practice, there may be some flexibility needed. For example, larger projects may need to be handled in smaller steps with more individuals involved, and projects involving less structured organisations may require fewer approval steps.

The numbers attached to each of the project manager's responsibilities refer to the numbered steps in the following sections. Whether each of the steps are followed, or followed in the order suggested, is a matter to be decided in relation to every project.

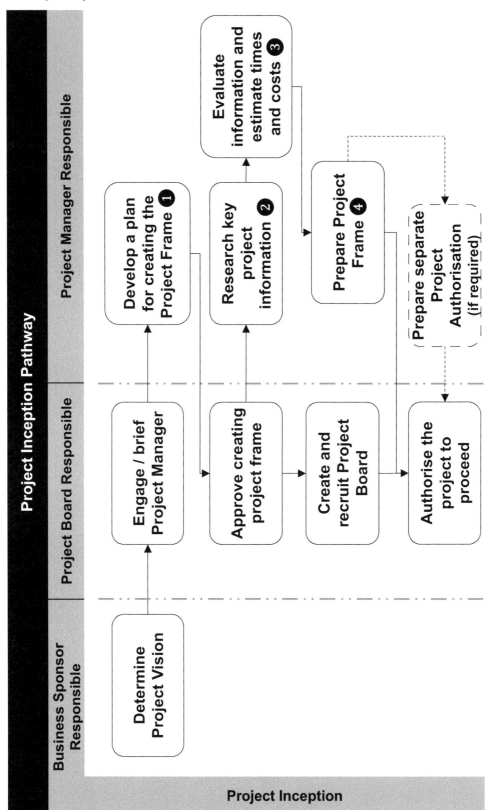

Figure 2.1 Project Inception Pathway

Introducing the Project Frame

Figure 2.1 shows that the main output from the inception stage is a project frame document, possibly with a parallel and separate project authorisation document if the client organisation requires it. We can think of developing the project frame as being like a using an artist's sketch book, in which we gradually focus our first thoughts, impressions, research, concepts, and conclusions into a final picture that will enable us to:

- Clarify the organisation's views about what should be achieved, i.e., its vision for the project and the factors that will influence our ability to achieve it.
- Start to build a consensus around the capacity of the project to deliver the organisation's vision.
- Communicate information about the likely basic shape of the project, including a range of likely costs and a potential time frame for delivery.

The importance of the project frame in setting the project off in the right direction cannot be underestimated. There may be pressure from the business sponsor and other stakeholders to "just get on with it," but additional taken time during inception will pay dividends later because we will have tightly defined the project's vision and objectives and obtained an understanding of potential costs, timescales, and constraints. To achieve this, we must first build our own knowledge and understanding of the project and start to draft out a route map to delivery. We are not talking about detailed plans, budgets, and business cases yet, as these will come later. The project frame brings together sufficient information to demonstrate that the project as we have pictured it could be viable and to justify the approval to move forwards to deliver the next project stage, preparation. It is in the preparation stage that we will develop a detailed business case and project plan. The approval of these later documents will give us the agreement to commence the "heavy lifting" of actually creating the full project and delivering the required solution.

The project frame needs to consider the following areas:

- **Vision mapping**. How we can draw out the client organisation's (possibly quite vague) vision into a series of project objectives. Ideally, we should be able to define all the project objectives so that they will fit with the SMART principle – i.e., that they should be:

  - **S**pecific.
  - **M**easurable.
  - **A**chievable.
  - **R**ealistic.
  - **T**ime-based.

- **Outcomes and end state**. What the project will physically deliver and what the new changed end state of the organisation will be. For example, a new building being used as the global headquarters, an integrated IT system supporting distributed working replacing antiquated paper-based methods, or even (as in our case study) a world-leading sewage treatment capability operating in five continents.
- **Success**. What will affect the perception of the success of the project.
- **Constraints and dependencies**. The limits on what we can achieve with the project. This includes corporate policies, legislation, finance, and also dependencies on other projects, e.g., acquiring and adapting a new building before beginning a project for replacing IT hardware.

- **Risks**. What risks we think may have to be addressed in delivering the project.
- **Timing.** How long we think it may require to deliver the project and what the key intermediate stage dates will be.
- **Resourcing.** What staff and other resources we will require to deliver each stage.
- **Outputs.** What the project will deliver in each stage.
- **Locations.** Where each stage will be delivered.
- **Costs**. How much the project will cost, both in terms of one-off capital spend and ongoing revenue costs.
- **Benefits**. How strategic and other benefits, both tangible and intangible, cashable and non-cashable, will be assessed. These benefits are the reason the project is being created, and if they are insufficient or insubstantial, we need to say so now.

Step 1: Develop a Plan for Creating the Project Frame

Drawing up detailed project plans will come later, but at this point we should set out what we have to do and how long we estimate it will take to develop the project frame. This is usually our first analysis where we have to think through how we are going to achieve any project goal. In preparing this limited plan, we should produce a schedule of the tasks that we will have to complete in order to reach the point at which the project frame can be approved and an indicative timeline for their completion. This scheduling can be a pencil, paper, and calendar exercise. It is not a task that usually needs to involve project planning software and complex Gantt charts – of which much more when we look at issues around planning in Chapter 4. When the timeline is complete, we need to add additional detail about each task that we have identified, including the resources we will require to complete it. This now becomes our stage plan for the inception stage of the project (again, more on these in Chapter 4).

The rationale for such an initial plan is so that we can understand our own current time commitment to the project and also set the expectations of others, including the business sponsor, about the initial route map we will be following. It should also answer their key questions, including:

- How much effort will be required to reach an agreed-upon project frame?
- Who will need to be involved? This should include the full range of different individuals who should be engaged in the development and agreement of the project frame.
- When things are likely to happen that will require involvement of these individuals?

Once this inception stage plan is agreed, we can proceed to develop the project frame, as set out in the following steps.

Step 2: Research the Project

Building our knowledge is what allows us to complete the project frame. Unless we are subject matter experts and have delivered similar projects before, we are unlikely to be able to prepare a robust project frame in isolation and will need input and support of others.

It's time to think about our fictional case study project again, which for politeness from now on is going to be called "bio-energy from effluent" rather than constantly referring to sewage.

The lift has just delivered you back to your floor. On the way down, you have come to the understanding that you are not doing this alone. Now, you have to reflect on the types of people that you need to help you deliver this project, which is going to dominate the world of "bio-energy from effluent."

Obviously, you need to speak to the head of the research and development team responsible for the current research work, and, if the boss has not already told them about you, you will have to update them on what has just happened. The sooner the better – and a call or face-to-face conversation will likely be a better start to this relationship than an email.

But who else are you going to talk to who can help fill in the background to the project? And who can tell you what you need to do to actually create a project within your company? Are there any mandatory project creation processes that you have to go through or people that you have to involve?

In real life any company scoring high on a matrix of project management maturity will have all the project management processes mapped out, manuals prepared, and often guidance provided online. In other companies, like this one, it is time to start speaking with our internal network of contacts and building everything from scratch.

But who would you put on your list of people to contact?

The Importance of Informal Enablers

Throughout the following chapters, we will be considering the very formal ways our client organisation works and designing the ways our project will mesh with them. In particular, we will be designing how documents are reviewed, approved, and signed off. But, before we go any further, it is also worth considering the ways the organisation hosting our project works informally. Even if our new project is within an area with which we are familiar, we will be working with different people and in different ways from a line management role. These differences will be magnified if we will be working in a new division or even a new organisation.

We might start by scrutinising the organisation charts, working through descriptions of business processes, and reading about the decisions on previous projects. All these help us understand the way the organisation is supposed to formally function. They are not *wrong*, and, unless we adopt a guerrilla approach to project delivery, we will have to respect them in the way that we design and operate our project to meet the expected norms of the organisation. The formal side, however, is never the only way any organisation operates, and to be successful, it helps if we understand the way that the informal organisation works as well. We will need to build a picture of this informal organisation, which will help us in designing effective communications plans, inviting the right participants to engage in the project, and, when necessary, lobbying for support from the most effective champions.

By way of explanation of the importance of the informal ways any organisation works, consider three people: Jan, Naz, and Chris. They don't occupy key management roles high up in the company organisation chart. In fact, it's quite the reverse; they all sit somewhere in the deepest recesses of the organisation. They are all pretty friendly and usually helpful to everyone they meet.[3] These three are the unsung heroes of successful project delivery, and engaging well with them will mean the project, and also our lives, will run more smoothly.

Jan has been attached to our project as an administrator. Jan has been around the organisation in many different jobs for a long time and is well respected. Therefore, the single best source of information when we need to find out who to talk to about something across the organisation and how they are best approached. A request from Jan for a meeting, or a room, or a projector will carry weight with most people. Jan is also the secretary of the staff social club and through that is on familiar terms with the chief executive, who also sits on the club committee. The chief executive, being an outgoing person, frequently chats to them about their work – which means that Jan is a backdoor way of finding out how our project is going and learning about what gets left out of formal reports. This channel is two-way, and through Jan, we have a priceless way of briefing and consulting the chief executive, providing we understand and respect that Jan always tells the truth from Jan's own perspective.

Naz is a smoker and is part of the project's communications team. Morning and afternoon, rain or shine, Naz mixes outside with the smokers who are the most informal group in any organisation. Whilst they smoke, they chat for a few minutes. Mostly about their work and their colleagues. This means that Naz has a finger on the pulse of the organisation and can provide feedback on its mood and feelings generally, as well as thoughts on any particular places that may be sources of potential opposition to the project. Naz is also a drinking pal of the person running the IT help desk, which is an added bonus when we hit IT problems or need a new laptop or workstation fix in a hurry.

Chris is the "go to" person for building management. Based in an office just behind the reception desk, Chris is known to keep one eye on people entering and leaving the building. If we need to know where additional space for a new team member might become available soon, or would like to get a desk moved or a whiteboard put up,[4] Chris is the person to ask. Chris also knows how to mend the printer and the photocopier and has a spare (and probably illegal) kettle that we can borrow when the team is working late or out of hours when the staff coffee bar is closed.

Through these three individuals, we can have effective but informal access to resources, contacts across the organisation, a channel to the top, and feedback loops on the impact we are having – for good or bad. Investing time in finding the equivalent of our Jan, Naz, or Chris for every project, and then in nourishing and nurturing them, does not appear on any to-do list for a project manager in any major methodology, but it always repays the effort required many times over.

Later we will consider interaction with the project's stakeholders, but from the outset, we also need to build up a picture of the interests, drivers, methods of working, preferences, and relationships of the key decision makers who will have influence on our project's success. Understanding all of these will make our interactions with them far more successful. Compare, for example, the way two business sponsors would prefer to make themselves available for an informal conversation.

Ash, business lead for a major project in a government agency, would be prepared for a discussion over a coffee and a round of toast in the canteen first thing in the morning – before the day starts to pile too many things up. The result of one of these early morning discussions was finding the solution to a problem that had stalled a national approach to multiagency collaboration for over six years.[5]

Alex, chief executive of a rural English local authority, preferred to meet at the end of the day and relax over a (non-alcoholic) beer in a nearby pub before driving home. One such informal chat about progress on a business transformation project resulted in an agreement to evaluate an additional area for inclusion in said project which would focus on income generation. Part of this extra evaluation required the direct observation of the effectiveness of the collection of waste from business premises.[6]

The point is that, as project managers, we are usually driven by very short time frames and often need access to key individuals very quickly. In these cases, rather than asking a secretary for a slot in an already packed diary, getting meetings at short notice is much more successful if we have taken the time to understand the individual and can ask, "Do you fancy a quick coffee in the canteen first thing tomorrow/meeting for a drink this evening? I'd really like your views on . . . ." Depending on the organisation's refreshment policy, and the tastes of the individual we want to meet, the following can also work: "Can I pop in for a quick chat over coffee/tea? I'd like to ask for your advice on something. I have custard cream biscuits . . ."[7]

In every area of our responsibility as project manager, it is important to recognise how far our own skills, knowledge, research, and time availability can take us and be able to identify when we need to call on others to help us. We can seek this help in a variety of different ways from other staff, stakeholders, subject experts, or professionals with expertise in key areas. Researching information is often our first opportunity to engage key staff and stakeholders in the project, and building sustainable and open relationships at this stage can be critical to achieving eventual success. Wider engagement also starts our transformation from being the new project manager to being the "go to" expert on the project.

Gathering Information – More Than Just Talk

Our research for the project frame is about gathering as much information as we can, using our seven honest serving persons as a structure. We should seek to speak with as full a range of experts in the business and subject area as possible, engaging stakeholders, staff, and customers. We should also review any similar projects within the client organisation or other comparable organisations. But remember, we always need to prioritise how we will spend our research time. This includes making sure that we understand the right people with whom to engage and the most effective way to engage them. At the project framing stage, our first decision is to differentiate between those individuals and groups with whom:

- We *must* engage.
- We *should* engage.
- We *could* engage and may do so if time permits.

Later on in the project, we often have to call on not only these people but also others for more of their time, support, engagement in key delivery tasks, and sometimes help when events are driving the project off track. One UK government agency's IT and business change project engaged with 104 external individuals representing nine government regional offices, one devolved national administration, two national professional bodies, and the representatives of over four hundred local authorities in addition to groups representing the agency's administration and professional staff. This level of engagement was maintained from developing the initial framing of the project's solution through to user testing and final delivery.

Various methods are useful for gathering information, each of which can have different strengths, but taken together they can help us rapidly build the knowledge that we need to complete the project frame, as well as being useful throughout different stages of the project. These methods include:

- Directly engaging with others, rather than trying to become experts ourselves.
- Researching the market to find out what is available.
- One-to-one interviews, often only with key stakeholders and managers.
- Facilitated workshops to draw out the widest possible range of views and then using these to build a consensus.

Directly engaging others to help us. During the information gathering stages of our project, this engagement may also have to include detailed discussions with experts in cost estimation and financial modelling to help us come to grips with preparing our cost and time estimates. In the case of a project which lies fully outside our personal experience as the "draftee project manager," such as in our bio-energy from effluent case study, our support may have to include the widest range of local knowledge and subject matter experts, plus expert marketeers, researchers, engineers, cost estimation experts, and stakeholders. When we become a project manager, learning how to harness the expertise of others and synthesise their views is essential. There are some basic rules which will support our success in doing this:

Listen a lot, learn fast, don't be afraid to ask questions,[8] and always take copious notes.

Researching the market. If delivery of the project's solution will involve a procurement at this stage, it is often extremely useful to undertake initial market and supplier research. Our objectives in doing this are to:

- Understand the scope of possible solutions currently available in the market and the options for delivering them.
- Gather information on how comparable projects have been delivered in other organisations, including any areas that required particular attention.
- Gain an understanding of what factors will drive costs – for example, numbers of users or volumes of materials – and any indications of potential overall costs for a solution of the size that we are envisaging.

Revealing the name of the organisation considering a project to potential suppliers can result in an inordinate number of future contacts by the suppliers' sales teams. Such contacts always have to be carefully managed to avoid any hint of bias in a future procurement. Anyway, at this stage, if they are offered, suppliers' presentations or solution demonstrations to stakeholders and the project team are much less helpful than informal discussions with the suppliers' existing clients.

One-to-one interviews. A programme of one-to-one interviews both enables us to understand the particular views of individuals and provides an extremely good opportunity to build the relationships which will help us successfully deliver our project. Such a programme will, however, often make very large demands on the scarcest of our resources – our time. Rationing this time can therefore be necessary, and it is useful to identify key decision makers and influencers and focus on them first. Interviews need to be structured to ensure that our key questions are answered and that we are able to show respect to the interviewee by keeping to the allotted time. As the clock ticks towards the end of our timeslot, it is always useful to ask two very open closing questions which should be tailored variations of: "In your view, what three things must the project avoid?" and "By contrast, in order to be seen as successful, what three things must the project achieve?"

This gives our interviewees the opportunity to set out their wider views, which we may not have covered earlier, but leaves us in control of the structure of the interview. Limiting the number to three negative and three positive things is quite deliberate and focuses the interviewees only on those things that are most important to them. The result of questioning in this way is that sometimes interviewees may ask for permission to include four things, hardly ever five, and sometimes they may only list one or two. The question order is also deliberate, and we should always finish on the positive.

Facilitating group exploration. Quite often, group exploration sessions are useful to tease out the views of others, but these can also be very dangerous. If left to open discussion, we can find that the session is dominated by those with the loudest voices or only reflects the views of the senior staff present, with more junior staff not wishing to contradict them. Skilled facilitation employing a variety of techniques may be required to ensure that the views of all participants are brought out and given equal weight. Some particular techniques which may be useful in information gathering include:

- **Action cards/sticky notes**. The facilitator will ask each participant to complete "answers" on a card to any of our seven key questions (What, Why, When, How, Where, Who, and How Much). The completed cards are then put on the floor or sticky notes placed on a vertical surface to permit the facilitator to lead the "grouping" of these answers into a common viewpoint, including resolving any conflicting views.
- **Round Robin brainstorm.** Everyone speaks in turn, and there is no cross-table discussion. The facilitator allots a time span for each speaker and rigorously enforces it. There will usually be several speaking turns during which ideas are challenged and refined. The problem with this approach is that it does still not fully compensate for the situation where more junior members of staff may not want to contradict more senior ones.
- **Progress sheets.** These simple sheets are completed either jointly or by every participant at the end of every session. They allow participants to record and assess progress, which helps them recognise their achievements, thereby providing motivation for further engagement. Sheets can also be circulated between participants for comment if desired. They provide a record for the project manager and help set the agenda for subsequent discussions. A suggestion of how such a progress sheet could be set out is shown in Table 2.1.

Table 2.1 Discovery Progress Sheet

Discovery Progress Sheet	
Issues raised	
Resolved	**Unresolved**
Closed and agreed to take forwards *Issues and activities closed because agreement has been reached that they should be addressed by the project.*	Issues and activities requiring further discussion *More discussion needed because agreement was not achieved or discussion revealed new items that needed further research or consideration.*
Closed archive *Issues and activities raised and closed because discussion has revealed them as not helpful or not required.*	
Next steps	
Suggestions for future activity.	
Prepared by	**Date**

There is a more detailed discussion of facilitation techniques in Appendix 1 – Facilitating Discovery.

Step 3: Information Evaluation and Estimation

When we have completed all of our research activities, the data we have gathered has to be organised, evaluated, and used to create an overview of the solution our project will deliver. We use this overview to form the basis for building our view of the potential costs and timelines of the project. As indicated in our end-of-interview open questions, one area we will need to address specifically is the perception by decision makers, stakeholders, and others of what a successful project will have to deliver for them. At the end of our initial research, we should be able properly to finish the sentence, "This project will be a success if . . ." We can then list the criteria by which the project will be judged by different groups of people. These criteria are called critical success factors, and delivering successfully against them is our principal aim if the project is to be perceived as a success. These critical success factors should set the agenda for everything that we do and every decision that will be made about the project.

We will find that part of our role in achieving successful project delivery will be to ensure that everyone we engage on the project also understands what success means to our stakeholders and how each project team member should focus all their actions upon it. They need to have their "eyes on the prize" at all times, and the critical success factors that we determine from our research tasks are the way we can represent the "prize" to them. Additionally, understanding our critical success factors means that when project changes inevitably occur, we can always judge our actions, and those of others, by whether they move us towards achieving them, away from achieving them, or even compromise our ability to achieving them at all. We should also regard critical success factors as forming a "golden thread" through all the stages of the project. We should eventually be able to trace them through all our documentation: reporting and processes from the project frame, through the business case, requirements specifications, contracts, designs, and benefits plans, right up to our handover for live operation, project closure, and the post-implementation review.

Our first task in this process of organising our data is to use our critical success factors to develop a baseline view of what the solution needs to include so as to meet our decision makers' and stakeholders' expectations of what needs to be achieved. This task may require resolving any conflicts that exist between the views of different types of stakeholders and, if necessary, agreeing on a hierarchy of desirability. We may then need to set up another round of panel sessions and feedback interviews to validate our conclusions.

Our second task is then to differentiate among the core features that the project must deliver, additional features that could also be useful, and those that should probably be outside the scope of our proposed solution. For example, when building a house:

- **Within scope.** Three-bedroom family home of at least 180 square metres on a plot of at least 800 square metres, with a large kitchen, utility room, lounge, dining room, one family bathroom, and one bathroom ensuite to the main bedroom.
- **Useful addition.** Double garage of 30 square metres, with electric door.
- **Probably outside scope.** Separate one-bedroom guest bungalow.

Determining the scope of the eventual solution is an iterative process, and we may revisit this first view of the solution later when we have completed our next tasks. We will also revisit the question of scope later during the preparation of the business case.

The first of our next tasks will be preparing our first estimates of how long the project will take and what it will cost. We will have to prepare estimates throughout the life of the project, starting now with the project frame by estimating overall costs, timescales, and resources at a very high level. We drill down into more detail when analysing options and preparing our business case. Later, during the implementation stage, when we start producing actual figures, we have to repeatedly update our estimates to show new "at completion" projections of costs and timescales. We need to understand that as project managers, whenever we estimate duration, costs, staffing resources required, benefits, and so on, we are essentially making educated projections about the future based on past experiences which may not be directly comparable. The real problem with estimation is that there is only one answer that will turn out to be right and there is an almost infinite set of answers that will be wrong; therefore, the skill, science, or art of estimation is to:

- Develop our "educated guesses" into formal estimates that are as close to a reasonable view of reality as possible.
- Understand what influences those estimates and the degree of certainty that we can have in them.

Eventually, however valid, wide of the mark, or just spuriously accurate our estimates turn out to be, we will have to be able withstand scrutiny on how we have prepared them. Consequently, we need to be certain that the estimation processes that we used are robust, clearly understood, and defensible.

Dealing with Uncertainty

The one thing that is certain when we start framing any project is that our projections and estimates are never totally correct. At this initial stage, we are all, *always*, wrong in some way. We can use all the appropriate estimation techniques discussed here, but we are still often wrong simply because much of the information we need is still unknown, and at the start of our project, some of it is genuinely unknowable and some of it is uncertain. For example, we cannot be certain of the costs of a building until the site is selected and the ground conditions have been assessed. A simple, and literally a concrete, example of the unknowable came when replacing an old domestic driveway. During the removal of the existing surface, the specialist contractor discovered that the depth of concrete in much of the driveway was far outside their range of experience. After removing three and a half very large truckloads of spoil, the team and client had to agree that concrete ranging up to 13 inches in depth represented an unexpected level of over engineering, probably more appropriate to a wartime military bunker than a domestic driveway. Hence, of course, the estimates of time and resources to remove it had to increase – as did the final costs to the client. In the same way, we cannot fully understand the potential costs of a computer system, fighter jet, railway line, or container ship until we know what the client's detailed requirements are and how they will be addressed by the project.

Consequently, right from the start of our project, we have to deal with uncertainty. We can reduce this by making reasonable and explicit assumptions not just about estimates but also about other factors that may influence the deliverability of the project. Ideally, these assumptions will be fully documented by us, shared with the stakeholders, and agreed on by the project sponsors. So, we might state as assumptions for building a three-bedroom house that the site that will be finally selected is level, utility services are available at the site boundary, and the site has access for large delivery vehicles.

Our assumptions will become more certain and our projections more accurate as we do more research and should improve still further later when the project itself starts generating data upon which we can base revised projections. So, when the client for the house finally selects a sloping site two kilometres from the nearest road with a peat bog at one end and accessed only by a very narrow and tortuous track, not only will we be able to understand how much to change our projections, but we can point out clearly why they have changed. At whatever stage of our project, whenever we produce estimates, they should come with a very large health warning, along the lines of, "These estimates represent an agreed-upon view of the future and are based on the evidence available at the point in time when they were prepared."

Sadly, most estimates do not normally carry such a warning, and public perception and selective recall can be a real problem for us. Although we, as project managers, may have publicised all of the caveats around the estimates, and regularly listed every one of the assumptions on which we based them, it is always only the headline figures that get remembered and reported. Selective memory and reporting will still apply no matter how much we explain how the estimates were produced and the degree of confidence that everyone agreed we should have in them. In our bio-energy from effluent case study, if we were to estimate a cost of $70 million to establish global domination in the bio-energy from effluent market within three years, these are the figures against which we and the project will be judged. The judgement against us will be harsh if delivery takes us longer or costs more than the estimates. It will apply even if we had a list of twenty assumptions and thirty caveats about costs and timelines, all of which were understood and agreed to by the business sponsor and the project board but turned out not to match what happened during delivery.[9]

Using Estimating Techniques to Reduce the Guesswork

There are many techniques that have been developed to help us reduce the uncertainty gap between our view of the future (the estimate) and the eventual project results (reality). Minimising this gap applies to all projects and is not a new problem. Incidentally, there are academic analyses showing why we tend to be over-optimistic when we are estimating. (Meyer, 2016)

To help in this area, some industries may have specific standards that should be used when preparing estimates, and there are whole bodies of professionals upon which we can call. Equally, there are also some generic techniques that we can apply to most projects, depending upon the level of information available at the time. Even where we are not undertaking the estimation process ourselves, it is helpful to understand some of the basic methods and techniques that may be employed on our behalf. These include the following four common techniques:

- Decomposition.
- Analogous estimation.
- Parametric estimation.
- Three-point estimation.

Decomposition. This is the breaking down of the project from the top into smaller tasks and units, often called the work breakdown structure, or WBS.[10] The basic concept is that smaller units or tasks can be defined more closely and so can be estimated with more confidence. For our estimation task, take as an example being asked to estimate the cost a project to build a new 10-kilometre rail line. We could say that based on previous experience of building rail lines, the average cost is £1 million per kilometre, so we should just multiply this average figure by 10. We have stated our assumption and made an estimate. It is a subjective view of likely costs, but it only works, of course, if our line conforms to the "average" specification. Our top-level estimation will fail if the proposed line has a different pattern of tracks, tunnels, embankments, bridges, and signalling than this average specification. If, instead of taking the line as a whole, we break the line down into its components as per the WBS – i.e., we decompose it and produce estimated costs for each component – we are more likely to obtain a realistic estimate for the overall project.

When we look at a project, however, we do not stop at one level of decomposition of tasks but continue on downwards until we reach a level where we are confident that we have a component that can reasonably be estimated. At the lowest level, each task will have attributes requiring estimation, such as cost, effort, duration, and quality. We need to understand the factors that influence each estimate, such as size or specification for materials, qualifications, and levels of seniority for staff. As ever, everything, including all of our assumptions about these underlying influences, should be recorded for subsequent review if required. Essentially, instead of one big subjective guess covering the whole project, we create a structure to enable us to make a series of smaller, more manageable, but still often subjective "guesses" and then build these into an overall estimate for the project.

A limitation to this process of decomposition, especially at an early stage in the project, is the level of information that is available to us to support the breakdown.

The WBS is actually a tremendously useful way of defining the scope of what we are going to do in delivering the project. We can take each component individually and define its limited scope as a discrete part of the overall scope of our delivery. Later, as we progress through the different stages of the project, we can turn this statement of scope into a full specification for that component – in the railway line example, we could produce the specification and design for each of the lengths of track, tunnels, embankments, bridges, and signalling. In turn, we can use these more detailed specifications to refine our estimates of effort (person days) or staff resources (skills, individuals, and their costs) and materials costs. Figure 2.2 shows how a simple WBS can be constructed.

Analogous estimation. Under this technique, we take a component of the WBS and look for an analogy in a previous project that has provided a similar component. We use this analogy as the basis for our estimate. In some cases, the analogy can be quite close – for example, we need an identical piece of equipment to a previous project, and we have the exact information on what it cost then. All we have to do is allow for any price changes since the last one was purchased. Now we have cost and effort estimates in which we can be reasonably confident, but we still have to check up on the supplier's lead times and to see that the technology has not changed.

Other adjustments that we may have to make when using analogous estimating include changes because the analogy does not align precisely with the current project element:

- **Sizing.** If our currently specified component is larger or smaller than the one used as an analogy, our estimates need to be flexed to take account of this.[11]
- **Specifications are similar but not identical**. If the current project has different outputs to the analogy, the input components may also be different and have different attributes. For example, if the last time a project included an extra doorway being made in an office building was to give better access to the janitor's basement cleaning store, the door will probably have a different specification to the doors that we need if our current project is about reconfiguring the executive suite on the 52nd floor.
- **Knock-on effects**. Different component specifications will not only have different direct costs but may also have unexpected knock-on effects elsewhere. For example, if the piece of equipment that we order for this project is bigger than the one for the last project, we may need to survey the access route and location, possibly strengthen the floor where it will be placed, increase power supplies, and add air conditioning. Maybe we will have to use a larger truck, a

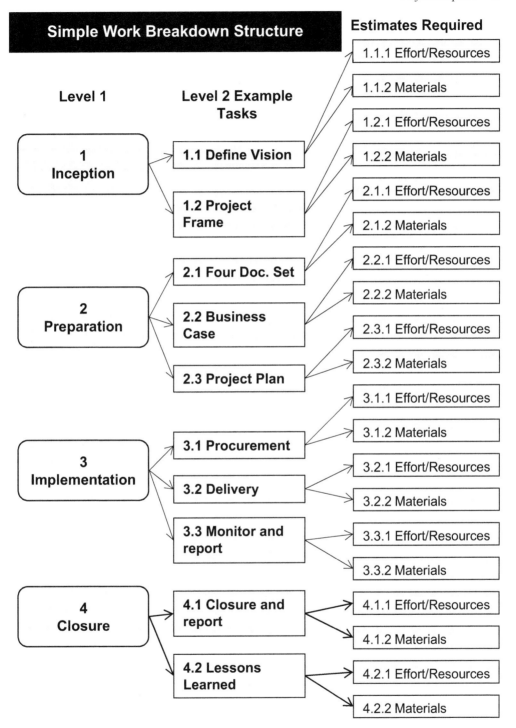

Figure 2.2 Work Breakdown Structure

crane, and a bigger team to put it in place. In our office door example, as well as allowing for a higher price for the increased specification of the door, we would be wise to budget some extra cost and effort for installing the executive suite doors over the one to the janitor's closet. This is because:

- We might need to protect the executive suite lift from damage.
- We might need a craftsman with higher skill levels to fit the door itself, which implies higher hourly costs.
- The new door may need to be fitted out of hours to avoid disturbance, also further increasing hourly costs.[12]

Parametric estimation. Some consider parametric estimation to have a higher level of accuracy than analogous estimation. Again, the method starts with decomposing the project into a WBS. This time, we have to reach a level where every component can be measured in appropriate units, and this can be tricky at the outset of the project. In our example of refitting the executive suite, the cost to the project of each of our doors will be a unit cost from a supplier price list, plus any delivery charge, plus the labour cost, which will be the amount of effort required multiplied by the unit cost of the labour with the required skills, plus any applicable taxes.

Our lack of detailed knowledge about the project can be a limitation to using parametric estimation. Consider the difficulty at the inception of the project of using parametric estimation in relation to building a warehouse. Here, the first stage of the project may be preparing the design of the building, and thus it is difficult to prepare unit-based estimates as only those units related to the site and the overall size of the building are available and none of the design or site details are known. Other factors of which to be aware include:

- Where effort is involved, we need to base the estimate on the projected hours for different skills and their respective labour rates.
- Material costs have to be determined based on projected usage. Whilst sometimes these can be based on industry norms – for example, a warehouse of floor area X would require Y tonnes of concrete for a solid floor – this is not always the case.

Three-point estimation. Three-point estimation is an arithmetically based approach to reducing subjectivity in our estimates. In essence it is straightforward. Instead of one estimate, we prepare three estimates by flexing our assumptions to represent the best, worst, and most likely scenarios. The more certain that we are of our estimates, the narrower the range is between the best- and worst-case figures. In the simple version of the approach, we add the three figures together and divide by three. A more advanced version of the techniques weights the answer in favour of the most likely case. The arithmetic then becomes

- Multiply the most likely scenario by four.
- Add the best scenario and worst scenario figures to the previous answer.
- Divide the new total by six.[13]

It is possible to amend these weights still further if required.

Irrespective of the way the estimates are derived, three-point estimation can be a good way to test and to moderate them.

Understanding Our Estimates – Now and in the Future

For each value that we have calculated in the estimating process, we need to understand and record:

- How we reached every estimate, including which technique we used and the assumptions we have made. Remember that we can use multiple estimating techniques depending upon the task and the information available to us.
- Our assumptions about the factors that may influence each estimate – for example, inflation rates and labour availability.
- The degree of confidence that we have in the estimate.

Our prime need for recording this information is not as a bureaucratic process but as an aid for ourselves. During the life of the project, and probably afterwards, our estimates will be revisited many times, and not only by ourselves as project managers but also potentially by project quality reviewers, clients, auditors, lenders, or grant funding bodies and as part of post-project reviews. Having the information to hand, in one place, greatly aids these activities and reduces the need for scrabbling around for notes or having to try to remember just how we derived our figures six months or six years ago.[14]

Looking to the future, having a comprehensive record of how we reached a figure also helps us justify any necessary alterations in the future when the information or assumptions underlying an estimated figure change. Table 2.2 shows the basic level of information that it is often useful to retain in some form. It may be that we can use a format such as that suggested in Table 2.2 as a top sheet summarising the estimate, with detailed supporting information being included in separate evidence sheets.

Final Words on Estimation

Even when following all the guidance on detailed estimation work, there are a number of factors that may unduly influence the final result if we are not alive to their impact:

- It has been put forwards by psychologists that, in general, we humans focus more on the positive than the negative. This bias can affect our ability to estimate accurately. One study of construction projects showed that not only was optimism bias influencing project planners and decision makers but they were also unaware of the impacts that such bias could have (Chadee et al., 2021).
- Within organisations, there is also a tendency, as information is aggregated and reported, for uncertainties to be absorbed, or at least not reported as fully as the data. This process is called uncertainty absorption and works as follows:
  - The person (us) preparing the figures notes any issues with them, documents these fully in a set of assumptions and caveats, and gives a judgement about the level of confidence they have in them.
  - The person at the next level up, who is anxious to meet a deadline instead of questioning us about them, says, "These figures are basically OK, but be cautious with some of the details," and neglects to pass on our carefully worded caveats, assumptions, and confidence levels.
  - The person at the next level up fails to mention anything about caveats or confidence levels at all.

Table 2.2 Estimation Record

Estimation Record			
Project			
WBS component			
Duration (elapsed days)	**Effort (person days)**	**Cost**	**Quality**
Source of base information	Source of base information	Source of base information	Source of base information
Influences	Influences	Influences	Influences
1	1	1	1
2	2	2	2
3	3	3	3
Assumptions	Assumptions	Assumptions	Assumptions
1	1	1	1
2	2	2	2
3	3	3	3
Uncertainty factors	Uncertainty factors	Uncertainty factors	Uncertainty factors
1 X%	1 X%	1 X%	1 X%
2 X%	2 X%	2 X%	2 X%
3 X%	3 X%	3 X%	3 X%
Estimate range	Estimate range	Estimate range	Estimate range
Worst case	Worst case	Worst case	Worst case
Best case	Best case	Best case	Best case
Most likely	Most likely	Most likely	Most likely
Three-point estimate	**Three-point estimate**	**Three-point estimate**	**Three-point estimate**
Date		**Prepared by**	
Version			

- Understanding the way that staff costs are calculated is always an area that needs exploration and documentation, as their treatment can vary wildly between organisations. There are many variants of allocating staff costs to projects in increasing order of cost. They can include variations on:

  - The time of existing staff is not charged against the project, but the employment *costs of staff engaged solely for the project* are charged to it.
  - The gross direct employment *costs of any employee assigned full time to the project* are charged as well as those engaged solely for it.
  - The gross direct employment costs of any employee assigned to the project full time *plus their apportioned share of costs of corporate services including accommodation* are charged (sometimes called fully loaded costs).
  - *Recovery of all fully loaded staff and management*[15] *costs* spent on the project are charged on an hourly basis, frequently based on a rate table.

  Remember, whatever basis we use for calculations of staff costs, we also need to consider the basis for evaluating the staff time-related benefits that it will deliver.

People do not always tell the whole truth, or sometimes any of it. This is understandable if our research involves talking to suppliers. Suppliers, in the hope of influencing our eventual procurement choices, may deliberately indicate lower potential costs or shorter timescales or higher quality than we will ever achieve in reality. Internal staff and stakeholders may have other covert motivations. A real-life lesson in hidden staff agendas came during a £1 million public sector project to radically change and upgrade services to clients. The major part of the costs would be in alterations in a series of buildings in different locations across England, and the project ran an extensive engagement with professional staff in these locations in order to frame the project. It later transpired that these staff were suspected of having colluded in pushing a hidden agenda of resisting the changes in services. On reading the team's project framing paper, the chief executive physically ripped it in half in front of both them and the business sponsor, saying, "They have lied to you. Go away and do it again." The resulting work took account of the unspoken staff resistance and showed a very different approach to the project, which was then accepted and implemented.[16]

Step 4: Complete the Project Frame and Obtain Authority to Proceed

The temptation is to prepare a large document as the project frame, and some organisations may demand it. However, it is a useful exercise to summarise the whole project frame on a single page (although admittedly it will probably usually be an A3 sheet or equivalent) and use this as a way of communicating the shape of the project with stakeholders and other interested parties. This "project frame on a page" should show how all the information about the project fits together as a coherent and considered whole view of what we are seeking to achieve. Table 2.3 sets out the information that a fully developed project frame could contain.[17]

An underlying set of similar pages, probably one per stage plus our explanatory notes, may also be required to record and share more detail. This could mean that the project frame document we present for approval will be longer than the "project frame on a page." It should still be a tightly written and succinct argument to support investing the time required in preparing the business case and the project plan in the preparation stage.

Table 2.3 Project Frame on a Page

Project Frame on a Page			
Project	**Prepared by**	**Date**	
WHAT are we going to deliver?			
Senior management vision		**Project high-level objectives**	
Outcomes and end state description		1	
		2	
		3	
HOW will the organisation know that we have done well? (critical success factors)			
1		3	
2		4	
WHAT might stop us?			
Constraints	Dependencies	Risks	
WHEN are we delivering? (stage dates)			
1 Inception		3 Implementation	
2 Preparation		4 Closure	
WHO is involved in each delivery stage? (project team, suppliers, stakeholders)			
1 Inception		3 Implementation	
2 Preparation		4 Closure	
WHAT are the outputs/deliverables by stage?			
1 Inception		3 Implementation	
2 Preparation		4 Closure	
HOW/WHERE is each stage delivered?			
1 Inception		3 Implementation	
2 Preparation		4 Closure	
HOW MUCH will it cost us by stage?			
Stage	**Capital**	**Revenue**	**Total**
1 Inception			
2 Preparation			
3 Implementation			
4 Closure			
5 Continuing costs			
WHY are we doing this? (benefits)			
Stage	**Cashable**	**Non-cashable**	**Total**
1 Inception			
2 Preparation			
3 Implementation			
4 Closure			
5 Continuing Costs			
Other project benefits			
1			
2			
3			

In many organisations it is unlikely that we will be allowed to further commit time and resources to the project in the next stage without submitting a formal request for approval. When the project has been framed, we often have to also prepare and submit a separate request for authority to proceed to the next stage. This request for approval is likely to draw very heavily on the project frame. The request for approval can be called many things in different organisations and methodologies – for example, project charter, project mandate, project feasibility, outline business case, or even just authority to proceed. This is a clear example of when and how we need to mesh our project activities and outputs with the requirements of the client organisation, and it is, of course, often very useful to base our submission for authority to proceed on a previous one, if it exists.[18]

The question of who gives authority to proceed is one that will vary among organisations and may change depending on the size of the next step. In some organisations, it may be the business sponsor who can commit to the next stage of a sizable project; in others, it may be that the full board of directors wishes to consider even very small projects. In the public sector, a whole range of different stakeholders and sponsors may have an influence. It may also be that it is not just the size of the project that is the deciding factor but its visibility to stakeholders and the wider public.

Our project frame (and if one is needed, an authority to proceed document) is the first example of how work in one stage of the project is going to be built upon the work that has gone before. In the next stage, preparation, these documents will become the basis for the full-scale business case and project plan. As the inception stage will end with the acceptance of these documents, we will need to follow the end stage processes set out in Chapter 13, including creating an end of stage report. We will also need to submit a plan for the preparation of the business case and project plan – essentially our next stage plan.

A Final Word on Deadlines and Managing Agreement

So now we have possibly prepared four documents:

- Our project frame.
- Any required request for authority to proceed.
- A report for the end of the inception stage.
- A plan for the preparation stage.

Obviously, we can now relax, as all we have to do is email them to our business sponsor and the project board and then we can sit back and wait for them to tick the box that says GO. Well, they might approve it, they might just metaphorically (or physically) rip it up in front of us and send us back to do it again, or it might just sit on their desk fighting for their attention with everything else they are doing. We can never fully avoid the risk that we just have not gotten decision makers' attention at the point where we need to gain approval. But there are actions we can take in advance to reduce the possibility of a refusal derailing us completely or a lack of priority delaying the project.

No surprises rule. Engage the decision maker(s) before finalising the submission. This could mean a series of interviews with the authorising parties in which we discuss the emerging results and gain their views on the conclusions we are reaching. This enables us to deal with any issues they would raise in advance of our submission of the required documents. A meeting to discuss the "project frame on a page" provides exactly such an opportunity.

Value their time. Understand the impact of timing when asking the decision makers for approval. If the decision has to go to a committee or board, make sure to know when their deadlines are and submit in plenty of time to allow the members to read our paper. We should never, never, never expect to get approval if we submit an unseen, undiscussed 150-page document for approval on the morning of the board meeting.

Pick our moment. We need to try and plan so that we are not submitting complex papers for approval at a busy time for the board or the business sponsor. If the September board meeting always has a full agenda, we should try and plan to avoid it or work even harder with individual members of the board in advance so that our paper can be approved quickly with little need for discussion.[19] We should find out when the business sponsor is going on leave and ask when they would like our submission. Equally, if we know that in October all the managers and directors are busy because there are staff appraisal interviews to conduct or next year's budgets to fix, we should try to avoid asking them for meetings or submitting lengthy documents for them to read during that time. We cannot always achieve this because the dynamic of the project may dictate otherwise, but it does mean that we once again have to work harder at respecting the ways our decision makers want to work.

Reflections

Using the same project as in our previous reflections, if possible, spend some time considering the following, and if working up the bio-energy from effluent project, be creative.

- What could a work breakdown structure look like, and do all components need breaking down to the same level of detail?
- What is the most appropriate way to prepare initial estimates, and does the same method apply to all elements of the project?
- Having prepared estimates by whatever means, go on to prepare three-point estimates to reduce the level of uncertainty and then consider how much they do or do not reduce the level of uncertainty.
- What knowledge do you need access to in order draft a frame for the project, and from where can it be obtained? This should include potential data sources, outside help needed, and support from subject matter experts.
- What are the three major things the project has to avoid doing?
- Similarly, what are the three major things the project needs to deliver in order to be seen as a success? Try to quantify these in some way. Perhaps being the market leader in bio-energy from effluent would mean that each year we win 50% of the contracts advertised to supply or operate treatment works in a particular market.
- What information is still needed to complete the project frame, and from where could it be sourced?
- Now, be highly critical and ask, if you were the business sponsor or project board and were presented with the information from the project frame in a suitable format, would you authorise the project to commit the required resources to develop a full-scale business case and project plan?

Remember, the object of these reflections is for us to think through the project manager's tasks and experiment with the results in a safe space. Don't worry too much about getting the "right" answer.

Notes

1 I once worked with a very small, incredibly efficient, formidable, and frankly feared person who ran the manual filing system for a municipal engineer's department. This system ran to several tonnes of paper. Everything was held in a central store, and files were collected from all offices every Friday and updated. She was one of the people whom staff needed informally on their side or else they might find that the files or drawings they required for their projects were not readily available – ever.
2 Spoiler alert: this is coming up in a few pages' time.
3 In reality, they are composites of many people we might encounter when delivering different projects, and they represent key individuals in the informal organisation.
4 There really is no adequate visual replacement for a whiteboard in a project office for logging and sharing issues of the moment. A bit tricky when our team is split over three continents, but it still makes a great backdrop for a video call.
5 But it did take two rounds of toast that day. Last I heard, over a million major transactions had used the solution discovered that morning.
6 Or to put it another way, spending the day riding round the English countryside on a refuse collection lorry. All part of life's rich tapestry when you are in the research stage of a project.
7 For our American friends, these are, I believe, known as sandwich cookies. Of course, the choice of tasty bribe is up to you, but remember that sugary jam doughnuts carry a serious risk of you losing all personal professional credibility when eating one in public.
8 There are no stupid questions – the only stupidity is in not asking them.
9 By the way, if you check back, you will find that Pharaoh's overseer in our first virtual site visit gave himself 100% contingency on time and was actually uncommitted on budgets. Good luck to any project manager trying to get away with either of these now, let alone both.
10 This is an unavoidable use of initials. I had to start using them sometime. I am very sorry, and I will try to keep avoiding them wherever possible.
11 Experience cynically dictates that increasing the size of anything results in a greater than proportionate increase in cost or effort, and similarly decreasing its size results in a less than proportionate reduction in cost or effort.
12 It was once alleged to me by a rail industry colleague that there was a knock-on effect on the operation of the cross-Channel Eurostar trains stemming from estimating the capacity of their foul water tanks based on French experience. This meant that without the additional capacity required to deal with the beer drinkers of the UK and Belgium, the trains' tanks had to be emptied daily instead of every three days. I don't know if this is actually true, but it's still a good story to illustrate knock-on effects.
13 If you really need a formula to feel comfortable, our three-point estimate (TPE) becomes:
TPE = ((Most Likely Case $\star$ 4) + (Worst Case + Best Case))/6.
14 Sometimes in the maelstrom of project delivery activities, it is difficult enough to recall what happened the previous Tuesday, let alone how we calculated both the total area needed and the cost per square foot of wall tiling late one night in August two years ago when we were trying to meet a reporting deadline.
15 And whilst we can control and manage most staff effort, we never know how much of their own time senior (i.e., very expensive) managers will dump onto our project code.
16 Accepted and implemented are not always the same thing.
17 Before scratching your head too much about cashable and non-cashable benefits mentioned in Table 2.3, take a sneak preview at the explanation in Chapter 3.
18 It is obviously useful to check that the organisation was happy with the previous example that we intend to use as a model and not to use one that was rejected!
19 Spoiler alert: Chapter 7 includes a virtual site visit which demonstrates what happened when individual project board members were not adequately prepared.

Bibliography

Association for Project Management, 2015. *Competency Framework*, Princes Risborough: Association for Project Management.

Chadee, A., Hernandez, S.R., and Martin, H., August 2021. The Influence of Optimism Bias on Time and Cost on Construction Projects. *Emerging Science Journal*, Vol. 5, No. 4.

Meyer, W.G., 2016. *Estimating: The Science of Uncertainty. PMI® Global Congress 2016*, Newtown Square, PA: Project Management Institute.

UK Infrastructure and Projects Authority, 2020. *Principles for Project Success*, London: UK Infrastructure and Projects Authority.

3   Preparing for Launch Part 1

Compiling the Four Document Set

Aim of this chapter: To show how we bring the project to the point where information gathering is complete and conclusions agreed, ready for the creation of the project plan.

Learning outcome: To be able to prepare the information to support a business case, including a high-level plan for the delivery of the project.

In the preparation stage, we have to commit significant time and resources to refine our picture of the solution, develop a business case to justify the investment in delivering the project, and prepare the full-scale project plans to show how the required outputs and solutions can be achieved. It is also the stage which in many projects can seem to progress very slowly, but during it we will be building our foundations for a successful project delivery. This chapter is concerned with the first steps in our preparation stage, and the following two chapters cover preparing the business case and the detailed project plan. These form the foundations of our project, and failure to spend sufficient time completing them now may be a cause of regret later.[1]

Before we go further, it's time to think about yet another emerging problem that faces you as the project manager for the bio-energy from effluent project.

The news is getting about that you have been given the role of project manager, and now you have received the following email from the head of marketing:

Re: Diversion from Marketing Strategy

Dated 13 September

I have just received a copy of the notes from the last board briefing session held two weeks ago, and I was disappointed to learn that a global marketing initiative is about be launched under your direction – I had not been previously informed of this. You should be aware that there is a Three-Year Global Marketing Strategy in place that was agreed to by the board last June and to which my staff are fully committed. I would be grateful if, without further delay, you could brief my senior staff and myself on this marketing initiative including:

- *How many of my staff resources you intend to divert from the global strategy work?*
- *When you will be demanding our support?*
- *What other expenditure will you be incurring that I will have to fund, not necessarily now but in the future? I have to be aware of this, in order to prepare a five-year forwards projection for the Global Marketing Spend.*

DOI: 10.4324/9781003405344-4

Generally, how do you intend to ensure that you integrate your demands with our existing heavy workload over the life of your project?
Regards,
Karen

Ouch. Obviously, she was missed when thinking through who needed to be talked to about the project. Currently her concerns cannot be answered as there is only a very broad outline of the project's vision, and approval for starting the detailed project design work has only just been received.

What are the different approaches that could be adopted to get her on your side?

One end of the continuum of possible approaches is to explain the current situation of the project to her and apologise for not previously contacting her directly but say that you thought she was being briefed by her board director as it is a main board initiative. Perhaps also invite her to a one-on-one discussion and, after checking her preferences with her secretary or assistant, also invite doughnuts, chocolate biscuits, or cake to attend!

At the other extreme is just simply ducking and passing the blame upwards.

This is your choice, and it will depend upon your personal management style and your assessment of relationships within the organisation.

PREPARATION STAGE STEP BY STEP

Typical high-level activities included in the preparation stage are set out in Figure 3.1.

In Chapters 4 and 5, we will look in more detail at compiling the business case and preparing the project plan. In this chapter, we consider how we need to build upon the detail in our agreed-upon project frame until we can state definitively and clearly:

What it is we are setting out to achieve.
How things will look when we have delivered the project.
How we should reach the desired end state by analysing our different options.
Why one option is preferred.
What will be delivered.
How Much it will cost.
Who will be engaged in the delivery process.
When we will deliver it.

These questions underly the four key areas that we now have to document in detail and justify during the selection of the preferred solution. These four areas can be documented as:

- End state design.
- Options analysis.
- Project delivery method.
- High-level plan.

Taken together, the thinking that is required to prepare this set of four documents will do most of the work of preparing the business case. The advantage to the project is that all four documents in the set can be discussed and agreed on in advance of our business case itself, which will then contain no surprises for the decision makers and so should have a smoother run through the approval process.

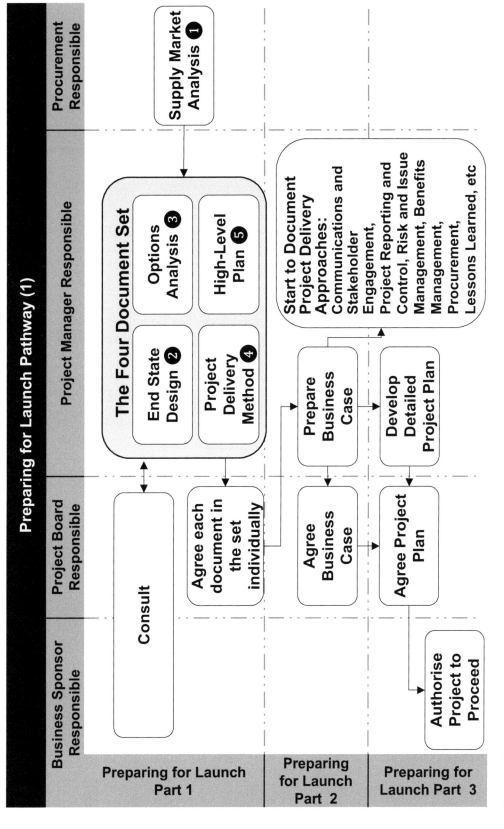

Figure 3.1 Preparing for Launch

Whether we choose to include the research, thinking, and preparing of these four documents as part of the business case preparation task or as separate tasks in their own right is largely academic. On balance, however, it can be useful to separate their preparation into different tasks, each with defined outputs on which we can focus and, where necessary, to which we can assign specialist resources to help us. When complete, we can submit each separately to the project board for approval.

Step 1: Supply Market Analysis and Supplier Engagement

As soon as we start to understand the shape of the possible solutions, we might choose to deliver our project, we need to start considering how it is going to be delivered and what we may need to purchase from the market.

We may have started our supplier analysis when preparing the project frame, but at this stage, we may require more intensive research into the possibilities available in the market, including solutions that may have been implemented by similar organisations. Our objective in continuing this early supplier market engagement is to:

- Explain our desired outcomes from the project.
- Expand our understanding of the risks and issues we may face.
- Allow the suppliers to suggest how our outcomes might be achieved.
- Get supplier input on required timescales to implement a solution.
- Get suppliers to indicate the range of budgets we might need to fund.

This is essential research and can have a major influence on the four document set that we are preparing. It can also help us if we engage now with the client organisation's procurement experts to support this process.

Even if we do not end up selecting an option requiring engagement with the market, there is a very strong rationale for this research step. This is that there are potential limitations to the success of the project if we are working only from an internally developed understanding of the client's critical success factors and requirements, in terms of the currency of knowledge currently available to the client organisation. This is because the client organisation and its staff may not be aware of the latest developments in solutions available and the range of the newest underlying technologies and service offers that are currently being deployed or in development. For example:

- External provision of services currently provided in-house.
- Suppliers' abilities to engage short-term specialists to help to deliver the project.
- Range of supplier partnership models currently available.[2]
- Current solutions, and those planned for imminent launch by suppliers, may offer much wider functionality than those currently in use – for example, information systems change projects, where the newer system that we are seeking may offer a much broader range of functions and options than the system being replaced.
- In construction, a lack of knowledge of newer construction materials that offer radically different building possibilities such as improved insulation, better drainage, or improved handling and installation.

Early engagement with suppliers can therefore have major benefits for ourselves, both directly in increasing our understanding of how the current market can help us deliver the project and for potential suppliers who gain a longer period to consider how they might respond to our needs when any tender invitation is forthcoming. Consequently, in many projects we will need to consider how to work with our procurement experts in a carefully controlled and managed way, in order to engage in a two-way early exchange of information with potential suppliers.

When carrying out any market analysis, it is important to be transparent with suppliers to ensure that no supplier can claim to have been advantaged or disadvantaged in any future procurement. We should always make clear to them that:

- The client's intention to undertake a procurement does not commit the client to actually do so or to do so within any particular time frame.
- Any information given by the client is for the supplier's background information in relation to the supplier market analysis only, and any procurement process will give all suppliers wishing to tender equal information.
- Any statements made by the supplier, including pricing information, will be similarly treated as being background information only and will not commit the supplier in case of any future procurement.

Undertaking a supply market analysis can include:

- Preparing an analysis of published information from suppliers in the market.
- Reviewing reports of suppliers in the market.
- Holding open supplier information days.

Two activities, however, should only be undertaken with great caution. Firstly, meeting selected suppliers individually at this stage can lead to allegations of unfairness in the procurement process. Secondly, individual solution demonstrations can be useful in theory, but if they are attended by anyone whom we may want engaged in the subsequent technical evaluation process, then such demonstrations risk leading to unconscious bias (positive or negative) in our evaluation. In order to mitigate this, it is best if every supplier engaging in the market analysis is asked to deliver similar demonstrations to the same group of attendees.

Step 2: End State Design

The end state design, sometimes called a blueprint, is our view of what the future may look like once the project has been successfully delivered. To achieve this future view, we need to take the client organisation's vision for the project's outcomes from the project frame and extend it into a working model, including detailing all the processes, practices, information, and technology required once the project has been delivered.

In real life the required deliverables from our project are not often black or white, on or off, single entities with single impacts. Instead, our solution, and hence our end state design, will have to be based on multiple and often subtle choices about what we are going to deliver and how it will integrate with the remainder of the host organisation's operations once it has been implemented. During our research the details of the organisation's requirements for the solution that we collect will often involve trade-offs and compromises that have to be reconciled. For example, "It would be great to have it in black" and "We have to ensure that we do not overheat in direct sunlight" could be contradictory requirements. We usually work with stakeholders to flesh out not only their requirements but also how these may be prioritised and use the results to shape the end solution that we will be delivering. This process is often best achieved through a facilitated discussion during which we can build consensus around the choices being made.[3]

A widely adopted tool to help to make the What choice is MoSCoW prioritisation. Put very simply, we list all the required features of the solution put forwards by stakeholders and other consultees and assign them to one of four simple categories:

Must have – solution is not viable without it.
Should have – solution can be delivered without it, but this might require some work arounds.
Could have – nice to have, but only if there is time and budget available.
Won't have this time around – is still desirable, but there are more important things to do first.

The end state design will necessarily be different in scope and complexity depending on the project. The more complex a project is, and the greater the level of operational change involved, the more likely an end state design is required. At one end of the spectrum would be a construction project where the resulting building or infrastructure is handed over to a third party on completion and the provider moves on unchanged to the next project. A clear design for the solution is still required, but it requires no changes to the organisation; i.e., effectively no changed end state design is required. At the opposite end of the scale, a project involving business change, new systems, processes, offices, and staff most certainly will need a comprehensive end state design that clearly sets out the solution requirements and the proposed new arrangements for management, people, processes, and technology. In most cases, one of the themes to address in the end state design is the continuing management, monitoring, and reporting of the benefits obtained from delivering the project.

If we need to prepare an end state design, a useful analysis technique often employed to set out its scope is POTI, standing for:

Processes – including business models of operation, costs, and performance levels.
Organisational structure, management, staffing, etc.
Technology buildings, IT systems, equipment.
Information and data required for operations and performance monitoring.[4]

If we are delivering different parts of our solution at significantly different points in time, we can also include these as intermediate steps in our end state design – for example, when we have delivered our sewage treatment works in the first market in year 1, across one continent in year 2, in two continents in year 3, and finally world leadership in year 4.

Step 3: Options Analysis

Bear in mind as we go through analysing our options that the eventual decision is not going to be ours but will be taken by others, often a project board or maybe a company board. Consequently, we need to document, explain, and possibly defend each step on our path to a final recommendation. Much of the output from this work naturally flows straight into the business case.

As we have been working on the project for a while, we will have received opinions from virtually everyone on the How of delivering the project. Some people will be experts in the subject, and others may hold strong opinions but have less direct knowledge or experience. Both sets of views should be taken into account and may need to be included in our range of different potential paths to glory – the first set of views because they act to increase our technical knowledge and the second because they may be indicative of human factors, such as the personal prejudices of key individuals and the organisation's informal social structure or historical thinking that will need to be addressed by the project.

Our starting point is to identify a longlist of potential ways of delivering the project's goal. At first sight this may seem simple, but as we develop the longlist, we will find that many of our options come with sub-options, and even they have sub-sub-options, and so on. This is because when looking at potential solutions, we must look as widely as possible and need to take account of opportunities for innovation in the project – for example, looking at ways of joining with or collaborating with other organisations or teams.[5] Our options analysis should also include a range of different ways of providing our required solution. These different methods of provision will vary depending on the project – for example, in one case outsourcing or even offshoring production might be feasible or in another using the option of customer online self-service instead of increasing staffing in regional offices. There may also be several different timing options available, perhaps an initial pilot operation, a staged implementation, a geographic rollout, or a mix of all three – or if we are brave, a "big bang," one-time implementation.

As an example of thinking through the different options, consider the fairly simple situation of an ageing IT accounts and payroll system and the tree of different options and sub-options that it generates, as shown in Table 3.1.

Table 3.1 Option Identification

IT Accounts System Replacement Project Option Identification			
Main Option	**Sub-Option Level 1**	**Sub-Option Level 2**	**Sub-Option Level 3**
Do nothing (this is usually an option).			
Buy time on an existing accounting system.	One from elsewhere in the company/similar public body.	Potential source 1	
		Potential source 2	
Buy new hardware.	Install on own premises.	Continue with existing system on new equipment.	
		Develop in-house system.	
		Implement a commercial package solution.	Package 1
			Package 2
			Package 3
	Install in commercial data centre. Which one? There are many including onshore and offshore with different service offers. A separate sub-option level 1 is for required for each data centre service provider/service option available.	Continue with existing system on new equipment.	
		Develop in-house system.	
		Implement a commercial package solution.	Package 1
			Package 2
			Package 3

IT Accounts System Replacement Project Option Identification			
Main Option	**Sub-Option Level 1**	**Sub-Option Level 2**	**Sub-Option Level 3**
Buy capacity from a commercial "cloud" service provider.	Which provider? There are many including onshore and offshore. A separate sub-option level 1 is for required for each cloud service provider/ service option available.	Continue with existing system on new equipment.	
		Develop in-house system.	
		Implement a commercial package solution.	Package 1
			Package 2
			Package 3

(For display simplicity this table does not show the sub-option level 1 entries required for each data centre or cloud service provider being considered.)

Let's Play Simon Says

We are now faced with choosing our preferred approaches and supporting the host organisation through to making its final decision on the solution. There is a huge body of academic work on decision making. According on the leading academic in decision making theory, Herbert Simon:[6]

> Now the salient characteristic of the decision tools employed in management science is that they have to be capable of actually making or recommending decisions, taking as their inputs the kinds of empirical data that are available in the real world, and performing only such computations as can reasonably be performed by existing desk calculators or, a little later, electronic computers. . . . Decision makers can satisfice either by finding optimum solutions for a simplified world, or by finding satisfactory solutions for a more realistic world.
>
> (Simon, 8 December 1978)

The point that he is making is that when deciding between options in the real world (in contrast to the simplified model world of economists like Milton Friedman), we are looking not for the optimum solution but for the first one that is likely to do the job given the information that we have and our ability to analyse it. Simon explains this in reference to needing a sharp needle to sew a seam. We must find one needle from many needles in a haystack. Before we mend the seam, he asks, is it most sensible to spend our time finding every last one of the needles, then comparing them all and selecting the sharpest, or do we check each needle as we find it to see if it is sharp enough, and as soon as we find one that is sharp enough, mend the seam? Simon calls finding the first solution that will deliver our needs "satisficing."

The Long and the Short of It

Moving on from Simon's "satisficing," we can use comparative options analysis as a tool to help us make the choices necessary in analysing our options. This tool enables us to extend beyond Simon's views to seek to make a choice not of the first needle that "satisfices" but of the one that we think is most likely to succeed from a broadly representative range. The process that we usually follow is to start with a list of all possibilities that we can identify (the longlist) and then to filter these down to those which have a likelihood of succeeding (the shortlist). We are trying to find

Simon's satisfactory solutions in our real world. If you are a golfing or other sports fan, you will recognise this weeding-out process as being exactly what happens in tournaments where golfers have to achieve a high enough position in the early rounds to "make the cut," which is when the number of players is reduced for later rounds. This is similar to football competitions where teams have to qualify in the group stages before getting to the knockout part of the tournament.

During project framing and subsequently, our research will have revealed a number of different options for delivery, and this can be supplemented by further workshops or interviews when we prepare our longlist of possible solutions. Our next task is to weed this longlist out and get down to a number of options that is more manageable – already we are not searching for the perfect solution, i.e., Simon's sharpest needle, but seeking to establish a manageable number of potentially viable options for which we can marshal data, evaluate information, and draw comparative conclusions.

Some of our longlist of options, although they may be technically feasible, may not be acceptable to the organisation. In our IT example prior, if the new system is for a public body, they may have a policy of supporting employment in country or locally, thus ruling out any options of providing hosting and support "offshore" or from another part of the country. A different organisation may have its own large data centres and follow a policy of in-house hosting of all its IT systems, or another might have moved completely to the cloud and only buy processing capacity on a revenue basis.

We should already have developed our list of critical success factors, and during our end state design work, these may have been subsequently enhanced and prioritised. These factors now provide the lens through which we examine the longlist to identify those options worthy of further analysis. First, we need to:

- Identify which options may fully meet the critical success factors.
- Park any decisions on options which do not satisfy all factors completely.
- Reject all of those failing to meet any one of the critical success factors.

This shortlisting approach maps onto a simple very bold visual presentation which can be used to both rapidly identify the judgements made on each option and communicate it simply to others. To illustrate how this process works, and using the options in the previous IT example shown in Table 3.1, the different options are reviewed against the following short series of (unprioritised) critical success factors as follows:

- High availability of the system, measured as 99.99% during normal office hours and 95% at other times.
- Short timescale to implement the chosen solution, with a maximum period of six months for an initial core system to be operational.
- Preference to support the UK economy, with a UK employed team.
- Limited requirement for software development beyond customisation of reports and interfaces with other internal systems.

At this stage a wider level of tolerance can be accepted because we are not attempting to select a preferred option but merely to weed out those that are less likely to succeed. In Table 3.2, the visual representation of our findings uses white cells for critical success factors not met, grey cells for parked queries, and black cells for those fully met. In the example, only two options fully meet the critical success factors, and six meet all but one, which has a parked query.

Table 3.2 Options Shortlisting Analysis

IT Accounts System Replacement Project Shortlist Analysis							
Options				Critical Success Factors			
Main Option	Sub-Option Level 1	Sub-Option Level 2	Sub-Option Level 3	High Availability	Short Timescale	UK Employed team	Limited IT Development
1 Do nothing.					■	■	■
2 Share/buy time on an existing accounting system.	1 Elsewhere in company or similar public body.	1 Potential source 1			■	■	■
		2 Potential source 2				■	■
3 Buy new hardware.	1 Install on own premises.	1 Continue with existing system on new equipment.		■	■		■
		2 Develop in-house system.			■		
		3 Implement commercial package solution.	Package 1	■	■	■	■
			Package 2			■	
			Package 3			■	■
	1 Install in commercial data centre. Which one? There are many, including onshore and offshore, and each may need a separate sub-option.	1 Continue with existing system on new equipment.		■	■	■	■
		2 Develop in-house system.				■	
		3 Implement commercial package solution.	Package 1	■	■	■	■
			Package 2	■		■	■
			Package 3	■			■
4 Buy capacity from a commercial "cloud" service provider.	1 Which provider and what sort of service? There are many, including onshore and offshore. Each needs a different sub-option.	1 Continue with existing system on new equipment			■	■	■
		2 Develop in-house system.				■	
		3 Implement commercial package solution.	Package 1		■	■	■
			Package 2	■		■	■
			Package 3	■	■		

When all of the identified options have been examined, if the shortlist does not contain sufficient options (normally five or six is a comfortable maximum), then we have to go back and review the parked options to consider if one or more can be carried forwards to the shortlist. In our example, the most appropriate next step would probably be to resolve the parked question about availability by undertaking further research. Assuming that the information is positive, we could proceed with a shortlist of four viable options.

If the shortlist still contains too many options to be manageable, it may be necessary to "bring forwards" some of the techniques discussed in evaluating shortlisted options in order to distinguish between them.

Evaluating the Shortlist

Our next step is to develop each of the accepted items on our shortlist into a much fuller picture including both the delivery process and the project outcomes. This fuller picture will need to include how well each option might meet our critical success factors, together with perceived costs, benefits, risks, timetable, phasing, and resources required to deliver. Our objective is to identify which of the range of shortlisted options is most likely to deliver the project's objectives. This approach is actually a step up on Simon's approach of just finding the first solution that will meet our requirements.

Whereas when reducing our longlist of options to our manageable shortlist we can be moderately broad brush in our approach, during the comparative options analysis of the shortlist, we need to be more targeted in our appraisal techniques and more precise in our data. The key to being more targeted is developing assessment criteria to a much lower level of detail – i.e., increasing their granularity and wherever possible including objective numerical assessments. Later in the project this increased granularity will be incredibly useful as it will support to the linking of our "golden thread." As set out previously, this "golden thread" should run from the initial project frame through options analysis, planning, procurement, and contracting into testing, delivery acceptance, and handover.

We have already engaged with one world-class economist, who helped us understand that we are looking for a reasonable solution that will deliver, rather than wasting time seeking perfection. Now, before considering some of the tools to help us choose which option we prefer, we have to call on a 14th-century monk for guidance. William of Ockham came up with a law of parsimony, usually called Occam's razor. Basically, he believed that the simplest explanation is usually the correct one. This "razor" has been interpreted as meaning that the solution that makes the fewest assumptions is the most likely to be the right one. In our case, it is a timely prompt for us to remember that when choosing between options, for each one we are making assumptions about an unknown future. The more assumptions we make about an option, the more opportunities we create for introducing uncertainty, risk, and error. Also, the more assumptions we make, the more that flexing one or more of them, sometimes even by very small amounts, can change the results of our selection process.

A real-life example of flexing assumptions occurred when choosing between companies to implement an IT system that would radically change the way a government agency would operate internally and interface with its customers. A comparative options analysis exercise was used to choose between alternative solution providers, and after a long day in a joint team and user meeting, there was a clear winner. The business sponsor arrived at the end of the day to review the conclusions reached. He decided that the business risk attached to the winner (a small overseas-based company) needed to be increased – i.e., the project needed to flex an assumption about the future performance of this potential supplier. The resulting changed scoring meant that the previously preferred solution was now in third place.

Options Analysis Techniques

There are a number of useful techniques for helping evaluate the different options on our shortlist and trying to identify a priority order for them. In theory, we could also use the same techniques to help to reduce our longlist, but if we have been precise enough in our definition of critical success factors, this is usually not necessary.

A useful tool for helping to prioritise our options is SWOT analysis.[7] This stands for the fairly self-explanatory assessment of the strengths, weaknesses, opportunities, and threats of each option. A SWOT analysis is often represented in a two-by-two matrix, as shown in Table 3.3.

Table 3.3 Example Option SWOT Analysis

IT Accounts System Replacement Analysis SWOT Analysis		
Option	2.1.1 Retain existing system, but implement on new hardware in a commercial data centre.	
	Positive	**Negative**
Internal	**Strengths** Features of the project that give it an advantage over other projects.	**Weaknesses** Features of the project that would place it at a disadvantage relative to others projects.
	1 Minimise staff training needs.	1 Lack of functionality of more modern systems.
	2 All data can easily be ported across to the new installation – no migration issues.	2 Fails to integrate with other software in use.
	3 Can be implemented at any time during the year.	3 Reporting is poor, and an extraction facility to a user reporting tool is required.
	4 Can parallel run the old and new systems.	
	5 No need to update existing computer room facilities.	
External	**Opportunities** Positive elements in the external environment of which the project could turn to its advantage.	**Threats** Negative elements in the external environment which could adversely impact the project.
	1 Opportunity to test current systems in a commercial data centre.	1 May not get immediate service in the event of system issues.
	2 Data centre could provide shared 24/7 cover at a rate below the cost of using dedicated internal staff.	2 Software supplier has said that the existing package will be unsupported in three years' time.
		3 A marginal increase in hardware requirements discovered as the project progresses could lead to a step change in costs.

Whilst there may be some objective data in the SWOT analysis, often it will include softer, more intangible information. Consequently, if we can include a group drawn from stakeholders, the project team, and the project board in working through the analysis, it is more likely our conclusions will gain acceptance.

Competitive analysis. The options for delivering some projects, especially those in the commercial sector concerned with transformational change or bringing new or improved products to market, will require an extra dimension in order to evaluate their potential for success. This is called competitive analysis. A full consideration of the different methods that can be used for appraisal of the potential commercial success of a new product, service, or other transformational change is outside the scope of this book.

There is, however, one commonly encountered model, Porter's five forces, that illustrates the types of additional issues that could be addressed when evaluating the different options for delivering such commercial projects (Porter, May 1979). Porter's views have been subject to academic criticism, but they do provide any project manager of these more commercial types of projects with a series of practical questions against which all potential options may be reviewed. These questions are:

- What is the current state of the potential competitors and their offers in the markets where we will be competing?
- How much bargaining power do our suppliers have? In an extreme case, there may be only one supplier for something that we require and they can increase prices, change specifications, or withhold supply at will. Conversely, if we are buying commoditised products or services with many suppliers, then each supplier should have little bargaining power. It might be, however, that whilst there are many potential suppliers, the cost and time required to change to a different one are unacceptable.
- How much bargaining power do our customers have? If our target market has a single or few customers for our proposed good or service, then we would be in a weak position, especially if they can switch suppliers easily. As the numbers of customers increases, their individual "customer power" diminishes.
- How easy is it for customers to find substitutes for our proposed product or service? If the market provides for easy switching to alternatives, then our relative power diminishes. If, however, our proposal will deliver new solutions, improved benefits, or lower costs for the buyer, then substitution may be more difficult.
- How easy will it be for competitors to enter our chosen market? Porter calls this force the threat of new entrants, but we could equally apply it to an existing player improving their offer to compete more effectively with us. If we are providing a new product, we may have a "first mover" advantage, and offering an improved one may also gain us an advantage in the market for a time. The period for which such advantages may be enjoyed, however, will depend on the time, effort, and resources that a competitor has to commit in order to offer a new or improved product.

These questions can either help supplement a SWOT analysis or be prepared as a standalone analysis. For those projects where it is applicable, competitive analysis is, of course, not only useful whilst preparing our shortlist but, with increased depth of research and analysis, can inform the final selection of the preferred option.

Beyond SWOT

Whilst SWOT analysis is a useful tool which may enable us to rapidly identify some options which will not deliver, it is often not sufficient alone for preparing our shortlist of options. Additionally, we need an approach which at least:

- Records data and assessments of different options on a directly comparable basis.
- Moves on from subjective text in a SWOT analysis to a numerically based evaluation, even if some of the numbers used are still subjective.
- Allows us to differentiate between evaluation factors based on their level of importance to the success of the project.

Consequently, wherever possible, it is useful if we can now start to decompose each critical success factor into a number of lower-level factors. We then need to identify a gradation scale against which we can form a rational opinion of how completely each of these lower-level factors is being met by each option. Taking the previous IT accounts system replacement project as an example, one of the critical success factors is "limited IT development," which could be expanded to include factors such as:

- Legally and tax compliant core accounting and payroll functions available immediately for each jurisdiction in which we operate.
- Prebuilt web interface able to be accessed securely from the organisation's intranet.
- Standard interfaces to our other systems, including a single sign-on process.
- Ability to use standardised printing process for reports but retain secure printing where required – for example, for cheques where needed and for confidential information.
- Banking system interface.
- Human resources system interface.
- Prebuilt report generator, including spreadsheet interface.

In addition to our expanded critical success factors, there are a number of other areas that we usually need to consider in detail whilst appraising options for every project:

- Risk.
- Non-cashable benefits.
- Financial impacts, including cashable benefits.

Risk

It is never too early in a project to start thinking about the risks associated with it, and we will be returning to risk in greater depth later. Risks will occur at many levels, from strategic risks to the organisation if the project fails to the risk of individual options, which may vary. The latter need to be taken into account during our options analysis. For example, if we chose a procurement route, suppliers may find our contract unattractive and either not bid or bid much higher prices than anticipated. Or, we could choose a wrong supplier. Or, during a long project delivery, technology may overtake us, resulting in extra costs. An example of this was in a long-running project requiring citywide mobile communications to ticket devices (which we will be visiting later) where not only did the communications technology move from 3G to 4G (and 5G was planned for rollout during the operational life of the solution) but also bank card verification technology moved from chip and pin to contactless, opening up new service design options.

Consequently, for each option it is advisable to identify any particular risks that are associated with it and to always include these as narrative descriptions as either weaknesses or threats in the SWOT. However, we ultimately require more than just that narrative description; we need to be able to differentiate between the options based on their total perceived level of risk. The approach used in risk management for comparing risks has a fairly straightforward foundation, and we can adopt the first steps of it here to build an evaluation matrix to help us compare our options.

For the first three stages, it may be useful to run a facilitated working session, as discussed in Appendix 1, to prepare a simple risk evaluation as follows:

- List the risks against each option.
- Consider the *impact* on the solution if that risk becomes a reality and put into one of three categories: high, medium, or low.
- Assess whether the *probability* of the risk occurring **in that option** is high, medium, or low.
- Assign a score of 1 to a low, 2 to a medium, 3 to a high.
- Multiply these probability and impact scores together.
- Total the results for each option.
- Rank the options.

In thinking about the impact of each risk, it is helpful to consider specific areas such as the potential costs of dealing with each risk, which are either:

- Costs of prevention – how much would countermeasures cost to avoid the risk occurring?
- Costs of treatment – how much extra cost would the project experience if the risk actually occurs?
- Impact on the project of delays, including lost or delayed benefits.

Table 3.4 shows how risks may be compared between different options, in this case drawing on our example of an accounting system replacement project.

One last area to consider is the inclusion an overall "risk" allowance as a further sum above our estimates for each option that may be required to meet costs arising from unforeseen events or from underestimation. This amount is called a contingency allowance, and the amount required for each option will need to reflect the particular risk profile of that option – for example, the costs of developing a novel IT solution always should carry a higher level of risk than implementing a commercially available package, where, given a set of operating parameters, costs can be determined reasonably accurately in advance. On a procedural note, it can be useful if there is an extra discipline around these contingency funds. Instead of just being added to the budget controlled by the project manager, they can often be retained as a funding reserve and released to the project by the business sponsor or project board based on a justified request.

Non-Cashable Benefits

Non-cashable benefits, put simply, are the benefits arising from the project which cannot be directly reflected in the client organisation's financial results. Consider the impact of new working methods which save the client organisation the equivalent of 40 hours per week of staff time. If that time is the role of one person, headcount can be reduced, and we can show a cashable benefit from reducing our costs. If, however, the time savings are generated by saving 6 minutes each by 400 staff, we are unlikely to be able to translate the savings into actual cash. Non-cashable benefits are valuable attributes of each option, but not all non-cashable benefits are equally valuable, and not all can be converted into a numerical value for comparison purposes, like our 40 hours staff time saving in the previous example. In simple cases, just listing the non-cashable benefits for each

Table 3.4 Comparative Risk Analysis

IT Accounts System Replacement Project
Comparative Risk Analysis

Main Option	Sub-Option Level 1	Sub-Option Level 2	Delay			Increased Costs			Reduced Benefit			Total Risk
			P	I	P × I	P	I	P × I	P	I	P × I	
P = Probability, I = Impact, Risk Score = P × I												
Do nothing (this is usually an option).			1	1	1	3	3	9	1	1	1	11
Buy time on an existing accounting system.	Elsewhere in company or similar public body.	Which one? There could be several.	2	1	2	2	3	6	2	2	4	12
Buy new hardware.	Install on own premises.	Continue with existing system on new equipment.	2	2	4	3	2	6	2	1	2	12
		Develop in-house system.	3	3	9	3	3	9	3	3	9	27
		Implement commercial package solution.	2	2	4	1	3	3	1	3	3	10
	Install in commercial data centre.	Continue with existing system on new equipment.	1	2	2	3	2	6	2	1	2	10
	Which one? There are many, including onshore and offshore.	Develop in-house system.	1	1	1	3	3	9	1	1	1	11
		Implement commercial package solution.	1	1	1	3	3	9	1	1	1	11
Buy capacity from a commercial "cloud" service provider.	Which provider? There are many, including onshore and offshore.	Continue with existing system on new equipment.	1	1	1	3	3	9	1	1	1	11
		Develop in-house system.	3	3	9	3	3	9	3	3	9	27
		Implement commercial package solution.	1	1	1	2	2	4	1	1	1	6

option in a priority order may be sufficient, but in other cases, it is often necessary to distinguish between options based on:

- The relative importance of the non-cashable benefit to the project outcome.
- The degree to which it is anticipated that the option will achieve the benefit.

Essentially this approach splits a single subjective judgement about the "value" of the total non-cashable benefit of an option, which we should have described in our SWOT analysis, down into a series of smaller subjective evaluations. The method works by assigning a value to the importance of the benefit – its weight – and then scoring each option on its performance. The resultant two numbers are then multiplied to give a weighted score. Weighted scores can be added to give a total value for the non-cashable benefits which can then be brought into the evaluation of the option. The resulting table is called a weighted decision matrix. Make a note of it, as we will be using it in different forms later. See Appendix 1 for one method of arriving at agreed-on weightings.

A typical comparison of the evaluation of the non-cashable benefits for each option could look as shown in Table 3.5.

Table 3.5 Non-Cashable Benefits Weighted Decision Matrix

IT Accounts System Replacement Project Options Analysis – Non-Cashable Benefits							
Benefit			**Option 1**		**Option 2**		**Option 3**
	Weight (a)	**Score (b)**	**Weighted Score (a × b)**	**Score (b)**	**Weighted Score (a × b)**	**Score (b)**	**Weighted Score (a × b)**
Aggregated savings in staff time – non-cashable performance gain	40	7	280	5	200	9	360
Reduction in elapsed time of process	10	8	80	6	60	5	50
Greater reliability	20	5	100	9	180	8	160
Improved management information	15	5	75	7	105	3	45
Improved customer experience	15	5	75	3	45	2	30
Total score	100		610		590		645

Key is gaining agreement both on the weightings and the scoring, and it may be necessary to facilitate workshop sessions with key stakeholders and the project board represented in order to achieve this.

Before We Play the Financial Numbers Game

Before we start playing with the financial numbers for each option, we need to be aware that there are multiple dangers in using the usual tool to do so, the ubiquitous Excel spreadsheet. Other spreadsheet tools are available, but most of the advice in this section applies to whatever software you chose to use, although any formulae may differ. We are about to use an infinitely flexible tool to ask the organisation to invest possibly a great deal of money, and large amounts of management and staff time, in a project stretching years into the future. Consequently, it seems common sense that this requires much greater caution in using the tool than pulling out a quick analysis off of last month's sales figures or reporting the previous week's UK coronavirus cases. Here the UK government managed, on at least one occasion, to fail to carry all data forwards from one spreadsheet to the overall summary!

In addition, we will have to be able to justify, defend, and, on occasion, even recall how and why parts of the spreadsheets were put together. This means that it is essential to adopt a rigorous standard of spreadsheet hygiene. The following suggestions about the use of spreadsheets are drawn from real events and represent lessons that often came at the expense of physically and mentally scarring various project managers.

Flexibility. The infinite flexibility of the spreadsheet introduces danger. Despite how people want to push, we need to try and keep the financial story simple and clear. One example we should avoid copying was a financial analysis prepared to support the business case for a major project for a UK public sector body. This analysis started with a very useful Excel workbook containing just four spreadsheets: one sheet for capital costs, one for revenue, one for benefits, and a summary sheet. But life got more complex as the organisation wanted to compare the options relating to different potential providers (six) and different payment mechanisms (three), all of which also required extra comparative sheets and summary sheets. A new sheet was inserted so that all the standard variables were contained in one place, making the constant changes easier to manage (a great tip) rather than embedding interest or exchange rates, or labour rates, in multiple formulae in different places. By the end, the body's financial model was composed of 77 spreadsheets. The UK Treasury, who were monitoring the project, then asked for a similar analysis using a slightly different economic basis, which required everything to be duplicated. So eventually, 154 spreadsheets were required in the financial evaluation model. It became essential to document not only how each spreadsheet worked but also the interaction between them. An example of how this can be documented using a more modest set of spreadsheets from a much larger project is in shown in Figure 3.2.

Clarity. It is a good practice to separate out your spreadsheet into different parts or even, as in Figure 3.2, into separate sheets. Separating input information, variables (e.g., rates of tax or inflation), and reporting into different places means it is much easier to manage alterations when stakeholders, the project board, the business sponsor, and everyone else wants to see the information reported in a different format or with changed assumptions. In the example of doubling of the number of spreadsheets, it actually required only minimal work to change some of the underlying assumptions and create the second 77 spreadsheet model.

Moving and reusing formulae. Formulae are also a pitfall. Sometimes even the simplest move, copy and paste of a formula, or tweak so that it works better can have unforeseen ripple effects downstream in other cells when the calculations suddenly do not quite behave how we would expect. This type of error will become increasingly likely if the model is not fully documented and if changes are made without stringent change control. Even the original author may not recall clearly and exactly how a set of interacting formulae was supposed to work. The more complex our model has become, the greater the danger of this type of fault occurring and remaining undiagnosed. A review of the model by another pair of eyes is always useful to help us spot such problems.

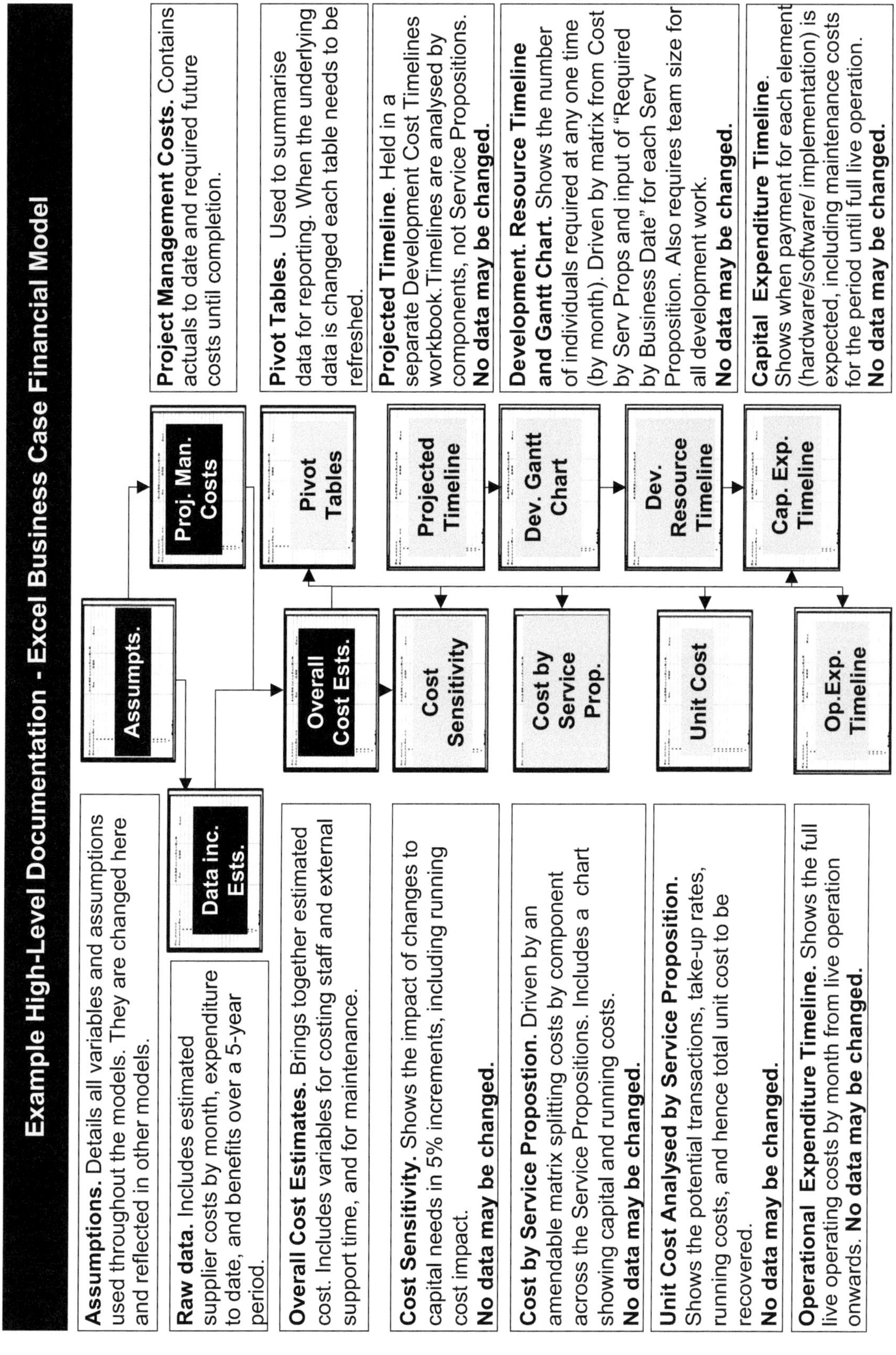

Example High-Level Documentation - Excel Business Case Financial Model

Assumptions. Details all variables and assumptions used throughout the models. They are changed here and reflected in other models.

Raw data. Includes estimated supplier costs by month, expenditure to date, and benefits over a 5-year period.

Overall Cost Estimates. Brings together estimated cost. Includes variables for costing staff and external support time, and for maintenance.

Cost Sensitivity. Shows the impact of changes to capital needs in 5% increments, including running cost impact. **No data may be changed.**

Cost by Service Propostion. Driven by an amendable matrix splitting costs by component across the Service Propositions. Includes a chart showing capital and running costs. **No data may be changed.**

Unit Cost Analysed by Service Proposition. Shows the potential transactions, take-up rates, running costs, and hence total unit cost to be recovered. **No data may be changed.**

Operational Expenditure Timeline. Shows the full live operating costs by month from live operation onwards. **No data may be changed.**

Project Management Costs. Contains actuals to date and required future costs until completion.

Pivot Tables. Used to summarise data for reporting. When the underlying data is changed each table needs to be refreshed.

Projected Timeline. Held in a separate Development Cost Timelines workbook. Timelines are analysed by components, not Service Propositions. **No data may be changed.**

Development. Resource Timeline and Gantt Chart. Shows the number of individuals required at any one time (by month). Driven by matrix from Cost by Serv Props and input of "Required by Business Date" for each Serv Proposition. Also requires team size for all development work. **No data may be changed.**

Capital Expenditure Timeline. Shows when payment for each element (hardware/software/implementation) is expected, including maintenance costs for the period until full live operation. **No data may be changed.**

Boxes: Proj. Man. Costs; Pivot Tables; Projected Timeline; Dev. Gantt Chart; Dev. Resource Timeline; Cap. Exp. Timeline; Assumpts.; Data inc. Ests.; Overall Cost Ests.; Cost Sensitivity; Cost by Service Prop.; Unit Cost; Op.Exp. Timeline

Figure 3.2 Finance Spreadsheet Model

Suppressing errors. Sometimes Excel very helpfully informs us when we have done something less than sensible and it puts an error message in the cell rather than the results of a calculation. The following harrowing experience showed that there are different ways of trapping this sort of error message which should always be used.

As a rule, when we pick up our underlying data (e.g., from cell D12) into our report area, instead of just putting

   +D12

in the data-receiving cell, we should instead try using the formula:

   IF(ISERR(D12),0,+D12)

This will put a 0 in the cell if there is an error condition in the data cell D12, and we can suppress the display of the 0 in the Options, or alternatively we can use a slight variation of this formula by using quote marks with a space between them:

   =IF(ISERR(D12)," ",+D12)

This will put a space in the receiving cell instead.

This lesson comes from a project which supported a team preparing a national report for the UK government. All UK public sector employers of a certain type of staff had to submit information on their projected requirements for these staff over a period, given different assumptions on training and other variables, in a set of data-collection spreadsheets. The submitted information was summarised into a set of national models and the data passed on to the project team to write up. The team, however, reprinted the models and included them in bound copy of the report. This was presented to the then UK prime minister personally by the head of the team. As the spreadsheet model was not intended for publication, the presentation had not been controlled and error messages suppressed. The first row of every column on each one of the summary pages read:

   #DIV/0!

The error was repeated ten times across every page no less. Hopefully the box of Singapore orchids, with which she was also presented, distracted her from reading the spreadsheet pages.

Getting the sums right. We have to make sure, especially when adding new data, that all the numbers we need are included in our totals. This is even more true when inserting new rows or columns. If you have a heading line with the data starting immediately below it and insert a row, data in the new row is not automatically included in the sum. The same is true for a column insert. In early versions of Excel, inserting a line immediately above a total also had the same effect. An example of what can happen was in a government grant application by a UK local authority for a road, which also required a large and expensive retaining wall to be built. This wall was the last item on a list in the grant application detail and even at first sight was obviously not included in the total grant being requested. The result was that the local authority had to fund the extra millions of pounds themselves. On another occasion, the business case for a project at some point had added a couple of items into a list, worth over €160,000, which were not included in the total cost. Was it just possible because a sheet was on manual recalculation and no one had pressed the F9 key after adding in the lines and had not bothered to check with a calculator? It is not onerous to spend five minutes with a calculator checking to see that the costs of €160,000, or several million pounds, have been included in a grant request. Maybe only the earlier versions of Excel had problems in this area, but it's always worth sanity checking. It is certainly better than reputational damage when someone else notices the mistake first.

Make it easy for the reader. We usually produce lists of numbers with totals at the bottom probably because historically columns were added downwards manually and this has been carried forwards into the present day. Consider, however, that the information most readers require most often is the total. This means that changing centuries of tradition and putting a column's total immediately below the column heading addresses the needs of readers much better. If they wish to for more information, they can always look down the column and examine the details. We can also allow plenty of room for the detail items below and maybe even colour the range that is being totalled, which makes it much easier to check later on. As ever, beware the impact of inserting an extra data row just above, or a column to the left, of your data.

Using spreadsheets – summary. There are, then, a few simple rules to make sure that our spreadsheets are safer to use for supporting major decisions:

- Try and keep the way the spreadsheet works as transparent as possible, perhaps using different coloured shading for different types of cells, data, variables, and calculations.
- Endeavour to keep everything as simple as possible and resist the demands to build a model that becomes a leviathan.
- Always have a copy locked away for security in the event that the model becomes corrupted in any way.
- Makes notes – firstly so that we can explain it to others and secondly because in three months, we will not remember the fine detail of how everything hangs together.
- We may understand that an error result showing in a cell is not important, but not everyone will, so suppress them.
- Always consider having an independent review of the model and its outputs before sharing the results. The more important the model, the more important this is.
- Always check personally that the results look sensible – even to the point of getting out a calculator and checking. It's not that Excel often makes mistakes, but we can, and we do.

Playing the Financial Numbers Game

We have to get just a little technical and mathematical because it helps if we understand the general principles and have at least enough knowledge to busk through the numbers and use a spreadsheet model to do the heavy lifting. A word of warning: towards the end of this section is a financial formula, and if such things seem scary, just look away for a second and then read the next section. There will be another warning.

A full discussion of the complexities of comparative financial analysis and the range of different possible techniques is well beyond the scope of this section. Be reassured because, if necessary, we can always follow our rule of humility and ask for help as this is an area where support is readily available from experts. Be warned, experience suggests that it can often be better and quicker to ask the expert to do the analysis rather than have them explain how it should be done.

When comparing between options, our problem is to find which option will deliver the highest level of financial return – that is to say, the tangible, cashable benefits minus the attributable cost. Managing benefits comes later, but in the context of our financial models, the first rule (another one) is that a benefit must be genuinely cashable or we can get into all kinds of arguments. Put simply, if we will be able to see the difference in the organisation's accounts, then it is a financial benefit. If everyone has a warm fuzzy feeling about how good it is but it doesn't make a difference in the accounts, then it isn't a financial benefit. As an example, remember the case of our new IT accounting system where we have identified that implementation would demonstrably save four hundred people around six minutes per week, which is equivalent to a full-time person's salary. Whilst this is a theoretical performance gain, in reality it would evaporate around the water cooler or coffee machine and deliver no financial benefit unless you cut salaries to reflect those six minutes. Whereas if the performance gain of four hundred minutes was related to one person, then we could reduce our headcount by one and hence and validly count our staffing cost savings as a cashable, tangible benefit.

Different rules can apply to public sector investment where the benefit is felt by the community but not by the organisation. If it is necessary to include valuing the increase in public good to validate a project, it really is time to call in the experts and probably don a helmet and flak jacket. The controversy surrounding how to justify the UK's second high-speed rail line (HS2 for short) is a prime example of the difficulties and emotions that can be involved.

Let's look at a couple of real-life examples:

Example 1. A project to replace manual ticketing in a concert hall box office with an online booking system was justified mainly by staffing reductions. The staff actually argued for their hours to be cut as it meant that cashing up each evening would be much quicker and they could get home earlier, especially after a concert. This was a real cashable benefit and could be included in the option evaluation. The new system also:

- Reduced ticket printing costs, which was also allowable.
- Cut about two hours of work each morning by the box office manager updating and analysing figures for up to 40 concerts, allowing him to report takings to concert promoters much more quickly each day. This time savings was not a cashable saving and could not be taken into account in the quantitative analysis. It was, however, a major benefit to concert promoters, who, when required, were able to take action much earlier in the day to increase publicity, even meeting deadlines for the local evening paper for the first time.

Example 2. In transforming a government agency from a paper-based to an online system, the only cashable benefit allowed by the business sponsor was a postal savings which he maintained was solid and unarguable, whereas staff cost savings might not be achieved. As the postage charge savings related to around fifty thousand packages of paper a year at a minimum of four pounds each, these savings were sufficient to justify the project on its own.

Financial evaluation is more straightforward if we are just looking at one year, but additional issues creep in when we have to look further ahead and work out what this really means. The general principle for evaluation is that money tomorrow is worth less than money today. So, we apply a discount factor to reduce the value of money in the future. Think of this discount as a sort of compound disinterest.

Comparisons can get complicated when we are comparing different patterns of expenditure and benefits over a number of years, but there are recognised techniques that are used to show the effect of different patterns of income and expenditure – applying the principle of reductions in future value. These include return on investment (ROI), discounted cashflow (DCF), benefit-cost ratio (BCR), discounted benefit cost ratio (DBCR), payback period (PBP), internal rate of return (IRR), and net present value (NPV). Different organisations will prefer different selections of these as part of our financial appraisals.

Net present value is one of the methods commonly used for evaluating options and has the advantage that it evaluates the future values of money and produces a single figure for our comparison between the options. NPV also has the distinct further advantage that it, in common with some of the indicators like the IRR, is a standard function when we are building our model of costs and benefits in Excel. So, rest assured that we do not have to hand build an Excel calculation based on the following formula:

LOOK AWAY NOW.

$$NPV = \sum_{t=0}^{n} \frac{(1+I)^n}{R^t}$$

Where t is the number of time periods, i is the discount rate, and R is the net cash (i.e., benefits–costs) in each period.

OK, IT IS SAFE TO LOOK BACK.

It is really not as hard as it looks. There are only a couple of tricky questions that we have to answer that will affect the outcome of our evaluation. The key factors that we have to consider are:

- The number of time periods (usually years) to be considered (t).
- The rate at which future values are to be discounted (i).

Some organisations will have rules on what time periods are acceptable for comparison and how much future values have to be discounted, but in many cases, it can be down to us to make recommendations.

Choosing a time period for comparison. The first decision we have to face is the appropriate period of years that we should be considering. The answer, of course, is that it depends on what we are doing, what the money is being spent on, and the rules of our organisation. As an extreme example, consider the London to Bristol and South Wales railway line in the UK, which was electrified at a cost of around £2.9 billion. This project was completed in 2020, and according to press reports, this outturn figure was a spectacular almost £2 billion over the original budget of £1 billion set out by then UK prime minister David Cameron, despite reductions in scope which cut sixty miles of mainline to be electrified from the project and "indefinitely deferred" other linked lines. Please use this as an example to avoid and not one to emulate.

The question is over what period we should value such types of investment and if all elements have to be valued over the same period. For a moment, consider neither the more recent electrification equipment, which will eventually have to be replaced as it wears out, nor the track and signalling, which will have been replaced over the years, but just focus on the actual construction of tunnels, bridges, embankments, and cuttings. Over what period should the original investors in the project have evaluated the benefits of their investment? These assets on the part of the line between London, Bath, and Bristol are demonstrably still in daily use and have been so continuously since 1838, when the first part of the line opened, and especially from June 1841, when the line was finally completed.[8]

Closer to home, consider our example of the investment in the new accounts system. Some options involve purchasing hardware, which is likely to need to be replaced after perhaps five years to maintain reliability, thus requiring new capital investment. Other options involve paying for shared hardware capacity annually. The software, however, will be subject to ongoing support from the provider and could easily last ten years or more, but with support and licence payments due each year. So, what time period do we choose to evaluate our options? The impacts of different time periods and expenditure patterns can be subtle. For example, capital options may ultimately be cheaper, but most of the expenditure comes early on and so has little discounting, whereas for revenue options, those costs incurred in future years are being increasingly discounted year on year. But in our IT example, by buying hardware, we also get the choice to "sweat the assets" – i.e., ignoring the recommendation to refresh the hardware in year 5 and continuing to use it for a number of additional years whilst accepting an increased risk of breakdowns. This, of course, disadvantages the revenue-only options.

The following tables (Tables 3.6 and 3.7) demonstrate how the NPV calculation works, using the Excel NPV formula to illustrate the difference between funding a scheme with large initial payments and one with equalised funding over a number of years.

Table 3.6 NPV Example 1

NPV Calculation Example 1								
Option 1	**Year 1**	**Year 2**	**Year 3**	**Year 4**	**Year 5**	**Year 6**	**Year 7**	**Total**
	£000	£000	£000	£000	£000	£000	£000	£000
Capital costs	150	50			100			300
Revenue costs	5	7	7	7	7	7	7	47
Benefits	25	75	100	100	100	100	100	600
Net benefit/cost in year	−130	18	93	93	−7	93	93	253
Discount rate	3.5%							
Net present value	199							

Option 2	**Year 1**	**Year 2**	**Year 3**	**Year 4**	**Year 5**	**Year 6**	**Year 7**	**Total**
	£000	£000	£000	£000	£000	£000	£000	£000
Capital costs								
Revenue costs	53	53	53	53	53	53	53	371
Benefits	25	75	100	100	100	100	100	600
Net benefit/cost in year	−28	22	47	47	47	47	47	229
Discount rate	3.5%							
Net present value	192							

The comparison shows that although the actual cost of the annual revenue-funded scheme is higher than the capital one, the impact of discounting the value of future money means that revenue funding gives a lower NPV.

Selecting what discount rate to use. Selecting a different discount rate in any financial options analyses may alter our recommendations, and consequently our choice of rate needs to be justifiable. In our example, increasing the discount rate in our example comparison will reduce the NPV of the revenue option and hence increase the gap between it and the capital-funding option.

A review of various government requirements illustrated the differences that are possible when selecting a discount rate:

* In the UK the government used a discount rate of 3.5% (originally fixed in 2003 and limited to the first 30 years, with a schedule of declining discount rates kicking in for year 31 onwards).
* Ireland used a flat rate of 4%.
* Australian government institutions have mostly used 7% since the late 1980s.

We could also choose the rate at which our organisation would have to borrow money as an indicator of the rate to use for discounting. In Australia the 7% rate was originally fixed when rates were around 6.8%. There was later discussion around its validity in the subsequent periods of low or even negative rates of interest. The impact of using Australian rates as opposed to UK ones is shown in Table 3.7, with the discount rate increasing from 3.5% to 7% in the IT replacement example. No other figures have been altered. The change of interest rate alone would mean that our recommendation should change from being in favour of option 1 to being in favour of option 2.

Table 3.7 NPV Calculation Example 2

NPV Calculation Example 2								
Option 1	**Year 1**	**Year 2**	**Year 3**	**Year 4**	**Year 5**	**Year 6**	**Year 7**	**Total**
	£000	£000	£000	£000	£000	£000	£000	£000
Capital costs	150	50			100			300
Revenue costs	5	7	7	7	7	7	7	47
Benefits	25	75	100	100	100	100	100	600
Net benefit/cost in year	−130	18	93	93	−7	93	93	253
Discount rate	7%							
Net present value	156							

Option 2	**Year 1**	**Year 2**	**Year 3**	**Year 4**	**Year 5**	**Year 6**	**Year 7**	**Total**
	£000	£000	£000	£000	£000	£000	£000	£000
Capital costs								
Revenue costs	53	53	53	53	53	53	53	371
Benefits	25	75	100	100	100	100	100	600
Net benefit/cost in year	−28	22	47	47	47	47	47	229
Discount rate	7%							
Net present value	161							

Remember, the discount rate and the time periods are just more of the assumptions that we use, and when the models are completed, it is worth conducting a sensitivity analysis – flexing the rate up and down, altering the time periods, and then reviewing the effects on our different options.

Step 4: Choosing a Project Delivery Method

A key component of this stage is designing and then getting agreement on how we are going to deliver the project – our project delivery method. The first thing that talking about a project delivery method is going to do is cause a major three-way-split in understanding what we mean, and this may be based on the project manager's background.

- The first group are project managers, often in software development, who will be thinking that we are going to consider the Agile Manifesto and its children, such as DSDM, SCRUM, Kanban, and even SCRUMBAN.
- The second group are project managers in the construction arena, who will think that we are going to consider design-bid-build, construction management multi-prime (CM MP), construction management-at-risk (CM@Risk), and design-build (DB).
- The last group is everyone else, who are by now scratching their heads and looking totally perplexed.

Before drafting our project delivery method document, it is worth taking a brief look at how our approach can be informed by some of the thinking around these different perspectives. We are not going to cover them all in depth, just pick up on some of the thinking and fundamental principles that underly them.

The traditional relatively sequential approach to project organisation is the most widely understood way to deliver a project successfully, but it has some shortcomings and dangers.[9] The concepts, some of the techniques, and the underlying issues that the newer approaches seek to overcome contain lessons for all project managers. Some of these approaches can be used as adjuncts to other methods, such as the PRince2 methodology;[10] any of them may have advantages for different types of projects; and each may have features which can be extracted and used elsewhere.

What Is Agile

Arguably all project management should be about agility and managing towards our objective by flexing our responses and plans as situations change and evolve. In the context of the approach to structuring the project's delivery, the term *agile* is used in a very specific way. It was first used in relation to software development. According to the Agile Manifesto website, it came about like this:

In 2001, at a ski resort in the Wasatch mountains of Utah, seventeen people met to talk, ski, relax, eat, and try to find common ground about how to develop new software more effectively. What emerged was the Manifesto for Agile Software Development. This manifesto set four priorities which it called pillars. These pillars changed thinking about what should be important in software development projects and provide lessons to think about for any project. The Agile Manifesto lays out these four pillars in the following statement:

> We are uncovering better ways of developing software by doing it and helping others do it. Through this work we have come to value:

Individuals and interactions over Processes and tools
Working software over Comprehensive documentation
Customer collaboration over Contract negotiation
Responding to change over Following a plan

That is, while there is value in the items on the right, we value the items on the left more.

(Beck, 2001)

There are times when adapting our project approach to adopt some or all of these four core pillars is beneficial. This is especially the case where we consider that there may be a comprehension gap between the stated requirements for the project at the outset and the requirements as they emerge or evolve during the life of the project. In a traditional project, the evolution of requirements might be regarded as scope creep or a failure of the opening stages of the project to define requirements properly. Evolving requirements, however, are a real-life pressure that all project managers need to acknowledge and manage and one that agile specifically addresses with its responsiveness to change.

The Agile Manifesto also has 12 underlying principles (Beck, 2001). Like the four pillars, these principles have implications that can often be carried over into wider – i.e., non-software – development projects; as the project manager, we should always consider these in relation to our own project. Of course, not all 12 principles will apply equally to all projects. The 12 principles are set out in Table 3.8.

Table 3.8 Agile Manifesto Principles

Agile Manifesto	
Principles	**Implication for Wider Range of Projects**
1 Highest priority is to satisfy the customer through early and continuous delivery of valuable software	How can benefits be delivered earlier?
2 Welcome changing requirements, even late in development. Agile processes harness change for the customer's competitive advantage.	At what point do requirements have to be fixed, and how can we cope with subsequent change: • Commercially? • In terms of deliverability?
3 Deliver working software frequently, from a couple of weeks to a couple of months, with a preference to the shorter timescale.	How can the project be structured to deliver visible results to the client at regular intervals?
4 Business people and developers must work together daily throughout the project.	How can we create integrated delivery teams composed of individuals from different parts of the organisation or from different organisations as appropriate? Can we create a single space within which the project team can become an entity in its own right, with members communicating on an equal basis?
5 Build projects around motivated individuals. Give them the environment and support they need, and trust them to get the job done.	
6 The most efficient and effective method of conveying information to and within a development team is face-to-face conversation.	

Agile Manifesto	
Principles	**Implication for Wider Range of Projects**
7 Working software is the primary measure of progress.	Are the outcomes of the project suitable for measurement during the life of the project? For example, we can categorise progress on a building by stage of construction, but how can this be applied to business change projects?
8 Agile processes promote sustainable development. The sponsors, developers, and users should be able to maintain a constant pace indefinitely.	Agile software development methodologies have planning processes to smooth otherwise uneven working patterns. Can these processes be replicated in different types of projects by structuring work allocations and our expectations differently?
9 Continuous attention to technical excellence and good design enhances agility.	In any project we should aim to reduce the amount of rework arising from poor performance at earlier stages, but how can we build in continuing control over the quality of working processes and outputs?
10 Simplicity – the art of maximising the amount of work not done – is essential.	In accordance with Lean principles, how can we plan to remove any work that does not add any value? Note – be warned, this does not usually include taking out a regular project manager's report to the business sponsor!
11 The best architectures, requirements, and designs emerge from self-organising teams. 12 At regular intervals, the team reflects on how to become more effective, then tunes and adjusts its behaviour accordingly.	The question for the project manager is how to facilitate self-organising teams, especially where they are multi-organisational teams? What behaviours are required to enable the project manager to: • Act as an "umbrella" shielding the team from extraneous influences. • Build an internal consensus within the team. • Present and defend the team's emerging views to sponsors and other stakeholders.

Before looking briefly at two of the more common agile methods, it should be noted in passing that according to a 2017–2018 survey, many project managers were using a mix of between three and four agile methods (The Scrum Alliance, 2018).

SCRUM

SCRUM is not an acronym. It was actually developed from a rugby analogy, put forwards in an article in the *Harvard Business Review* in 1986. The article drew on in-depth research into the operation of six new product development projects (three projects within Canon and one each from Honda, NEC, and Fuji Xerox). The researchers identified that a new method of product development had emerged in these projects, and from their examples in just four companies, they generalised that companies were now using a "holistic" method of project delivery. The authors compared this holistic method to rugby, where the team moves up the field as a unit and passes the ball between team members. The authors went on to compare the new method to the traditional approach, which they likened to a relay race with the baton being passed from one department to another in a sequential approach. In the rugby analogy, they saw the team as being multidisciplinary and made up of continuously interacting members. It is this interaction which then drove the product development process (Takeuchi, 1986).

The researchers observed that the successful product innovation in the six projects considered was based on a number of common concepts, which are now built into the SCRUM approach. Even without adopting SCRUM or another agile approach, these concepts contain major lessons for the organisation of people into teams for achieving project goals. The common concepts upon which SCRUM is based are summarised Table 3.9.

Table 3.9 SCRUM Concepts

SCRUM	
Concept	**Explanation**
Built-in instability	Management creates the overall objective for the project with deliberately challenging goals, such as to deliver product for half the current cost. The original article also included a quote from an unnamed Honda executive suggesting that the team's creativity will come from exerting extreme pressure on individuals in the team (this view may be less culturally or socially acceptable in other countries or jurisdictions).
Self-organizing teams	• Autonomy – the team sets its own agenda for achieving the management-set goals. • Self-transcendence – the team sets its own goals and evaluates their achievement during the project. • Cross-fertilisation – this only occurs when individuals widen their view and consider the goals of the team, rather than working from their professional perspectives.
Overlapping development phases	Removing end-of-phase checkpoints means that work on tasks normally allocated to a later stage can begin earlier because there is no decision-making delay.
"Multi-learning"	This has two aspects: • Multilevel – individuals pursue personal development of skills within their discipline. • Multifunctional – individuals develop their knowledge and experience in new functional areas.
Subtle control	Management needs to hold a balance between controlling instability to avoid chaos without harming the team's creativity.
Organisational transfer of learning	This could include when team members move on to other projects or by disseminating lessons learned by the project.

The SCRUM approach picked up this highly contrived rugby metaphor,[11] but, putting the name aside, it embodies these principles into an iterative and incremental framework for projects, software, and product development:

- The core of the method is an overall development cycle which is split into a series of short development cycles called sprints, each of these lasting no longer than one month.
- Dates are set in advance, and when the end date is reached, the sprint closes and the next sprint starts. Unlike in a conventional project plan, the dates are never slipped.
- Instead of a detailed plan for the whole project, at the start of each sprint, the project team selects the work that is to be completed within the sprint from a prioritised list of requirements (called the backlog) and commits the project to their completion.
- The work to be undertaken during the sprint is fixed in a sprint backlog, and any other work that emerges, either internally or externally, is added to the main backlog and prioritised for inclusion in subsequent sprints.
- Each day, the team reviews progress on these items in a time-boxed meeting (15 minutes) and adjusts the steps required to complete the remaining work to be delivered within the sprint.
- When the sprint ends, the work completed should result in a finished "product" or increment to a product that has been tested and quality approved and is therefore, at least in SCRUM terms, potentially "shippable" to the client – for example, a completed module of a computer system.[12]
- On the last day of each sprint, a two-part sprint review meeting is held. The first part of the meeting is a review and demonstration for the stakeholder or customer. The second part is a team retrospective. The output from these meetings is then fed back into subsequent sprints.

The basic flow of a SCRUM project is shown in Figure 3.3.

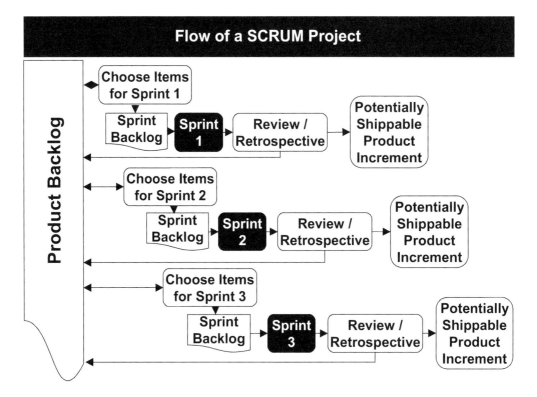

Figure 3.3 Flow of a SCRUM Project

DSDM

Originally called the dynamic systems development method, DSDM was developed in relation to the production of computer software. Underlying DSDM as its two guiding concepts are common sense and pragmatism, which are defined as:

- Common sense – the practical judgement that does not rely on any sort of training, intelligence, or expert knowledge.
- Pragmatism – actions and policies based solely immediate practical consequences and not dictated by any sort of theoretical approach or dogma.

DSDM is based on eight principles, many of which are directly transferable to any project:

- Business need as the prime focus.
- On-time delivery.
- Collaboration.
- Quality.
- Incrementalism based on firm foundations.
- Iterative development.
- Communication.
- Control.

There is a fundamental concept in DSDM's approach to project management. Time, cost, and quality standards for the project are fixed at the end of the project's initial stage. Any contingency is managed by reducing the features of the solution (scope reduction). This process operates by first prioritising each requirement, and when slippage occurs, the lower priority requirements are dropped from the project or deferred to a later stage.

Kanban and Project Management

The origins of Kanban lie in the Toyota Production System, which evolved in the company's Toyoda and Toyota manufacturing plants in Japan over many decades. Kanban is based on three core principles and six core practices, which have now been adapted from their manufacturing origins into wider industries, including software development. The three Kanban core principles are:

- Start with what you do know, including understanding current processes as they are really operated and respecting current roles, responsibilities, and job titles.
- Gain agreement to pursue improvement through evolutionary change.
- Encourage acts of leadership by people at all levels.

The six core work practices, which describe what should be done on a daily basis, are:

- Visualise the workflow. This is done using a basic Kanban board. At its simplest it needs to allocate tasks into one of three columns: to do, being done, and done. However, being done, for example, could be broken down into a number of different stages – e.g., design, develop, test, deploy. The benefit of this visualisation is that it can be made clear to everyone how work is progressing.[13]
- Limit work in progress (WIP). The theory is that overall productivity rises by allocating just enough work to each team member and avoiding overloading that can occur when allocating tasks beyond the team's capacity. The Kanban board is used to manage the WIP limits – i.e., managing the "being done" columns.

- Manage and enhance the flow. Incrementalism is the preferred path, making small changes to improve existing flows and learning from current performance – and mistakes – rather than risking wholesale change.[14]
- Make policies and workflows explicit to reduce rework and to focus efforts where they will be most effective.
- Feedback within and between teams.
- Continuously improve.

Kanban is less structured and more adaptable than SCRUM, has fewer rules, and can be applied to any process in place. Rather than being timeboxed like SCRUM, Kanban limits the amount of work in progress to drive change.

A software industry has grown up around these principles, providing tools for the project manager to use when managing Kanban projects.

Construction-Based Delivery Approaches

The delivery structure of construction projects differs from many other types of projects because there are a number of roles within a construction project which are traditionally delivered by different organisations with separate contractual relationships. They are not, in effect, different from the roles performed in most projects but use different names:

- The owner – commissioning the project.
- The consultant – responsible for design and management of the project. This role may break down into several different roles, including architect, engineer, project manager, etc., each of which may come from different specialist companies.
- The contractor – undertaking the physical work. There may be only one contractor, or there may be a prime contractor supported by a series of specialist sub-contractors.

Although there may be many variations in detail, there are five main delivery models that have evolved to organise these roles to deliver a construction project. Like the agile approaches, which were designed originally for new product development and software delivery, these delivery models are based on concepts which can be applied much more widely and are set out in Table 3.10.

Table 3.10 Construction Delivery Approaches

Construction-Based Delivery Approaches			
Approach	**Features**	**Benefits**	**Disadvantages**
Design–bid–build (DBB)	Traditional linear approach. Owner engages consultant and contracts for the build only after design stage.	Most familiar and widely used project delivery method. Lowest in price, at least up front. Maintains owner engagement throughout. Clear responsibilities in event of litigation.	Contractor and designer not collaborating during design phase – e.g., about materials and construction techniques, resulting in more change orders. Sequential nature of tasks lengthens delivery times.
Design–build (DB)	Design and construction are handled by one firm.	All elements of delivery in place at the start of the design phase under one contract. Integration between designer and contractor means high levels of communication at all stages and fewer change orders. Project delivery times are shortened. Design changes can be vetted, approved, and executed more quickly.	Owners carry heavier upfront load in preparing a detailed set of specifications and contract. Control is passed to the single supplier, and the owner may not be as engaged in approving designs and change control as in DBB. Disagreements with the supplier can lead to delays and potential for litigation.
Construction management-at-risk (CM@Risk)	Consultant acts as prime contractor with design and construction sub-contractors. Often fixed price.	Single contract and liability and risk lies clearly with the consultant. Owners work directly with designers and contractors from the start. Greater awareness of costs and greater certainty. Changes can be explicitly priced.	Pricing in risk by the consultant increases cost. Owner is liable for approving the completeness, accuracy, and details of the design plans. Disagreements can lead to schedule delays, increased costs, and litigation.
Multi-prime	Separate prime contractor for each phase: design, engineering, and construction.	Owners are acting as the project managers and have greater control over the project. Improved control over suppliers as each is managed separately.	Increased scope for omissions, duplications, and misunderstandings arising from imperfect communications. More difficult to monitor and control final project costs, especially where changes require input from multiple prime contractors. Increased problems in controlling and integrating work schedules from different prime contractors.

Construction-Based Delivery Approaches			
Approach	**Features**	**Benefits**	**Disadvantages**
Integrated project delivery	Owner, designer, and contractor share single (or linked) contract.	Increase in transparency between partners leads to early contractor involvement and reduced need for conflict resolution. Knowledge is shared across the project team, reducing misunderstandings, conflicts, and rework. Projects can sometimes be completed more quickly. Costs reduced through removal of staff in different parties needing a counterpart, i.e., having a design engineer in the consultant having to be paralleled by one in the contractor.	Trust between team members is not automatic and has to be developed. Some funders are not familiar with the approach and may be reluctant to participate. The same can apply to some designers and contractors.

Lean Project Management

Lean principles,[15] like Kanban, are derived from the Toyota production system, and there are many arguments about the similarities and differences between the two. There is held to be a difference in emphasis between the two in that Kanban is a visual method focusing on delivering value whereas Lean aims to reduce waste during the process. The underlying principles all have relevance to project delivery as well as to manufacturing, both of which are essentially "production"-focused processes.

The Lean Institute introduces the underlying principles of Lean as follows:

> The five-step thought process for guiding the implementation of lean techniques is easy to remember, but not always easy to achieve:
> - Specify value from the standpoint of the end customer by product family.
> - Identify all the steps in the value stream for each product family, eliminating whenever possible those steps that do not create value.
> - Make the value-creating steps occur in tight sequence so the product will flow smoothly toward the customer.
> - As flow is introduced, let customers pull value from the next upstream activity.
> - As value is specified, value streams are identified, wasted steps are removed, and flow and pull are introduced, begin the process again and continue it until a state of perfection is reached in which perfect value is created with no waste.
>
> (Lean Enterprise Institute, Inc, 2020)

Like the principles set out in the other approaches, the Lean principles can help us design the approaches we use when setting out how we intend to tackle the creation and management of our project.

Defining Our Project Delivery Method

We can learn from these wider perspectives, as well as from the traditional sequential project. We can adopt a mix-and-match approach in working out how our own project will operate. For example, we could adopt:

- Kanban's visual tracking.
- DSDM's iterative development.
- SCRUM's daily team meetings.
- An integrated project delivery team.
- Lean's removal of wasted steps.

One size does not fit all, and we need to create the most appropriate method of delivering our project given our objectives and the constraints within which we have to operate.

Throughout our description of the project delivery method we wish to implement, we must demonstrate not only what we propose to do but clearly show the reasoning behind why we are recommending it. We can draw on evidence from any methodologies we wish to adopt, the thinking about different approaches (such as those outlined prior) and, most importantly, research and lessons learned from other similar projects. If we can find no comparable projects, it is worth including data on industry averages, if they are available. As ever, we must always remember to be explicit about the assumptions we are making.

Project Management Structure

The first key aspect of the project delivery method is to identify the project's governance and project management structure and to describe each of the roles identified and their responsibilities. The following two examples (Figures 3.4 and 3.5), both based on real-life projects, reflect two different approaches to structuring a project team.

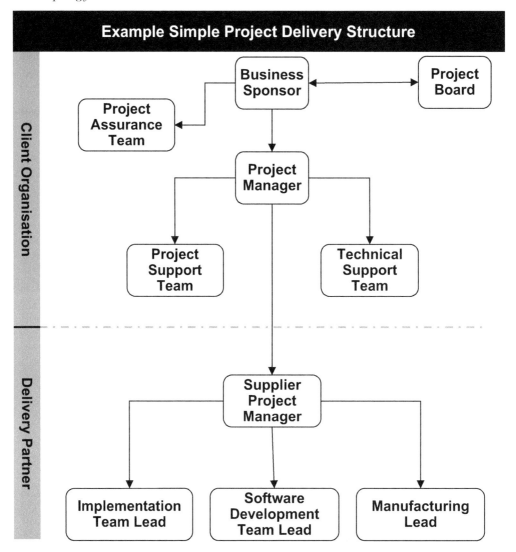

Figure 3.4 Simple Project Delivery Structure

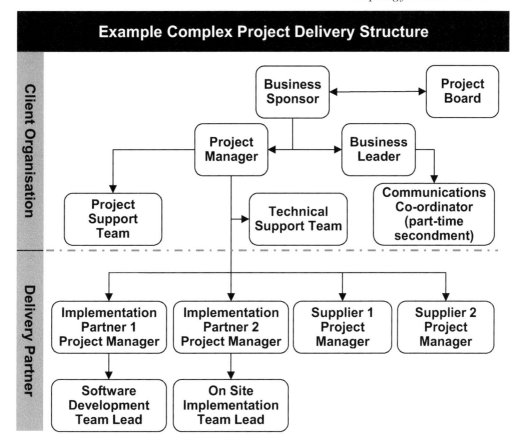

Figure 3.5 Complex Project Delivery Structure

It is worth looking at these two projects a little more closely. Both projects involved IT development and the installation of displays and other hardware in public areas.

- The history of the project with the simple structure was that there were originally more external organisations involved in delivery, but responsibilities were increasingly transferred to the single supplier. This project was more complex and had a much larger overall spend than the second one. It had struggled to surmount various technical and environmental issues. The independent project assurance team was included part-way through the project after having been determined to be a necessary requirement by an independent scrutiny of the project delivery approach. Sadly, this project ultimately failed and was cancelled, although this was not necessarily the result of the project's structure.
- The project with the more complex project structure had been initially begun by a local delivery partner but was transferred to the eventual client organisation, which had a national remit. Its scope was widened to include multiple geographic areas, and it overcame a number of technical and environmental issues. It delivered successfully.

The project delivery method needs to set out a brief description of the roles, skills, and experience required to deliver it. At this stage some individuals will have already been assigned to the project, usually at least the business sponsor, the project manager, and the members of the project board.

In describing the team's roles, we should include the way we intend to operate the project. So, in a sequential project, one of the duties of the project board could be written as follows: "At the end of each stage of the project, the project board will review the project's progress and the planning for the next stage. It will authorise the closure of the stage and progression to the next stage." By contrast, in an agile project, the project manager could have a duty as follows: "Before commencing a sprint, the project manager will review and prioritise the project backlog, identify those tasks to be completed in the sprint, and prepare the sprint backlog."

Whenever a project proposes to use external suppliers, the project delivery method must cover the proposed procurement approach, as this will not only influence the shape and timings of our project plan but also the management of the delivery of the project. It is often advisable to work closely with a procurement expert in selecting the approach best suited to the project's needs because the procurement process is governed by interlocking sets of local, national, and international rules. This is especially true of larger value contracts and for those in the public sector. The result of this collaboration with a procurement expert should be a rationale for the selected approach, which we should include in the project delivery method, and a procurement timetable which will sit within our high-level plan.

Procurement is only the first part of the story when considering external suppliers. We also need to set out how we propose to control the delivery by the suppliers. This should include describing how delivery teams should be organised and defining the responsibilities for the management, monitoring, and control of:

- Supplier delivery performance, including quality assurance and incentivisation.
- Planning.
- Risk.
- Costs.
- Change.
- Payment.

The other point to make is that our high-level plan and project delivery method may be co-dependent, and having decided upon one, we should always review its impact on the other and make any adjustments necessary.

Step 5: The High-Level Plan

Our high-level plan is usually a short document, perhaps even presented mainly as a graphic. It details how we will break the delivery of the project into manageable chunks over different time periods. We may be following a traditional sequential approach in which the timing of each stage depends on our estimate of how long it takes to complete each task within it, or we may be bringing in aspects of one of the agile approaches where we are timeboxing delivery and the content of each timebox is limited by the time available.

In either case the objectives of our high-level plan are:

- To record our approach and communicate it to others, including decision makers.
- To provide the basis for subsequent detailed planning for each stage or timebox.

It's time for another virtual field trip.

This time you are in a European city, standing opposite a modern but nondescript office on a narrow street. It is 11 a.m., and for a change, the weather is fine. In fact, it is pleasantly warm, so you take off the heavy coat and gloves that you had worn as a precaution. I promise to try to remember to give you a clue about clothing next time.

A tanned and well-dressed man of around 40 years of age with slightly greying hair in a dark check jacket and yellow open-necked shirt is exiting the front door of the building. He is carrying a small brown cardboard box. He trips, stumbles, and falls forwards in a peculiar slow motion. A few items spill out of the box onto the pavement. You rush over and help him get up and collect his spilt possessions: a picture of his wife and children, a coffee mug bearing the legend "World's Best Boss," and a few small mementoes of different conferences, each emblazoned with a supplier's name. Obviously, he is a man who has been told to clear his desk, and these were the only personal items he kept in the office. He is visibly shaken by tripping, and you offer to buy him a coffee whilst he gets himself together. Over the coffee, he starts to tell you his story.

"Until this morning I was a project manager for a local city agency. In fact, I should tell you, I was the latest of a long line of managers for the project. I'd been there probably nine months, and I was responsible for implementing a card system for the city region, costing over fifty million euro. No more paper tickets; the city's idea was for users to access services by just tapping a card on readers which would be widely located across the region. Fairly common stuff nowadays, but the city council had added a twist.

"I inherited the project, and I started a fair time after the supplier had been chosen, contracted, and started to deliver. The remote hardware and communications technology that we'd bought was well tried and has been implemented in many other cities. The central IT hardware was pretty standard stuff too. All of it was well understood and well proven. But the city really wanted the kudos of being seen as a leader in the field and had specified a new third element, an innovative system of accounts linked to a personalised card – engraved with the holder's photo, no less. I don't think very many organisations had done this before, and certainly not our selected supplier. This innovation had required them to develop a new complex central software system with fast links to every piece of remote equipment – which had to operate within a second or so, during which time the card had to be checked against the central computer system and the service authorised. Get the timing wrong and huge queues could build up at critical points, such as the electronic gates at stations during peak times. So, no pressure there.

"Again, long before I came on board, the city had structured the project to deliver both the tried and tested hardware and the innovative accounts system together in a single 'big bang' implementation. I was worried about this and told them so.

"Looking back, it was almost inevitable that the delivery of the new personalised accounts system would be delayed several times, with knock-on delays to the first operational date. With such a publicly visible system – everyone was waiting for the gates and card readers to appear – the delays led to widespread press and public criticism of the project. It was obvious that the supplier would not be able to deliver the complex accounting system, and so I had already had to terminate the contract with them. This morning it's been announced that the CEO of the agency and his boss, who is the local politician responsible, have both been removed from office. The CEO's last task was to tell me to pack up and leave as well."

Now, being keen to learn more, you ask him what he learned from his failing project and what advice he might have for you as an aspiring project manager.

He gives you a weary look and says, "I could blame the supplier, who were also the manufacturers of the hardware. They seemed only to be interested in manufacturing and then getting paid for their hardware. They just didn't seem to want to deliver a whole working system for us. But I think the fault really lay with the city, or whoever originally structured the project plan. The core problem was not the city's vision, which is great, but their pressure for the system to be seen as innovative from the 'go live' date and to launch everything at once. I wish I had argued much more strongly that we should follow the principle of keeping it simple, and go for two delivery stages. Stage 1 would implement the core proven systems – prepaid cards, readers, communications, and central IT system – just as other cities have. We could even have rolled this out to different parts of the region over time. Stage 2 would have been running in parallel, developing the new software for the personal accounts system, just as we have been, but clearly separating it out as a separate deliverable at the outset would have meant we could deliver it later when the core system was fully operational.

"The benefits from this would have been that we could show rapid progress with Stage 1 and get a large slice of the benefits from the system much earlier, and, because we could have been more certain of the dates for the initial rollout, we could also have managed the communications with our stakeholders and the public with much more certainty.

"Adopting this two-stage approach would have meant that when the accounts software development turned out to be taking longer, and in fairness to everybody, also turned out to be much harder than we planned, we would either have had the goodwill of our stakeholders and got continuing support or we could have reduced our scope. We could have pulled out of the personal accounts idea, saving some money, and still have delivered a major system for the public with massive benefits.

"In the well-known American saying, 'Pioneers are the people with arrows in their backs.' That's me now, arrows in my back. In general, my advice to you is that there's much less inherent risk in adopting proven solutions, moving swiftly forwards with the project, and then bringing the innovation in later. Taking baby steps towards your goal is less risky than trying to get there in one giant leap forwards. As project manager, you have to stand firm on this. Get some early triumphs in place. You'll gain kudos and recognition for delivering and be less likely to find that you, like me now, have extra enforced leisure time.

Thank you for your help and for the coffee. Here's my personal business card just in case you hear of anyone that might be looking for a project manager. Now, if you'll excuse me, I have to go home and tell my wife."

All or Nothing

The project manager on the field trip pointed out the reduction of risk that comes with phased delivery. This is a major and well-recognised issue that we have to consider before fixing on the high-level plan. In its Digital Scotland review, the Scottish government noted:

> There are many examples of projects that have failed in a big-bang implementation. That is where the whole project was due to become operational on one day, usually a critical or legislative deadline. Teams must consider the implementation approach and appropriately stage the project and go live.
>
> (Audit Scotland, 2017)

The same Scottish government paper quotes a similar conclusion by the Netherlands Court of Audit in 2007, which found that because of political, organisational, and technical factors, government IT projects are often overly ambitious and complex. They concluded that such projects should start small and proceed in small increments.[16]

Generating short-term wins can be a key part of the successful delivery of a change project, as they create a positive environment for the project to continue delivering. The problem is partly one of stakeholder psychology. If the stakeholders can see progress is being made, especially if the project is slipping, they are more likely to be supportive. Therefore, another first rule of planning a project should be that wherever we can, we should aim to deliver some benefits as soon as possible so that stakeholders can see progress, rather than waiting to deliver everything in one go. As we saw on the field trip, not paying attention to this rule can have dire consequences.

Applying the Scottish government's requirement to consider the implementation approach in preparing the high-level plan, our first questions to ask are:

- Can we sensibly get some early benefits by delivering this project to live operation in stages?
- What would be the benefits and associated risks with a phased delivery over a big bang?
- Are there any points that we can identify where the extra benefits delivered by a phased launch will exceed its additional costs?

Having answered these questions and thought about the impact of more agile approaches, we can now start to identify our high-level plan – i.e., when and how we are going to deliver the project. At this point the work breakdown structure comes into its own in helping us determine the timeline for the remainder of the project.

Since we prepared the WBS, however, we have gained more information and thought carefully about the delivery approach, so we may need to revise it to reflect any modifications. A good example of a necessary change would be the impact of the procurement route chosen on both dates and delivery methods. Depending on organisational policies, the value of the procurement, and often the jurisdiction of the procurement, we have to work within sets of procurement rules which will govern our timing. Another external timing constraint is meeting cycles. If we need a main board decision in regard to our project, we need to plan to hit the board's meeting cycle[17] – including deadlines for submission of reports to them. We need to reflect such constraints in the skeleton of our high-level plan.

At this stage we really are painting with a broad brush, but always bear in mind that when the project board gives its approval, we will have to construct and then obtain agreement to a much more detailed plan for each stage. Being over-optimistic now about what can be achieved in any one of the time periods identified will create unrealistic expectations about project deliveries, which we would have to work around and manage later on. Equally, bear in mind that we should not "pad out" the timeline and extend delivery beyond what can be reasonably expected, as this may increase costs by maintaining the project team for longer than necessary, delay benefits, and consequently adversely impact the viability or acceptability of the project.

Before completing the high-level plan, there is one last set of questions that we have to answer which will apply to every project where we will be going to the market to engage one or more suppliers to help us deliver. These questions include:

- What is required to ensure that our project, and the underlying contracts that it will need, can be structured to be attractive to potential suppliers?[18]

- What is the expected lead time needed by the suppliers for the supply of our required services or goods?
- How long would the supplier be expected to take to complete, based on the experience of others?

So now it is time for deep thinking on how to structure the project. We could call a working group together, or we can put a "Do Not Disturb" sign on the door, lock ourselves in, and start putting together different scenarios based on all the information collected so far. Whether we approach the plan by working backwards from the final deliverables or working forwards from the beginning is usually a matter of personal preference. There is now a temptation to jump straight in to using project planning software, but, before doing that, we need to think through our assumptions about how we are going to deliver, and we will need to show the project board how our conclusions were reached. Table 3.11 is a very simple working table that will help us record our thinking in producing the high-level plan. This example table assumes that our example IT accounts system replacement project is totally straightforward and has a single supplier taking full responsibility for everything.[19] The advantage of using a table, such as the example, is that it provides a record of the thinking behind the high-level plan and can be supported by notes as required. Multiple versions of this table can be used to identify how different scenarios employing different ranges of assumptions could work.

Table 3.11 High-Level Plan Working Table

IT Accounts System Replacement Project New Accounting Package Installed on Own Premises High-Level Plan – Working Table – Scenario 1					
Stage	Task No.	Task Description	Resources Used	Elapsed Days	Notes/Constraints
1	Task 1	Project frame	Project manager and support team	21	Completed.
2	Task 2	Four document set	Project manager and support team	21	Support from finance is required for the NPV analysis.
	Task 3	Business case	Project manager and support team	14	
	Task 4	Project plan	Project manager and support team	14	
3	Task 5	Solution design	Consultants and users	77	Single supplier responsible for all elements of hardware and software.
	Task 6	Procurement	Procurement team and project manager and users	28	Assumes that international trade rules will not apply; if they do, period could extend – see alternative scenario.
	Task 7	Agree contracts	Legal and project manager	14	Assumes using standard supply contract, no special terms.

Stage	Task No.	Task Description	Resources Used	Elapsed Days	Notes/Constraints
	Task 8	Mobilisation	Supplier	7	
	Task 9	Solution dev. part 1	Supplier	105	Split of deliverables to be agreed on in project plan, but part 1 must include core software delivery and hardware installation. Supplier to provide small development hardware system whilst main hardware is ordered/installed. As new hardware will be smaller and require less power/air conditioning than the existing hardware, no upgrades are needed to the IT suite.
	Task 10	Acceptance test and go live part 1	Supplier and users and test manager	14	Unit and system testing to be supplier responsibility. Cannot coincide with finance year end 31 March.
	Task 11	Solution dev. part 2	Supplier	70	Will include integration with other systems and reporting package.
	Task 12	Acceptance test and go live part 2	Supplier and users and test manager	14	Unit and system testing to be supplier responsibility.
4	Task 13	Handover and project closure	Project manager and user management	7	Including project closure report, formal review to follow after three months.

IT Accounts System Replacement Project
New Accounting Package Installed on Own Premises
High-Level Plan – Working Table – Scenario 1

Other assumptions

1	User departments make suitable staff available to attend requirements workshops, supplier presentations, etc.
2	Suppliers are able to offer solutions that meet the company's unique needs – including data integration requirements.
3	Go live dates will avoid periods of peak activity of the finance teams.
4	Historic data can be ported to the new system, and it will not be necessary to keep a version of the current system to access it.

The working table is not, however, the usual way of communicating the timing of a project; this is most often done using a Gantt chart.[20] If you are not already familiar with Gantt charts, they are a simple way of laying out tasks – or, in our case at the moment, our project stages – against a timeline. Gantt charts allow us to communicate our intentions in a very visual way.

Table 3.12 is an example of a Gantt chart into which the data in our very simple working table (Table 3.11) has been translated. The Gantt chart now shows our different stages and tasks identified against dates when they should occur. For display purposes, the example is formatted to show where work is scheduled to occur in one-month blocks, but these blocks can be revised according to the overall length of the project and the length of each task or stage. When we are undertaking detailed planning, block length could be revised to daily or weekly, but for project board reporting purposes, monthly blocks are usually sufficient. Although this is a sequential project, Gantt charts can be equally applied to agile methodologies, such as to the sprints in SCRUM, for example.

Table 3.12 Simple Gantt Chart

IT Accounts System Replacement Project Example Simple Gantt Chart																		
Stage	Task	Start Date	Target End Date	Nov	Dec	Jan	Feb	Mar	April	May	June	July	Aug	Sept	Oct	Nov	Dec	Jan
1	Task 1	29-Nov	20-Dec	▓														
2	Task 2	20-Dec	10-Jan		▓													
	Task 3	10-Jan	24-Jan			▓												
	Task 4	24-Jan	07-Feb				▓											
3	Task 5	07-Feb	25-Apr					▓										
	Task 6	25-Apr	23-May							▓								
	Task 7	23-May	06-Jun								▓							
	Task 8	06-Jun	13-Jun								▓							
	Task 9	13-Jun	26-Sep										▓					
	Task 10	26-Sep	10-Oct											▓				
	Task 11	10-Oct	19-Dec												▓			
	Task 12	19-Dec	02-Jan														▓	
4	Task 13	02-Jan	09-Jan															▓

Microsoft and others also provide downloadable Gantt templates for Excel which are useful at this stage and for simple project tracking.

Reflections

If you have not selected a real-life project, then you are going to have to really harness your imagination from now on, especially if you are already working up the bio-energy from effluent project. But as project managers, we are creative people anyway, so this is not going to be a problem for you.

If possible, using the same project as in your previous reflections, spend some time considering the following:

- Considering your identification at the end of the last chapter of three major things that the project needs to deliver in order to be seen as a success, extend this into a list of critical success factors (five is now the minimum) for your project and show how different options/sub-options could be mapped against them.
- What is the most appropriate approach to delivering the project? You should consider the contribution that could be made by different agile approaches at different points in the project's life cycle.
- How many aspects of the project can be delivered and implemented early to ensure that the project is seen to be progressing and the risks associated with "big bang" deliveries can be removed?
- When could the main stages of the project be delivered? Collate your information into a table of events, showing the major issues and constraints affecting each of the main stages and tasks that you have identified. When you are happy with the result, prepare a high-level Gantt chart to enable you to show clearly how the project will progress. Hint – remember that Excel has Gantt chart templates.
- Develop an analysis of the potential net present value of the project for two different patterns of expenditure, one with most payments during the project and the other with payments spread over the entire period of the planned operation of the solution. Hint, you will need to choose appropriate time periods and discount rate.
- Just how wrong can you be? Examine your assumptions and establish for each what level of change would make the project financially non-viable and how cautious you can be in your estimation whilst still justifying the project.
- Now, once again be highly critical and ask yourself, if you were the business sponsor and were presented with the high-level plan, what questions would you ask the project manager?

Notes

1 It is a promise almost always fulfilled that something skimped now will bite the project manager later.
2 See Appendix 2.
3 For details of a few facilitated techniques to help to achieve this, please see Appendix 1.
4 You may have noticed so far that I have tried to avoid the contrived and outrageous acronyms in which project management abounds, but I cannot get around this one.
5 If anyone says that we should "think outside the box" when identifying options, ignore them. The secret is that there is no box, and there never was one. Try checking out Edward De Bono and his approach of lateral thinking instead.
6 Apologies for taking not one but two quotes from his memorial lecture issued on accepting his Nobel Prize for Economics in 1978 and also for leaving out the really interesting bits in the lecture when Simon

was sniping, very politely, at the economist Milton Friedman, who had won the prize two years earlier and was accused by Simon of denying that Decision Making Theory was really economics.

7 Apologies, another unavoidable acronym, sorry.

8 When it opened at 1.83 miles, the Box Tunnel outside the city of Bath was the world's longest railway tunnel. The story that it was aligned so that the tunnel was fully lit by the sunrise on Brunel's birthday of 9 April seems to have been disproved – it is now claimed by some that it was more likely to happen on his sister's birthday, 6 April, instead.

9 This does not mean we cannot plan for multiple tasks at one time, but these are usually assigned to our different stages, which are sequential.

10 PRince is, of course, another contrived acronym, standing for PRojects In Controlled Environments. Someone should have been made to stand in the corner facing the wall until they were truly sorry for that one.

11 For non-rugby aficionados, the scrum is the group of slower, heavy, usually muddy forward players alleg-edly known for dubious tactics. These forwards repeatedly try to battle through as many opponents as they can until they can somehow squeeze over the try line to score, unlike the backs, who, when they get the ball, pass it out to a fast runner who deftly streaks up the field, neatly skipping round the opposition and touching the ball down between the posts usually before the forwards have extricated themselves from the last crash of bodies that they created. How should we aim to manage our projects – light, swift, and agile or relying on brute force to slam our way into the nearest group of opponents, hopefully with-out losing the ball – again?

12 Note that this is not necessarily something that has to be put on a truck or a ship.

13 Good luck finding a wall for mounting a large whiteboard in most open plan offices. Software is now often used for visualisation, which also makes it easier to share with dispersed or virtual teams.

14 The term *incrementalism* was coined by Charles Lindblom in the 1950s.

15 For further information on Lean, see www.lean.org.

16 This may be OK for IT projects but pretty difficult to achieve if the project is to build a suspension bridge.

17 Meeting cycles can be a killer for a time-critical project, especially if meeting slots are only available every two or three months. It's a measure of successful project and time management if someone (usually us) does not have to be working all night immediately before the 10 a.m. deadline for a submission to the board.

18 Contrary to popular belief by purchasers, not all suppliers want to supply us. They may perceive the pro-ject as too big or small for them, it may not appear profitable enough, or the balance of risk may not be acceptable. One professional services company was so risk averse that its legal team was nicknamed the Business Prevention Department. Obviously not to their face – it never pays to upset the lawyers.

19 We should be so lucky!

20 Not an acronym but named after the American, Henry Gantt, who modified it, rather than the Pole, Karol Adamiecki, who has a better claim to have devised it.

Bibliography

Audit Scotland, 2017. *Principles for a Digital Future: Lessons Learned from Public Sector IT Projects*, Edinburgh: Audit Scotland.

Beck, K., Beedle, M., Bennekum, A. van, Cockburn, A., Cunningham, W., Fowler, M., Grenning, J., High-smith, J., Hunt, A., Jeffries, R., Kern, J., Marick, B., Martin, R.C., Mellor, S., Schwaber, K., Sutherland, J., and Thomas, D., 2001. *The Agile Manifesto* [Online]. Available at: http://agilemanifesto.org/ [Accessed 11 December 2022].

Lean Enterprise Institute, Inc, 2020. *www.lean.org ©Lean Enterprise Institute, Inc Copyright All Rights Reserved. Used with Permission* [Online]. Available at: www.lean.org [Accessed 20 September 2020].

Porter, M.E., May 1979. How Competitive Forces Shape Strategy. *Harvard Business Review*, Vol. 57, No. 2, pp. 137–145.

Simon, H.A., 8 December 1978. *Rational Decision-Making in Business Organizations Nobel Memorial Lecture*, Pittsburgh, PA: Carnegie-Mellon University, ©The Nobel Foundation.

Takeuchi, H., and Nonaka, I., 1986. The New New Product Development Game. *Harvard Business Review*, Vol. 64, pp. 137–146.

The Scrum Alliance, 03 January 2018. *State of Scrum 2017–2018 Scaling and Agile Transformation*. The Scrum Alliance. Available at: www.scrumalliance.org: https://resources.scrumalliance.org/Article/state-scrum-2017-2018-report

4 Preparing for Launch Part 2

Writing the Business Case

Aim of this chapter: To understand the need for and set the context for a business case and to give guidance on the organisation of its contents.

Learning outcome: To understand the steps required to prepare an effective and persuasive business case which can be approved by the relevant decision makers, whilst being aware that the business case is a living document which is periodically updated throughout the life of the project.

This chapter is about the second part of the preparation stage of the project, in which we bring together all the information that we have amassed and included in our four document set. We now have to review it, structure it, filter it, and present it in a suitable business case format for the project board. Usually, the project board will take the decision on whether or not to proceed with the project and also on how it should be delivered if given the go-ahead. The client organisation may, however, have other levels of decision making up to board level, and in public sector projects, in particular, other stakeholders may be engaged in the decision. When agreed, our business case forms the baseline against which the performance of our project will be measured in the future.

According to the UK government's comprehensive business case guidance, the rationale for a business case is as follows:

> A business case should include a cost-benefit analysis and demonstrate strategic, economic, commercial, financial and management justification . . . . The business case should be developed over a number of phases and should be updated to reflect changes and reviewed prior to every gate or decision point to justify continuing the work.
>
> (UK Infrastructure Projects Authority, 2021)

A good business case, however, also explains the answers to the following basic questions to a wide audience, including decision makers:

- Where are we now?
- Where do we want to get to?
- Why should we make the journey?
- What is the best route?
- How are we going to get there?

In summary, it is in the business case that we have to persuade the organisation's decision makers and stakeholders that the project, as we envisage it, will meet their requirements in a timely and cost-effective manner.

DOI: 10.4324/9781003405344-5

How Should I Prepare an Effective Business Case?

The business case is the single most important document for which we as project managers are responsible because, unless we prepare it with great care, it can:

- Lead to a perfectly viable project not being approved.
- Set unrealistic expectations regarding the benefits to be expected from the project.
- Mislead the organisation in relation to delivery time and project costs.
- Lead to the project's goals being delivered in an inappropriate way.

In one document we have to:

- Relate the project to the context of the organisation and its strategy.
- Show that due diligence has been completed in considering different alternatives.
- Encapsulate the financial case for continuing the project.
- Show that there is capacity to deliver the project.
- Set out the way forwards to completion.

Figure 4.1 shows, in more detail, the simple steps we could follow in preparing our business case.

It's time to put on the mantle of the bio-energy from effluent project manager again.

The telephone is ringing — it's the finance director.

"I understand that you are progressing with preparing a justification for all this sewage nonsense. I have to tell you that I'm not convinced it is a good idea, and it is not what the company should be doing at all. Sewage, I mean, really! But putting that to one side, if the numbers add up, I'll support it when it comes to the board. My concern is that your numbers are shown to be robust and that the project will be financially sound."

Not a call that you expected, but here is an opportunity to gain a potential ally at board level, if they can be brought round. For now, reassurance is needed, as is a direct and immediate response. The project will need both the finance director's support at the board and engagement from finance staff in preparing financial projections. A quick answer is needed to show that this has been properly thought through.

So, how can this call be best handled to ensure that the professional credibility of the project (and the project manager) is enhanced?

Step 1: Review and Update Information from the Four Document Set

The good news is that because we have completed the project frame, possibly a separate justification for starting the project, and the four document set, we already have a lot of the information we require. Also, in consulting on these documents and obtaining agreement with them, we should have already built the relationships with our key stakeholders and project board members that we will need in order to develop our business case and win approval for it.

In even better news, the readers of our business case are going to ask us to provide convincing answers to the same *seven* questions that we have already asked and answered, and at the risk of repeating ourselves (please chant aloud if you wish):

Their names are What and Why and When,
And How and Where and Who,
(And How Much).

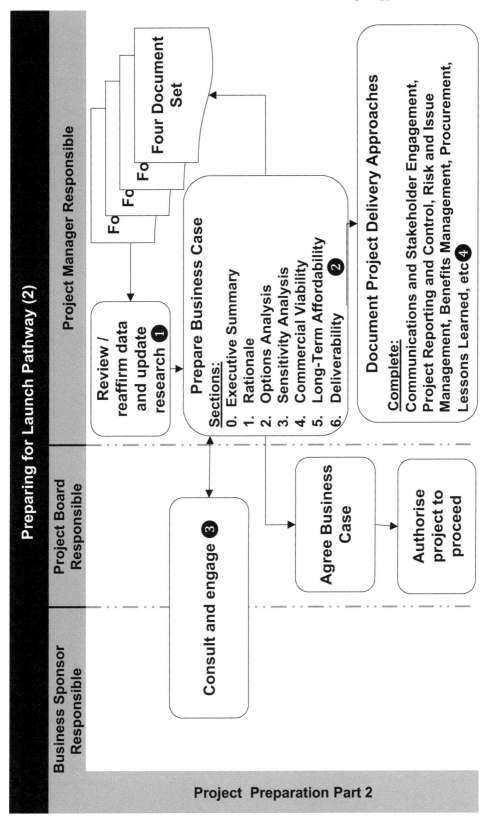

Figure 4.1 Preparing for Launch Pathway (2)

Although we have already done much of the research, thinking, and relationship building required for completing our project's business case, we need to confirm that it is still valid. This is doubly the case if there has been a delay in the approval process. Whilst some of our information will remain static – for example, performance and financial figures relating to older previous periods – there may be fresh data on the most recent periods, the current situation, and projections for the future. Equally, prices and costs may change rapidly, and if we are using third-party suppliers, their offerings may also change. Consequently, we need to take a short period to review and refresh the information upon which we will be building our business case. It can also be useful to meet again with key subject matter experts and stakeholders to reaffirm both the information and any fresh conclusions that are drawn from it.

Step 2: Building the Business Case

Taking a step back and considering how a business case can be presented, our old friend from Chapter 3, economist Herbert Simon, can help us explain what we are trying to do with a business case. In his Nobel Prize lecture, he also posed the following question: "Given a particular environment of stimuli, and a particular background of previous knowledge, how will a person organize this complex mass of information into a problem formulation that will facilitate his solution efforts?" (Simon, 8 December 1978)

The business case is a collation, a focus, of the mass of information that we have built up about the project. Our task is to organise it in such a way as to convince others of the justification for investing in the project, accepting the results of our options analysis as the best solution, and agreeing on our high-level plan as the most appropriate way of delivering it – i.e., it is a document primarily targeted at a particular, and usually very small, group of people: those who will approve the project. We can think of the business case as needing to be a very persuasive sales document aimed at explaining to the project board what needs to be done, satisfying any hesitation they have about the correctness of the approach, and overcoming their fear of committing resources to an uncertain outcome. It has been said that the main motivators to address in selling anything, including ideas, are fear, uncertainty, doubt, greed, and envy. Similarly, our business case needs to address each of these possible negative blockers in the mind of each member of the project board. And we will know all about them individually because we have been in constant dialogue with them.

This means that in deciding the contents of our business case, we have to address two issues:

- What information does our audience really need in order to make their decision?
- What information do we have?

If there is a gap between the two, we have more work to do, and if we have more information than is required, we need to edit our document very carefully so that readers can extract the information that they require without losing it in a mass of, what is to them, irrelevant detail. We are, after all trying to get them to agree to our recommended solution, and we need to write the business case in such a way as to help them do so. Keeping the document to the point and presenting only the information that the decision makers need is in line with our principle of valuing their time.[1]

There is much guidance on good practice for business cases available to us, so the decisions on how to organise Simon's "complex mass of information" is much easier. The G20 Infrastructure

Working Group has set out five principles that should be included in a business case, which are a useful starting point for even the smallest project; these are rationale, options, viability, affordability, and deliverability (G20 Infrastructure Working Group, 2018).

The size of the organisation and the relative size of the commitment required to deliver the project will help determine the level of complexity required in the business case. In the UK, the government Treasury Department issues a regularly updated "Green Book" running to over 120 pages of how to evaluate investment proposals in the public sector, plus supplementary guidance notes, and another book of 130 pages on "Business Case Guidance for Projects" (HM Treasury, 2018). This actually contains information and guidance on three different levels of business case: strategic, outline, and full. Other jurisdictions and large organisations will also have their own policies and procedures and standards for completing the business case. Not surprisingly, business cases can end up being lengthy documents.

Some organisations may require an outline business case initially and follow this later in the process with a full business case. More generally, organisations are highly likely to require very full and detailed reporting on specific items, such as procurement strategy and risk management. The golden rule for us is that if the organisation has rules for business cases, follow them exactly. Failure to do so may result in unnecessary delays whilst the business case is reworked to meet the client's standards or could even lead to the project being dropped entirely.

A useful first step is to research with the business sponsor if there is a recent, similar, and successfully approved business case that we can use as a model. Failing that, and if there is any uncertainty about the shape and contents of the business case that is required by any host organisation, the PRince2 methodology has a useful solution that we can "borrow," even where we are not following the full PRince2 approach. Before starting any main document (such as a business case), PRince2's solution involves preparing an initial document, called a product description. This initial document is probably only a few pages long and allows us to define the business case's content and obtain agreement from our project board to the:

- Information that the completed business case will contain.
- Sources of that information.
- Document authorship.
- Quality criteria.
- Responsibility for any required quality reviews.

This approach of using a product description fits very closely with our "no surprises" rule from Chapter 2.

Putting It All Together

Whatever the required level of content, our business case has to be written persuasively in the language used by our target audience. It should also tell a strong story, starting with the broad overview and increasingly leading the reader thorough to our unarguable conclusion of what needs to be done and how we intend to do it. We can prepare such a strong narrative by ensuring that we adhere to the G20's five principles by answering the following questions:

- Why are we undertaking this project?
- What ways of delivering a solution have we considered?

- How have we arrived at a recommended solution?
- Why would a supplier engage with us to deliver it?
- How much will it cost, and what will the benefits be?
- How will the project be delivered, where, and by whom?
- When will the project be completed?

These questions follow the recommendations of the G20 Working Group,[2] and the answers to them provide us with a useful breakdown of the required sections by which we can build a compelling business case to organise Simon's "complex mass of information." The table of contents of our business case could therefore be as follows:

0   Executive Summary
1   Rationale, including some or all of the following:

- The business context.
- Current situation analysis.
- Scope of the required solution.
- Critical success factors.
- Stakeholders.
- Constraints and dependencies.

2   Options Appraisal, showing the reasoning behind the approach used and the individual evaluations:

- Longlist of options.
- Shortlist of options; for each option show: benefits, including "soft" i.e., non-cashable benefits; costs; and risks.
- Recommendation.

3   Sensitivity Analysis
4   Commercial Viability of the recommended solution
5   Long–Term Affordability of the recommended solution, including benefits
6   Deliverability of the project, including:

- Project delivery method.
- High-level plan.
- Post-handover operational and benefits management.

However, to reiterate, we always have to follow the host organisation's guidance, rules, requirements, and templates when structuring a business case, and we may need to add further sections or content as required.

Part 0: Executive Summary

The usual guidance on what should go into a business case, at whatever level and from whatever source, seems to often miss one key point. The area of coverage of business cases and marshalling the information we have amassed means that they are often lengthy documents. But, sadly, not all our readers will be either motivated or have the time to read our tremendous and compelling masterpiece thoroughly. There are, however, two things that we can do to help them:

- Place *all* detailed tables and supporting evidence that we wish the reader to have access to in a series of appendices. This avoids losing our key narrative in a mass of detail.

- Prepare an executive summary. Although first on the contents list, compiling an executive summary is our very last task in preparing the business case. This should follow the structure of the main document and condense our main points into a short readable chapter.

Our objectives for the executive summary are to:

- Enable the reader/decision maker to understand the critical information quickly.
- Allow the reader/decision maker to easily delve deeper into the substance of the business case wherever they wish for further information. It is to facilitate this process that we should structure the executive summary to follow that of the main part of the document; usually subheadings in the executive summary should relate to chapters in the main document.

It is best to write the executive summary last.

Part 1: Rationale

The Business Context

We may believe that our audience is fully engaged with the client organisation and aware of the problems, issues, and opportunities it faces. This is usually not the case, especially where outside individuals will be engaged in making or supporting decisions about the project. These individuals may be non-executive directors, independent committee members, or drawn from funding institutions, professional bodies, and so on. As project managers we cannot be sure of what they do or don't know, so it is always worth including a moderately wide overview of the adopted strategies, goals, and objectives of the organisation to set the context of the proposed project.

This statement of the wider context should include reference to the business strategy that lies behind the project. In our bio-energy from effluent case study, for example, this could read:

> The company has achieved 5% growth per annum and delivered increasing shareholder value for the last ten years by expanding our range of contracted services in global markets. The company has identified that to maintain its growth plans for the future five years, it needs to adopt two strategies:
>
> - To grow organically by delivering new types of service to its existing client base and expanding existing services into new industries and new markets globally.
> - To seek out and develop new service areas of specific interest to new industries in each market. This will require both inorganic growth – acquiring existing high-performing companies – and delivering market-leading advantage through the development of novel technologies.

We can also draw on the vision of the project and can set out the business problem or opportunity the project will be addressing. So, in our bio-energy case study, the vision could become (and by now you should know what's coming next):

> Following the development of our world-first patented bio-energy from wastewater plant, which as its sole "waste" product produces a high nitrate fertiliser, we will become the leading player in the supply and maintenance of sewage treatment services globally. Innovative funding arrangements will mean that utility operators' customers will be able to

access our equipment on a payment-for-use basis, and they will not have to find capital investment funds. The ongoing cost to the customer will be more than met from their cost savings and from their energy and fertiliser sales income. This initiative will generate a continuing revenue income stream for us that is guaranteed for the life of each contract. Market penetration will be delivered through supply or partnering arrangements with an existing respected player in each market and potentially through their ultimate acquisition.

Current Situation Analysis

Moving on from the generic goals of the organisation, this section focuses on the Why of the project. Wherever possible, it is useful to attach numbers to the description – for example, in the IT systems replacement project example:

> The ageing accounts and payroll system is now suffering a mean of over 30 system faults per week resulting in an average system availability during office hours of only 75%. This has major knock-on effects on the operation of the business especially during critical points such as month end. On three occasions this year monthly-paid staff have had to be paid more than one day late.

We should also include an analysis of the problem and its root cause or define the opportunity that the project is seeking to exploit. Wherever possible we should set out the current measures of performance, which will be the baseline against which we will subsequently map and quantify the benefits of the project overall and of our recommended solution. In our IT finance system example, our analysis of the problem could continue:

> The existing accounts and payroll systems were licensed from company XX seven years ago, and we currently hold licences for their use by YYY staff. The systems were installed on new hardware which was purchased for them and has been renewed once, nearly four years ago. The current hardware is out of warranty, certain elements are no longer maintainable, and the entire hardware platform now needs to be replaced. In addition, the software supplier was acquired by a competitor three years ago and stopped developing new functions for our systems. The supplier has now announced that all support for our systems will end on 31 March. Our internal support team has recently had to apply multiple software patches from the supplier to keep the system running, including on three occasions this year having to switch the system to the backup site whilst undertaking major repairs. To undertake each reversion to the backup site takes a minimum of eight hours, with a similar time to bring live operation back online – if everything runs smoothly. Such switches between data centres need to happen overnight. The impacts of these failures and switching to the backup site in the last three months alone have been:

- A reduction in service availability percentage to 85% during normal working hours.
- Identifiable additional non-staff costs directly attributable to systems failures: £15,000.
- IT department staff time, including weekend, night, and overtime working: 2,000 additional hours.
- Delays in processing accounts resulting in additional overtime and weekend working by accounts and payroll staff: 1,250 additional hours.
- Total additional staff costs over three months: £85,000.

In our continuing example of the bio-Energy from effluent project, the opportunity could be stated as:

> Pioneering work by the research and development team has led to a ground-breaking development in the processing of raw sewage. This paradigm shift in sewage treatment creates a major advantage for us as a prime mover in the global sewage treatment marketplace. Using a novel biochemical process, housed in a sophisticated but easily transportable processing unit, it is possible to transform the sewage into an easily handled sterile fertiliser, clean water – which can either be used for irrigation or released as steam – and most importantly, the reaction process is exothermic and can be used to generate electrical power for resale. Each plant installed will deliver estimated return of 10% per annum above costs of operation and depreciation and will have a 15-to-20-year life depending on the operating conditions.
>
> Our prime mover advantage which we will obtain from leveraging this new technology will enable us to expand into new markets where we currently have no footprint and over time to build on our success by bringing in a wide range of our other proven services. We anticipate that the project will directly deliver opportunities for sustainable growth in group profitable revenue of 10% annually by year 5 and increase growth in other revenue streams by a further 5% by the same time.

Scope of the Required Solution

Now is the time to pillage our end state design for the right words to use here. It is also useful to add in any technical standards that the solution will have to meet. For example, in IT systems this could relate to information security, or in construction it could relate to health and safety or environmental considerations. In summarising the potential scope of the project, it is useful for us to introduce the reader to the different levels of options available that we have identified and will be reviewing in detail subsequently, including:

• The minimum scope to meet the immediate business need, including "do nothing."
• Any intermediate option.
• The full scope of the project and the proposed changes required to reach the end state.

In the IT finance system example, a statement on scope of the required solution could be written as follows:

> When the new accounting systems are installed and fully operational, users will be able to enjoy 99.99% available during normal office hours, with a 90% system availability at other times to allow for essential systems maintenance and other operations. The systems will operate under UK VAT and other accounting rules and will support tailoring to mirror our existing and unique chart of accounts but still allow for future flexibility. They will integrate automatically with other systems, such as purchasing, customer relations, payroll, and human resources. The new systems will support up to 90 users at any one time, including data entry, management, and reporting. These users may be operating either in central or area offices or remotely via a secure web interface. In order to preserve the operation of the finance department, the new system will operate with a failover to a backup site that is operational within five minutes of a main system failure and with no loss of data. In addition to inbuilt standard reports, flexible management reporting using a commercial business intelligence package will be available through prebuilt interfaces.

Critical Success Factors

We originally outlined our critical success factors when we completed the project frame, and by now these may have been further refined based on our increased understanding and updated for use in our options analysis process. In any case, following our "no surprises" rule, by now, the critical success factors should be well understood, have been discussed and agreed with the decision makes and other stakeholders, and by this stage should not be controversial.

Again, as the business case will usually have a wider circulation than the documents prepared to date and will form the overall baseline for measuring the success of the project, wherever possible we should always include relevant objective measures for each of the critical success factors.

Stakeholders

Depending on the nature of the project, there may be several different groups of stakeholders, and it is useful to describe the principal concerns and ambitions of each group, then note where there may be conflicts between them.

A simple, but probably extreme, example of different groups of stakeholders having different priorities arose in one project which required procuring a series of public hospital IT systems in the UK. Unresolved conflicts arose when deciding on priorities for the project which eventually surfaced when choosing the recommended solution:

- The technical evaluation team were prioritising an easily implemented, lowest cost, "off the shelf" solution from a North American company.
- The users' priority was for the most feature-rich solution possible provided from a European software company, even at a much higher price.
- The main board reflected the government's preference for a new, bespoke solution that could be developed in the UK at a projected cost between the other two options.

Constraints and Dependencies

We originally identified constraints and dependencies in developing the project frame, and they need to be clearly restated here and, if necessary, expanded as the result of our now increased understanding. Essentially, the constraints form the boundaries of the space within which we have to deliver the project. These may be our, by now, old friends of time, cost, quality, and scope but can also include the rate at which the organisation can accept major business change or contractual commitments, such as the time period for an expiring contract or other limitations on timing – for example, avoiding times of major staff commitments like peak workloads, holidays, and other annual cycles. Simple examples include scheduling highway projects near schools as much as possible during school holidays and avoiding replacing an accounting system just after an organisation's financial year end, which is often a busy period for finance staff.

Our project may also be dependent on the successful delivery of other projects – for example, an office relocation which depends on the refurbishment of a building, or the launch of a new product which is dependent on the completion of a new production facility. Equally, other projects may be similarly dependent upon the successful completion of ours. These crosscutting dependencies are why several discrete projects may be co-ordinated and managed as one programme.

Part 2: Options Appraisal

We have already completed all the necessary work on each option by this stage, including benefits, costs, and risks. Now it is a matter of presenting our options appraisal succinctly in the business case.

Longlist of Options

Ideally, we should include our reasoning and the information prepared to support our reduction from the longlist to the shortlist. We can help our readers by taking a visual approach (such as that in Table 3.2 in Chapter 3) to summarise both the data and assumptions used in reducing our options down to the shortlist. Where the information on each option is complex and detailed, it is extremely useful to pull only the summary data into a table into the main text and follow the rule of including our detailed work and possibly supporting spreadsheets as appendices.

Shortlist of Options – Comparison

If the approach of agreeing on the different elements of the four document set in advance has been followed, this section should only bring forwards the previous work agreed on by the project board in our options analysis paper. When, however, we are considering our different shortlisted options, it is always worth revisiting a couple of areas in particular before preparing our recommendation:

- The costs of each option, including contingency funds, and the effect of timing of delivery upon them. For example, a capital-funded option, where funds are required before live operation, will have a very different profile to one where costs are incurred over the life of the solution being delivered and may be less acceptable to the client even if the total cost over the life of the project is lower. A famous historic example of this approach was the business success of the Boulton and Watt company in the UK in the late 18th century. The company introduced a revolutionary new steam engine designed by James Watt which used a staggering 75% less fuel than previous engines. Their business model was also innovative. Instead of selling the engines, which would have had a high up-front cost for purchasers, they levied licences fees for the use of the engines. These licence fees were charged over 25 years and were to be funded by one third of the savings in the purchasers' fuel costs over using less efficient engines.
- The benefits for each option merit further careful review at this stage because there may be variations in types of benefits between options, and these may also influence the way their achievement will need to be managed, monitored, and reported.
- The nature and value of cashable benefits may vary between different options. One example was in a system change where one option allowed the reduction of posted-out mail in favour of electronic communications, reducing the direct costs of postage. There was also a non-cashable service improvement as time lags in the process were reduced.
- There may be variations in non-cashable benefits, and it can be useful to consider the benefits of each option using a wider approach, such as the balanced scorecard introduced by Kaplan and Norton (Kaplan and Norton, 1996). Their approach indicates that we can split benefits into four main perspectives, financial, customer, internal processes, and learning and growth.
- The timing of start points for achieving benefits may change. In one major government-to-citizen service project, the first option was to progressively replace a number of commonly used documents that citizens had to submit, and the agency had to then enter into their systems,

with a web-based submission of information. This meant that the agency could start to reduce the staff input required to key in data very early in the rollout process. The second option was to introduce the whole service in one go much later, meaning that staff savings would not have been achieved until some years afterwards.

Recommendation

This is the key part of the business case in which we draw together and compare information from our analyses of each option to produce a clear argument for a preferred solution. The winner will be the solution that can be delivered quickly, has the lowest risks and costs, and will deliver maximum benefits. Sadly, life is always a compromise, and our next step is to evaluate each option against these parameters. The most useful tool for performing this comparison against multiple criteria is, once again, the weighted decision matrix.

Tables 4.1 and 4.2 show how this would work in the IT replacement accounting system example.

In Table 4.1 we simply tabulate the results of our assessment of the performance of each of the shortlisted options against different variables. In this example, four factors are being used; the net present value, non-cashable benefits, risk, and time to complete implementation.

Table 4.1 Analysis of Option Performance

IT Accounts System Replacement Project Analysis of Option Performance				
Shortlisted Options	**Assessed Performance**			
	NPV £000	**Non-Cashable Benefits**	**Risk**	**Time to Complete (Months)**
1 Do nothing (usually an option).	65	0	100	0
2 Buy time on an existing accounting system.	90	30	30	6
3 Replace hardware and use existing accounting package/update own premises.	75	30	85	9
4 Install replacement hardware in commercial data centre/existing accounting package.	85	30	85	9
5 Install replacement hardware in commercial data centre/new accounting package.	67	95	20	9
6 Buy capacity from a commercial "cloud" service provide/new accounting package.	100	105	20	3

Table 4.2 takes our previous results and gives each variable a standardised score. In this example, the score given reflects a simple ranking of the six options being considered by order with the highest rank of 6 being given to the "best" result and 1 to the "worst" result. The next stage is to multiply the ranking by a previously agreed-upon weighting factor and then to rank the overall total results.

Table 4.2 Option Ranking – Weighted Scoring

Shortlisted Options	Ranking of Assessed Performance				Weighted Scores (Rank x Weight)					Final Rank
	NPV	Non-Cash. Benefits	Risk	Time	NPV	Non-Cash. Benefits	Risk	Time	Total Weighted Score	
Weighting					**60**	**30**	**15**	**10**		
1 Do nothing (usually an option).	1	1	1	1	60	30	15	10	**115**	**6**
2 Buy time on an existing accounting system.	5	2	4	2	300	60	60	20	**440**	**2**
3 Replace hardware and use existing accounting package/update own premises.	3	2	2	4	180	60	30	40	**310**	**5**
4 Install replacement hardware in commercial data centre and use existing accounting package.	4	2	2	4	240	60	30	40	**370**	**4**
5 Install replacement hardware in commercial data centre and use new accounting package.	2	5	5	6	120	150	75	60	**405**	**3**
6 Buy capacity from a commercial "cloud" service provider and use new accounting package.	6	6	6	2	360	180	90	20	**650**	**1**

IT Accounts System Replacement Project
Weighted Option Ranking

Table 4.3 shows a slightly more sophisticated approach in which, instead of a rank for the assessed performance of each option against a variable, a standardised score is allocated based on the assessed performance of each option as a percentage of the "best" performance. In the example shown in Table 4.3, the standardised score has been multiplied by 6 solely for comparability with Table 4.2. The next stage is to multiply the standardised score by the previously agreed-upon weighting factor and then rank the overall total results. This approach has the advantage over ranking in that the standardised scores are still related to actual performance and a "near miss" is not penalised in the same way as when rankings are used.

Part 3: Sensitivity Analysis

We already understand that our options analysis is based on the best information that we have at the moment, plus a series of assumptions, and that the extent of the assumptions may vary between different options. Before completing our recommendation on the best solution to adopt, we need to establish just how wrong our assumptions can we be before our recommendation should change. Or, when expressing the same thing to the project board, we can say that we need to demonstrate the robustness of our analysis of the options.

The tool that we use for this confirmation of robustness is sensitivity analysis. Depending on the number and intricacy of the assumptions that we have had to make for each option, this modelling can become a very complex analysis in its own right. We could rerun our options appraisal models many times, each time making one small incremental change to one of our main variables (the so-called one-at-a-time, or OAT, approach). The result is a large tabulation showing all the results, and we can report conclusions such as, "Option 1 will still be preferred if build costs go up by 7.9% or less, but Option 2 will drop from second to fifth place except when implementation time is less than six months."

We need to bear in mind our satisficing stance on decision making. We are not seeking to describe every possible scenario or find the ultimate perfect solution. Our search is for an option that is likely to fit our needs across the widest range of potential scenarios – i.e., Herbert Simon's theory that we should be looking not for the optimum solution but for the first one that is likely to do the job given the information we have and our ability to analyse it. Remember, in his terms we are searching for a needle that is sharp enough to sew our seam in the widest variety of realistic scenarios, rather than the absolutely sharpest needle in the haystack. Therefore, and at the risk of annoying people who love developing and running models for decision making, we do not need to vary every estimate, weight, and assumption and then consider every possible combination of these estimates, weights, and assumptions in seeking that optimum solution. We need to focus only on those changes most likely to impact our real-world project and our four key influences, which are, as always, time, cost, quality, and scope.

It may be, however, that the results of the testing of the options against different stresses can be communicated better to the project board and stakeholders if they are based around realistic possible scenarios that could affect the project. As we have previously identified the major risks, we already have the basis for identifying the scenarios to use. This approach also means that for each scenario, we will usually be amending multiple variables and looking at the impact of their

Table 4.3 Scored Option Weighted Ranking

Shortlisted Options	Standardised Score *				Weighted Scores (Standardised Score × Weight)				Total Weighted Score	Final Rank
IT Accounts System Replacement Project — Scored Option Ranking										
	NPV	Non-Cash. Benefits	Risk	Time	NPV	Non-Cash. Benefits	Risk	Time		
Weighting					**60**	**30**	**15**	**10**		
1 Do nothing (usually an option).	3.9	0.0	1.2	6.0	234.0	0.0	18.0	60.0	**312.0**	**6**
2 Buy time on an existing accounting system.	5.4	1.7	4.0	2.0	324.0	51.4	60.0	20.0	**455.4**	**3**
3 Replace hardware and use existing accounting package/update own premises.	4.5	1.7	1.4	1.0	270.0	51.4	21.2	10.0	**352.6**	**5**
4 Install replacement hardware in commercial data centre and use existing accounting package.	5.1	1.7	1.4	1.0	306.0	51.4	21.2	10.0	**388.6**	**4**
5 Install replacement hardware in commercial data centre and use new accounting package.	4.0	5.4	4.8	0.7	241.2	162.9	72.0	6.7	**482.7**	**2**
6 Buy capacity from a commercial "cloud" service provider and use new accounting package.	6.0	6.0	6.0	2.0	360.0	180.0	90.0	20.0	**650.0**	**1**

interaction on each of the options. In our IT finance system replacement example, scenarios could include examining:

- What would happen if options with new IT buildings/extensions which required planning consent are delayed because consent is refused by the local authority and we have to go to a planning appeal, possibly delaying the project by many months?
- How would an increase of 10% in building refit costs impact those options using our existing premises?
- What would be the impact if a key software supplier withdrew from the market?
- How would a major change in tax legislation affect the delivery and maintenance of each option?
- How much in additional cashable benefits would accrue for those options using existing facilities if delivery could be condensed to complete in four months?

As ever, it is always best that if we decide to adopt the scenario-based route for sensitivity analysis, we should consult and gain agreement in advance on the scenarios to be used and on how the results are to be reflected in the recommendation of a preferred approach.

Part 4: Commercial Viability

This section of the business case applies to projects which will require external suppliers – i.e., those that will need to be governed via one or more contracts. It draws on our previous market research with both suppliers and other organisations that have implemented similar projects. We need to demonstrate that we have considered alternative commercial arrangements and that those proposed are both feasible and sufficiently attractive to the market to generate interest from suppliers. These arrangements could include items such as:

- Allocation or sharing of risk of different types.
- Performance incentivisation mechanisms, including retention payments, and scaling payments for under- or overperformance.
- Contracting models such as partnership working or early contractor involvement.

Other issues that we can set out in this section could include, as appropriate:

- Our preferred procurement route – in many cases this will be dictated by company policy or legislation, but options may be available within this, including, for example, the use of existing supplier contracts or framework contracts, which are common in the public sector and may reduce procurement times.
- Whether the contract will be specified in terms of every detailed requirement in our solution specification or in broad business outcomes – e.g., in the replacement accounting system, we could specify every piece of hardware necessary for data storage and its backup or simply require that the solution provides for at least ten years of accounting transactions to be held available online. Another very simple example was in an accounting system procurement where a key requirement was stated as the value fields must have sufficient digits to cope with a twice-yearly remit of a maximum value in yen from a Japanese subsidiary, rather than specifying that the solution needed to have a value field of format 9,999,999,999.99.
- The length of any service and support contracts and any potential issues relating to length. For example, a short-term support contract may be less attractive to the marketplace and carry a higher cost than a long-term one, but that, in turn, risks tying in to a supplier who may not perform adequately.
- How the risk of non-performance or sub-optimal performance of the solution will be shared with the supplier.

Part 5: Long-Term Affordability

In our NPV calculations, we have already identified the costs (including contingency funds) and benefits of the solution that we are recommending. In this section, we have to set those costs and benefits over the whole life of the project in terms of the client organisation as a whole as benefits in one area can lead to additional, and sometimes hidden, disbenefits elsewhere. One example of this was a government agency that relocated the IT service from London to a remote area in South West England beyond feasible daily travel limits. As the new location had lower premises costs and wage rates, direct IT costs were reduced. However, the effect of the move was to increase the travel times, as well as travel and hotel costs, and reduce the productivity of all managers and staff in HQ departments and area offices who needed to attend meetings in the new IT offices.

As the impact of cashable benefits will inevitably lag behind the additional expenditure that the project requires, we also have to set out a cashflow statement showing the benefits, costs, and sources of funding required by the project and by our implemented solution over its lifetime. The information required to prepare such a statement is already contained in our NPV analysis, but discussion with the business sponsor and the organisation's financial staff may be required to identify how the project costs will be funded and any additional fees that may be required to cover lending arrangements and costs.

In order that the impact of the project can be monitored, we need to agree with the organisation's financial managers how project and post-project expenditure and benefits can be identified and reported. We then set out these arrangements in this section. These arrangements will need to recognise that existing financial management arrangements and systems are often departmentally based and so the financial impact of a project can be difficult to split out from normal operating incomes and expenditures. This difficulty is exacerbated where the work of many different departments is affected by the project and the solution it has delivered.

Part 6: Deliverability

The objective of this section is to demonstrate that we have suitable arrangements in place or planned to ensure that:

- The project can be delivered successfully under appropriate control and quality assurance.
- After handover of the solution at the end of the project, there will be suitable operational management controls.
- Ongoing management and monitoring will be in place to ensure that anticipated benefits are achieved.

If we followed the four document set approach prior to our preparation of the business case, project delivery should have been extensively covered by our recently prepared project delivery method and high-level plan. This will have been reviewed and agreed on by the project board, and they can now be slotted directly into the business case document at this point with little or no further work.

The final areas that need to be covered in the deliverability section are the post-project management of operations and benefits. These should have also been previously detailed in our end state design and reviewed and approved by the project board. This work can now be pulled directly into the business case and supplemented as necessary by any further thinking and the outcome of any more recent discussions with operational management.

Step 3: Consult and Engage

Throughout the business case completion process, it is advisable to be in regular discussions with the business sponsor and the members of the project board. Partly this is us following our "no surprises" rule, but mostly it is because their collective knowledge and views will assist us in shaping and selecting the options that we have to consider. Even after completing the final business case document, when it is tempting just to send it out to the project board, we should not sit back and wait for the meeting to approve it. If we have planned our timing properly, previously discussed the four document set, and consulted during the development process, we will have enough time to work through our final draft of the business case with the board members individually before we issue it to them formally for agreement. Even though the board members should have seen many of the elements of the document already, this step can be incredibly useful as a final sanity check. Our main objectives in doing this are to ensure that we have addressed any concerns they may have expressed previously and to guide them to the relevant part of the document where we have covered them, but we may pick up on surprising and previously unstated issues as well.[3]

Business Case as a Living Document

So, the business case is done, finished, and approved and we can now put it away for ever and get on with delivering the project, or at least developing our detailed procedures and plans. Well, not quite, remember the second part of our quote from the UK Infrastructure Authority about business cases, which is repeated by the UK Government Standard for Business Cases, includes the following phrase: "against which continuing viability is tested" (HM Treasury, 2018). All too frequently, business cases have not been used properly to help us control the project delivery and keep the community that we have involved engaged and supportive. Too often, the business case, once prepared, submitted, and approved, is put on the shelf and left to gather dust. The business case is not a document produced for a "one off" approval but as the standard against which the project and its various deliverables, outcomes, and end solution can be validated during the remainder of the life of the project and afterwards.

We already understand from our detailed information gathering and estimation processes that our knowledge of the project and its costs, risks, benefits, and timescales will evolve during the project. Consequently, the information upon which the business case relies will naturally change, and we should review and update it at agreed-upon points and critically examine whether the project is still viable as originally conceived. It is always great news if our intermediate project reviews show that we got the business case close to right and everything still looks to be progressing within the tolerances that we expected. If, though, when we review the business case, it is not such great news, we all have a psychological problem – admitting failure. Afterall, we were the ones who recommended how the project should be shaped and justified and its objectives delivered.

Time for another field trip. In fact, you are back in the role of project manager for our favourite bio-energy project.

No need for a coat or probably even a jacket. It is some months after our last visit; things are not going so well, and you are going back up to the boss's office to face a less than happy superior.

Boss: "Thanks for trekking up here, and for your review of progress against the business case. I have to tell you that it has not gone down well with the main board. I know that everybody happily agreed with all those assumptions that you made about the state of sewage treatment across five continents, but they still regard you as their, owner and they are disappointed. What they are saying is that we are facing a potential failure of the project with no visible means of resolving it."

Now, you could say to your boss, "Weeeell, it's not quite what we thought and looks like it might not be, you know, well, things have changed a bit. The researchers' plant isn't scaling up well. None of us foresaw that this would happen. And sort of, well, probably I think that with a bit more time and cash I can promise that we will get there . . ."

Or you could own up to the changed information and changed conclusions and try and make the most of the opportunity that the discussion presents.

Speaking Truth to Power

Fortunately, there is a simple assertive technique that can be very useful when we have to present bad news, such as in our case study:

- **State** the problem in clear and unemotional terms.
- **Express** your views on the issue.
- **Review** the requirements for moving forwards.
- **Explain** the results that you are expecting.

This is how this approach could play out in the case study scenario. So back to the boss's office, and this time we take firm charge of the discussion, starting with a metaphorical, and probably physical, very deep breath and saying:

> **(STATE)** *"The current situation has arisen because the initial small production-sized plant is not generating the predicted scaling up of power yields from the small research plant that was expected when the business case was approved. In fact, the test yields are currently running at around 70% of the predicted yield. This means that the solution as envisaged in the approved business case is no longer profitable."*
>
> **(EXPRESS)** *"I have looked back at how we got where we are, and I regret that the range of our sensitivity analyses only looked at a range of plus or minus twenty-five percent – i.e., they only extended downwards to seventy-five percent of predicted yield. This calculation showed that at this level, the project would be marginally profitable. Obviously if we had dropped the yield predictions by a further five to seventy percent, it would have shown that the project would be unprofitable."*
>
> **(REVIEW)** *"We now have a choice that either we close the project now and write off the expenditure to date or we change course. We can produce a second-generation research rig*

exploiting everything that we have learned and then jump from that straight on to a full-size production model using the improved technology. In parallel, the research team are also going to look at the mix of micro-organisms in the digester and see if we can tweak it to obtain better yields by preferring the most advantageous organisms. There is also a good chance that we could not only increase the overall yields but improve the hydrogen-to-methane output ratio to reduce carbon emissions. This revised plan keeps closest to the project timescale and maximises both our previous investment and our opportunities for the future."

(EXPLAIN) *"If we follow this route, I suggest we will only need to add six months to the project timeline and find an additional $Y million. There is a very strong likelihood that this approach will succeed and the eventual result will be a vastly improved power yield, but obviously until we have done the work there can be no guarantees."*

Doing well so far, so it's time to push for more:

"I also feel that I need to keep a much closer eye on the project and need to be able to move onto it more or less full time. I know that's not what we planned, but we could move Charlie up from the insurance team leader role into becoming my deputy. I would continue with support, coaching for Charlie and working together on the major decisions. Not only does it release more of my time to focus on the project, but it provides an opportunity for Charlie, who is almost ready for a move up to gain management experience. If I focus solely on the project, I can get an updated business case and modified delivery plan out to the board by next Friday."

Boss: *"OK, I'm going to speak individually with the board members so that they know it is coming. Can you brief Charlie?"*

Your internal sigh of relief comes just a little too early.

Boss: *"I also think that I may need to keep closer liaison with you on this. Consequently, going forwards I have arranged for you to have an office up here on the executive floor so that I can drop in easily and you can keep me updated."*

If you would have adopted something like the second approach anyway, congratulations. This was a difficult meeting well handled, and the result is buying some time to prepare the updated business case and for the boss to start getting used to the idea that the project timeline will extend and that it will cost more.[4]

Many projects will require some adjustments to meet expectations, and facing such an issue with the business sponsor or project board always requires that deep breath and an open, honest approach. This can be nerve racking, but the alternative of drifting forwards, always chasing targets and numbers that are now impossible, is far worse.

Step 4: Getting a Head Start on Other Project Documentation

Experience shows that as project managers, we frequently end up working very hard to meet external deadlines imposed on us for preparing the business case. We then have to wait for the decision-making process to review and question our information and conclusions and, we hope, agree with our recommendations. Consequently, the period whilst waiting for a decision can be less pressured. If this occurs it can give us time to start working in more detail on other areas that will help the next stage get up to speed more quickly. We can start on the development of the range of policy documents, often called strategies in project management methodologies, which will govern how we deliver our project. This work includes formulating the management and reporting processes for the project and drawing up high-level approaches for items such as managing risk, communicating with stakeholders, and ensuring that benefits are delivered. We can also start preparing any templates for items that we may not have already, such as our risk and issues registers and project management reports.

The temptation is always to use this time instead to press ahead developing the detailed project plan, but there is always a risk of the high-level plan or project delivery method being amended during the business case approval process. Any changes that the project board or others require are more likely to lead to having to redo work on a detailed project plan than to our project management process and strategies.

There are a range of areas that we will need to document, and these are mainly covered in detail in other chapters. The following are areas which we could start considering now if time permits:

- Project management processes.
- Software support.
- Delivery strategies, including ones covering quality management, risk and issue management, benefits management, and lessons learned.

Project Management Processes

As we have already established, every project will have its own unique features and exists within its own specific organisational environment. As project managers, we have to take these into account when designing how the project will operate. If we have not yet done so, now is the time to ask a few key questions:

- Does the host organisation have an infrastructure to support project delivery, such as a project management office? If so, it is necessary to contact the responsible team and seek guidance.
- Does the host organisation have particular requirements for the logging of information and the management of user and other requirements? If not, we have to create systems to ensure that the amount of information that we uncover, create, and store can be managed effectively.
- Are we required to adhere to strict disciplines, such as a mandated project management methodology, in preparing future strategy and control documentation and manging project implementation? Or is a more flexible approach allowed or even encouraged?

Gaining an early understanding of the systems and processes through which the project is to be delivered is a major factor in helping the project run smoothly. Irrespective of whether we are mandated to use accepted standard systems within the organisation or have to create them specifically for our project, we need to include, as a very minimum, the following set of arrangements, systems policies, and processes:

- Systems for requirements management – tracking and tracing of solution requirements through to design, development, testing, and delivery.
- Version control and storage of project documents.[5]
- Methods for sharing documents with other parties under strict control.
- Planning, monitoring, and reporting activity and finances.

A real-life example of poor system control over solution requirements resulted directly in a pro-tracted contractual dispute, which was resolved short of engaging legal teams – but only just. The dispute arose at the end of the design stages of a public building refurbishment programme, and it absorbed large amounts of both contracted designer and client management time to resolve. It also resulted in a substantial amount of design rework. Properly enforced requirements management using dedicated software (or even a decent paper-based system) could have avoided the seemingly trivial dispute, which, at its core, was over whether the light fittings in the final design should have been round or square. The dispute erupted when the designs had been finalised and were presented for formal sign-off by the client. Claim and counterclaim were made, but in truth, the argument should not have arisen. Whichever party to the dispute was right is immaterial. The failure was not in design or client specifications but in the project process to track and agree on requirements as early as possible in the task. There would have been no need for expensive rework if the shape of the light fittings had been documented as a solution requirement by the designers and signed off on by the client's project manager in a timely manner at the outset. This, together with all other solution requirements, could then have been clearly tracked through the project to the final design and subsequent installation.

Software Support

Increasingly, project management and administration tasks, such as planning, reporting, document control and sharing, resource management, project time sheets, and requirements management, are supported by specialist software tools. Even in a simple project, a planning tool is usually essential. The more complex and/or longer term the project is, the greater the need for more complex tools covering a wider range of functionalities. If there is no standard software tool set to use within the host organisation, now is an appropriate time for us to schedule our particular requirements and investigate the current level of software support available in the market.

Probably the best known of these project support tools are Microsoft Project and Oracle's Prima-vera,[6] but there are multiple and multiplying alternatives for project planning and control. There is a clear division in the software products between standalone "best in class" software packages that handle one aspect of project management and integrated applications that bring all the functional-ity to support a project into one package – but each feature may be less rich than in the standalone ones. Some of the packages have free-to-use versions, admittedly with limited functionality, but this may be sufficient to support smaller or less complex projects.

The types of features we may require from software to support project management and team operations may include:

- **Plan development and task management.** The plan, as we will see, is the core to deliver-ing our project, and the ability to develop a plan, and to display Gantt charts, is core to the required functionality from project software. Our requirements go beyond developing the original plan, and any software chosen should also let us assign project team members to tasks and schedule their tasks against their availability. Ideally, we would like the team members to log their time and progress against their assigned tasks.[7] This data is the raw material for a variety of progress reports and allows us to monitor the health of our project and, for example, identify if a team member is coming under pressure and tasks need to be reallocated.
- **Collaboration.** Teams are now often dispersed, and software functionality that enables team members, stakeholders, and suppliers to engage remotely and hold audio and video

discussions – for example, around shared documents or other files which can be jointly edited in real time – is increasingly essential. Other collaboration features could include discussion boards, direct messaging, and automatic email updates generated by the software – for example, if the project plan is amended.

- **Document management.** Strong control over documents generated or received by the project is an essential element of effective project management. Software can be used to ensure that documents are effectively organised and stored securely in a single place and that versions are controlled. Many large organisations will have such systems operating organisation wide but structured on a departmental basis, and consequently we will need to work with the system support function to ensure that the needs of our project can be effectively supported.
- **Email integration.** Email integration has a number of main features that we may want to consider, including: generation of email notifications, such as alerts by the system itself, for instance to a team member if a task allocation changes; use of email to generate new tasks or update tasks which are logged to the system; and the capture of project-related emails for storage as part of controlled document management.
- **Integration with host financial systems**. A method for feeding back into the system the budget plans and financial transactions can be invaluable for managing the project's finances. Experience shows, however, that caution is required because in projects we are concerned with spend up to today, rate of burn of money today against progress today, and also forecasted stage and project outturns – which may be one, two, or several years off. Many hosts' financial systems, however, work to monthly and annual cycles and have a mainly retrospective focus.
- **Mobile working**. Where project team members are working away from the host site, perhaps when working from home or travelling, the ability to access the project management software via a mobile phone can be a very useful function. This could be either via a secure web access or by a dedicated app provided as part of the software package.

Delivery Strategy Preparation

If there is still time waiting for business case approval, we can usually start developing some of our project's delivery strategies whilst waiting for project board approval, with a low risk of needing subsequent alteration if the business case has to be amended. All these areas have been addressed at a very high level in the business case but will need to be expanded into complete standalone project policy documents. These strategies can be updated as further information emerges or as new situations arise. They include:

- Quality management.
- Stakeholder communications.
- Risk and issue management.
- Benefits management.
- Lessons learned.

Subsequent chapters include consideration of each these areas. Like our business case, each one should be a narrative which takes the reader on a journey and explains the objectives that we are seeking to achieve and our route to achieving them. A good start is by addressing tailored versions of our seven questions, such as:

- Why is a separate strategy required?
- What are the main objectives of the strategy, and what processes will we use to deliver them?
- When will activities take place?

- Who will be responsible for ensuring that activities designed to meet the strategy are delivered in a timely manner? If people are to be involved from several areas, can we identify a single point for co-ordinating monitoring and reporting?
- Who will be responsible for monitoring delivery?
- Where in the project development and delivery process should we build reviews to demonstrate that outputs in this area continue to meet both the strategy and evolving stakeholder expectations?
- How will any required reviews, reports, and other events be triggered?
- How much investment should be made in this area?

Reflections

If possible, using the same project as in your previous reflections, spend some time considering the following:

- The business sponsor has asked you to prepare a 15-minute presentation for the project board that works through the main arguments in each of the sections of the business case.
- Prepare a short explanation of why the weights were chosen for each of the five critical success factors against which you evaluated the options for delivering the project.
- Two of the main board members have been highly sceptical of the project and the benefits that you expect to achieve. Plan out agendas for a meeting with each to work through your evaluation of the benefits and address their doubts.
- What are the main aspects that need to be included in a quality strategy for the project?

Notes

1 OK, if you just must put in those extra 20 pages of data tables of which you are so proud, please put them in an appendix. See Appendix 2!
2 It is worth noting the contrasting example of a view that can be prevalent with some, particularly smaller, clients. The CEO of an owner-managed business of around 25 staff and a greater number of contractors was resistant to preparing any business case at all for acquiring new customer management and accounting systems because they had already formed a decision to proceed. The instruction given was "Just get on with it!" Risky.
3 Spoiler alert: we are going on a virtual site visit in Chapter 7 that shows what can happen when a final document is not properly "socialised" with the board members before submission. The scars have still not healed.
4 OK, well done, and you will now also have an office on the executive 52nd floor. You'll find out later that you still don't get a window as it's a converted janitor's storeroom located between the lift and the toilets. This location makes it really convenient for the boss to "drop in" multiple times a day. Sorry.
5 As ever, it is truly aggravating after a lengthy, and possibly testy, discussion for it to slowly emerge not only that one of the participants is using a version of a document that is two weeks old but that the participant with the out-of-date document is us and nobody thought to give us the current version.
6 A personal admission to having cursed Microsoft Project many times over the years, particularly every time it automatically rescheduled a carefully timed and resourced project in an unexpected way. The fault, though, is personal, and not the software. It resulted from only having had very minimal training on an older version of the software and never having had updated training on its newer functionality. For major projects one employer always used Primavera P6, and it was a joy to be able to use a trained project planner to construct the plan on the client's behalf. The lesson learned is to that to get the most out of any software

that we decide to use, we should either invest in the necessary training to use it effectively ourselves or bring a trained and experienced user into our team.

7 Without a client organisation that already records such information or team members entering this information themselves, our project management office, which in a small project may just be us on our own, will get to do lots of exciting data entry.

Bibliography

Audit, N.C.O., 2012. *Lessons Learned from Government ICT Projects*, The Hague: Netherlands Court of Auditors.

G20 Infrastructure Working Group, 2018. *Principles for the Infrastructure Project Preparation Phase*, New Delhi: G20 Research Group.

HM Treasury, 2018. *Guide to Developing the Project Business Case Better Business Cases: For Better Outcomes*, London: UK Government Assets Publishing Service.

Kaplan, R.S., and Norton, D.P., 1996. *The Balanced Scorecard*, Boston, MA: Harvard Business School Press.

Simon, H.A., 8 December 1978. *Rational Decision-Making in Business Organizations Nobel Memorial Lecture*, Pittsburgh, PA: Carnegie-Mellon University, ©The Nobel Foundation.

UK Infrastructure Projects Authority, 15 July 2021. *Government Functional Standard GovS 002: Project delivery Version 2.0 – Contains public sector information licensed under the Open Government Licence v3.0.* Retrieved 05 December 2021, from www.gov.uk: https://www.gov.uk/government/publications/project-delivery-functional-standard

5   Preparing for Launch Part 3

The Project Plan

Aim of this chapter: To introduce project planning as a data-driven process and to underline the importance of the project plan as a working set of documents essential to the unfolding narrative of the project.

Learning outcome: To be able to prepare a project plan and understand its use as the major tool for monitoring progress throughout the delivery of the project.

Why Not Just Get On and Do It?

What we are trying to do with a project plan is to shape the future to be able to deliver our project and to logically arrange and order the work that will be required. The agreed-upon project plan also represents the baseline from which we will start to deliver our project and against which departures for any reason will be measured. Without a strong project plan, we will always only be able react to events rather than having any chance of influencing them. As we move forwards with the project, based on our plan, we assign resources, manage tasks, and take decisions to deliver our desired solution and the projected benefits to the host organisation. Our project plan will also allow us to track our progress to date and forecast how and when we will complete the project. From now on the project plan is core to our existence as the project manager.

Having got the business case approved, from now on as project managers, our job comes down to creating and delivering the project plan. The path for doing this is set out in Figure 5.1.

The Planning Process

When planning, we are essentially preparing a series of visions of the future at different points – these visions are perhaps one, three, or six weeks, months, or years ahead depending on the timescale of the project. The project plan is also the tool with which we can communicate the story of how the project will unfurl. Like a good novel, we are going to lay out a clear story line, mostly in charts; populate it with characters, the project team; add a good deal of narrative text, our work package descriptions which explain the plot; and finally arrive at a convincing conclusion. Our plan is, therefore, much more than just a schedule of tasks, and by now you can guess which six of the seven questions about each task the project plan is going to have to answer.[1]

When we assembled the information for the project frame, including estimating resources and durations of the major parts of the project, we gained a grasp of the dynamic flow of the project, and we have already set this out in our high-level plan. The development of this we can now regard as part 1 of the planning process – and without it we would not have been able to prepare our business case, but now we need to move on to part 2 and address the details of the project plan.

DOI: 10.4324/9781003405344-6

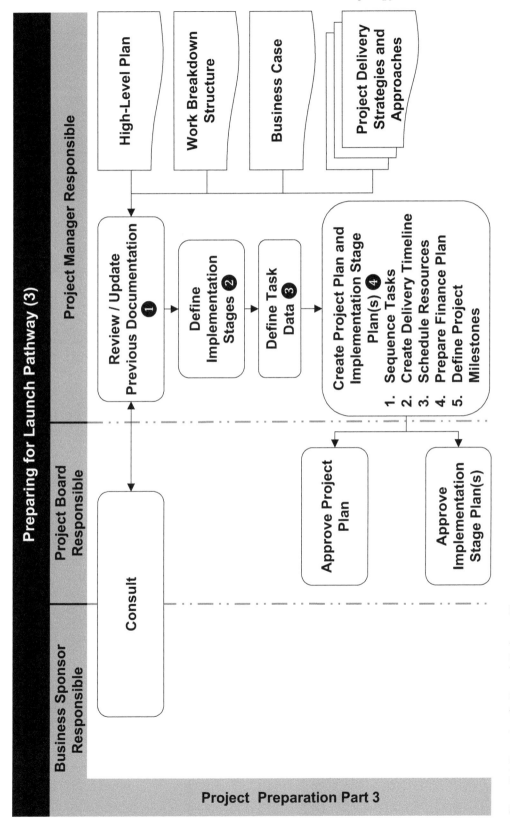

Figure 5.1 Preparing for Launch Pathway (3)

But before we move onto stage 2 of planning, please slip your bio-energy from effluent project manager hat back on for a few moments. We are just at the stage where the business case, and hence the high-level project plan, have been approved. How might the following email from the boss impact the detailed project planning process?

Re: The Way Forward

Dated 21 December

I just wanted to drop you a quick note that at the board Christmas "away day" last week the view was expressed – and no-one seemed to contradict it – that we should focus growing our business glob-ally in markets which were not fully developed. There is going to be a paper coming out soon called "The Way Forward" setting this out. I know that all our work on the project to date has envisaged the first rollout across Europe with the Middle East to follow, but now we are going to have to flex this and look at underdeveloped markets first. I should add that this is all commercially sensitive until the paper comes out, so please do not communicate it to anyone. Happy Holidays!

What would you propose to do? Consider which of these two possible courses of action might be appropriate – or is there a better way?

The first course of action is to immediately start by altering the high-level plan and commence rebuilding any work done on the detailed project plan. This is going to make the forthcoming holidays difficult.

The second approach is to treat this as advance intelligence of the way key stakeholders are thinking. Something that is interesting, worth thinking about in case it actually happens, but not yet something to rip up our current direction of travel of the project. It may even be that this project will fall outside the scope of the decision. So, although our thinking to date has been around starting in Europe and then moving into the Middle East, we may have to change priorities. However, as we do not yet have an agreed-upon project plan, and we have not started delivering against it, any changes necessary will be internal to the project anyway.

The advantage of the second approach is that we could enjoy the holidays and come back and sort it out in the New Year when the board may finally have come to and publicised a formal decision.

Of course, there is no definitive right answer. What would you do?

At about this point someone always says or writes, "Failing to plan is planning to fail." This is true as far as it goes. But the reality of a good plan is that it includes planning to fail, if by failure we mean not delivering fully as originally envisaged – which is exactly what a cautious delivery with planned "quick wins" facilitates. If circumstances change and the project is halted at some point in the future before we complete – and it does happen – we will have planned for, and we hope delivered, as many benefits as possible as early as possible rather than waiting for a single "big bang" at the end. This is one of the ways we can start to design risk out of our project from the detailed planning stage.

As an example of what can happen without an effective plan, in our field trip to a European city in Chapter 3, what the project manager didn't tell us was that the organisation had only ever had a partial project plan that covered delivery by the main hardware and software supplier and that both the supplier and the agency had resisted preparing a fully a comprehensive plan that covered all other activities such as staffing, training, communications, stakeholder management, rollout to the public, and so on. They had also gone for a "big bang" launch option. Truly a classic example of failing to plan being planning to fail.

Another tactic might be to look at reducing risk in our supply chain by engaging multiple suppliers or sources, especially for time-critical project elements. For example, if a construction method is selected that involves a need for a continuous pour of concrete (and cannot have a gap between pours of more than two hours), we might plan from the outset to have the option of supplies being delivered from multiple concrete plants. This was exactly the inherent risk of delivery schedule failure for the project team who constructed the lift-testing tower in Northampton, UK, which, at 127 metres tall, was built using four thousand tonnes of poured concrete over a two-week period with the building rising at up to 7.2 metres per day. Their constraint was that any delay that allowed the concrete to start setting at the interface between concrete pours would cause a weakness in the final structure. Consequently, the team planned on using multiple sources so that once concrete pouring started, they would be able continue the pour within acceptable timing tolerances until the building was finished.

It is worth recapping about agile projects discussed in Chapter 3, especially the SCRUM approach. Removing the problem of estimating time is an area where the agile approaches score well. These approaches have the underlying attitude that we set a fixed time – the sprint – and during that time will only do what we can with the resources that we have available. Consequently, we can be much more certain on the timeline than in a conventional, sequential project, but we cannot be definitive on the outcomes, as uncompleted elements will be returned to the backlog and may never be completed. By contrast, in a sequential project, the theory is that the outcomes are fixed and the period of delivery is extended, although there are any many examples where the scope of the solution also gets reduced.

Whatever approach is used to deliver the project, we do need to analyse the tasks that need to be done and estimate the resources, both type and quantity, that we will need to complete them. During the remainder of this chapter, we will look at planning using a sequential approach, but bear in mind that working in an agile way will still require a subset of the same planning activities in that we need a reasonable estimation of the work required for any task when it is being considered for selection from the backlog for inclusion in a sprint.

Now, it is time for another field trip. This time we are only going to be indoors, so remove the raincoat, hat, scarf, and gloves.

We are about to gate crash the first joint planning meeting for a major project. This project is going to revolutionise the way a national government-to-citizen service is delivered. The anticipated cost is already north of £200 million, and in the way of these things, that is only the starting point.

The clock reaches 4 p.m., and as agreed, a small, neat man in a brown suit – the business sponsor, who is also the organisation's deputy CEO – enters to receive a report on progress in the meeting. We slip in unnoticed behind him.

Look carefully around this room. It has dark wood panelling and is so old fashioned that it occasionally gets hired out for as a location for films and TV shows set in the 19th century.

Today, it's an early spring afternoon, the heating is still on, and it's very hot and stuffy in here even with a window open. The room is much too small for the 20 or so people who have been packed in for nearly two hours working out how the project is going to be delivered. There are representatives from every area of the business, from the very remotely located IT department to suppliers, public relations, operations, and the project team itself.

Ignoring the old oak panelling, someone has stuck a huge sheet of brown paper a metre high and at least five metres long along one wall. Future years are marked off each metre or so along the sheet.

Everyone has had their own firmly held point of view about how long their contributions to the project will take and the best time to launch changes on the unsuspecting public. There have been tough discussions and heated arguments, and probably at least two people have had to go outside for a smoke and cool off. Now, yellow, brown, and green sticky notes have fully populated the five-year timescale on the brown paper. This inoffensive sheet of brown paper has transformed itself into a five-year master plan for spending that £200 million and thereby revolutionising the work of the agency, as five years has been agreed by all the representatives in the room as the shortest feasible time to carry out all the work needed.

First, the deputy CEO, with barely a glance around the room, asks for a sticky note in a different colour and a felt tip pen. He writes something on a pink sticky and approaches the brown paper time-line. He places his pink sticky on the brown paper against 1 July of year 2 and reads out, "Project end date – delivery complete."

The silence that follows is prolonged, and everyone stares disbelieving, first at the deputy CEO and then at each other. The project director, who has been refereeing the meeting and is the owner of the brown paper timeline, looks towards the deputy, raises a quizzical eyebrow, and breaks the stunned silence with one word: "Why?"

"Because that is exactly one year before I retire, and I want to see this project completed and every-thing in place and operating for a full twelve months before I go."

The meeting breaks up in disarray, with many muttered comments and still more people slipping outside for a calming smoke.

You are keen to know how the project director is going to handle this, and so as the room clears, you go over and introduce yourself.

"Excuse me," you say. "Observing the meeting, I was impressed by the level of commitment, thought, and sheer detail that everyone had put into the process. The outcome of the meeting came as a complete surprise to me, as I suspect it did to everyone else, and I was wondering where you go now."

The project director considers for a moment. "Well, I have known John, the business sponsor, for over thirty years now, and once he has arrived at a decision, there will be no budging him. I find that when someone has arrived at a view by irrational means, no amount of rational argument is going to shift them. Anyway, as far as he is concerned, the timing of the delivery of the project is a business decision, and it is his role to take business decisions, and for better or worse he has taken this one. He will, of course, bear the consequences if it is the wrong decision.

"I actually have two strategies; firstly, I am going take the plan that we have up on the wall and work with a couple of the team leaders to prepare a version that meets John's timelines. This is what I formally have to do, and it will be done, and I will let it be known that John's timelines will be very challenging. Challenging, like brave, is always a such good word to use. Secondly, I don't have to do anything because our project administrator, Jan, who arranged your visit today and was taking notes in the meeting, is secretary to the social committee, which is chaired by the chief executive, and they are very close. He will ask Jan, informally, of course, about today's session, and Jan, who can be very forthright, will let him know, in no uncertain terms, how she feels about John's actions. His method of operation is that he will probably call me for an informal chat, probably over a coffee, maybe mid-morning on Friday, after he has had his weekly walk around the building and chat with staff. If the chocolate biscuits are out as well, that will be a strong signal that he not overly concerned and our discussion will be fairly relaxed."[2]

Over-Optimism in the Planning Process

In 2013 the UK National Audit Office issued its report titled *Over-Optimism in Government Projects*. In the introduction the author states, "All too frequently over-optimism results in under estimation of time, costs and risks to delivery and the over estimation of benefits. It undermines value for money at best, and in the worst case leads to unviable projects" (UK National Audit Office, 2013).

The first rule of project planning is that it always takes longer than you think. We need to take this into account when we are working through our list of tasks to be completed and setting expectations for delivery. We need to:

- Positively counter optimism bias.
- Be highly sceptical of all information that we receive.
- Strongly challenge those who are providing us with it.

On one large project, the project manager had carefully worked through the details of every task of a complex work breakdown structure with a panel of experts to arrive at an estimate of the time and resources required and added a generous 20% to counter optimism bias. The programme director took a different view: "I know these people; we have to take everything that they say and double it." And she did.

We are unaware of the potential dangers that await the project and often tend to discount those over which we can have no control. Even when we do, and build in other potential contingencies, precise estimation of timescales for a project is difficult.[3]

Step 1: Review and Reaffirm Prior Project Documents

When we start the detailed planning work, we have already completed a significant amount of the preparation required, which is contained within the WBS (see Chapter 2), our project delivery method (see Chapter 3), the high-level plan (see Chapter 3), and, of course, the business case itself (see Chapter 4). These are major inputs to our planning process, and we need to revalidate them before building their information into the project plan in order to check:

- That there are no changes to the project scope or deliverables that have occurred that the project plans will need to reflect.
- The impact of any subsequent decisions on delivery methods or phasing of deliverables.
- The tasks that we have identified do not need to be modified as the result of any further knowledge that we have gained since the documents were prepared.

Step 2: Define Implementation Stages

The second rule of project planning is that the further we look into the future, the hazier our vision of the individual tasks can become and the greater the number of assumptions that we will have to make. Imagine, for a moment, that our project will involve a procurement exercise for a supplier who will undertake most of the delivery work. Whilst we can work with a procurement specialist to detail each of the tasks required precisely, each supplier may take a slightly different approach to delivery and so we cannot finally plan the delivery stage in detail. We may also know in broad terms how we could roll out the project across the business, which is more completely within our control, so the latter part of the plan may have more detail than the middle section. To counter these issues, we can prepare a project plan which takes account of this uncertainty by breaking it down into a series of detailed implementation stage plans. Implementation stages are logical breaks in the implementation process which will depend on the nature of the project – for example:

- Preparation of detailed solution design.
- Procurement.
- Resourcing and contracting.
- Development/delivery, which may itself be split into delivery of different workstreams.
- Testing and acceptance.
- Transfer to live.

Ideally, these should already be reflected in the high-level plan.

We may be able to have a very clear picture of the details of the next implementation stage and how it should progress, some good ideas about the one after that, but by the next stage only a broad idea of what we should be achieving. This effect of future uncertainty is often referred to as the planning horizon. This is a slightly misleading term as the effect of future uncertainty is more comparable to looking into a fog where, as images are further away, the less distinct they are. Consequently, the individual implementation stage plans are prepared and agreed on only as they are needed, and this allows us to keep the overarching project plan at a higher level.

The following process steps will free us from needing to prepare vastly detailed plans about a future which lies beyond our planning horizon:

- Based on the WBS, create the project plan itself, usually at a higher level of detail. This should identify the different implementation stages and their key tasks and target dates – essentially this is an expanded high-level plan, supplemented by task descriptions.
- Create a series of templates for more detailed individual implementation stage plans, which will sit underneath the project plan, and plan the first implementation stage in as much details as possible.
- Create a template for work package descriptions. The work package description forms a mini-brief to the people delivering the work and ideally also will provide for a signed authorisation by the project manager for the work to proceed and a corresponding signed acceptance by the person responsible for its delivery. The template should provide full details of the requirements for the work to be undertaken, including a specification of the required output, resources to be used, standards to be followed, quality assessment and acceptance criteria, timing, and sometimes finances. Using such a description is not universal, but it is a useful discipline in all projects and is essential where the work is to be undertaken by a supplier or contractor.
- As soon as information about any of the necessary tasks in an implementation stage becomes available, start populating the relevant implementation stage plan and completing the work package descriptions. It is useful to review each one to make sure that it answers our full set of seven questions about the activity.
- As a part of our governance, the implementation stage plan for the next implementation stage always has to be completed as one of the documents that need to be signed off on by the project board by no later than the close of the previous implementation stage, otherwise that previous stage cannot be closed. Sometimes the full set of work package descriptions can be included as supporting documentation for the implementation stage plan.

Step 3: Data Drives Everything

The temptation is for us to rush straight into opening up the project planning software tool of our choice and feverishly start entering each of our identified tasks into stage plans and the project plan. Many project managers do just this, and hey, presto, they produce a magic schedule showing just how wonderfully the project can be delivered by three months from next Tuesday. It is, however, important to understand that project planning and scheduling is a data-driven process starting with understanding individual tasks. Whatever method we use to develop a project plan – on paper, in a spreadsheet, or using free software or a major corporate system in the hands of a planning expert – we first have to create and validate that task-based data. Moving forwards, this task-based data will be the basis of both our scheduling to produce plans and later on of our project monitoring and reporting. Figure 5.2 illustrates how the work undertaken previously and the data about individual tasks can be used to drive the project planning process and the outputs required – i.e., the project plan, implementation stage plans, work package descriptions, and task-related data.

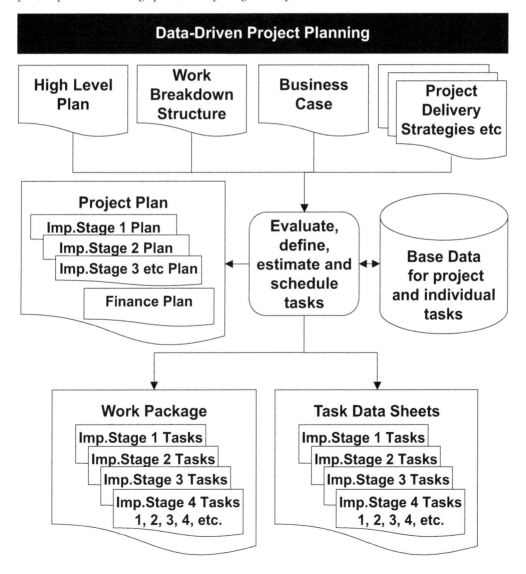

Figure 5.2 Data-Driven Project Planning

Whilst metaphorically locking our door, wrapping a cold towel around our heads, and labouring in splendid isolation to produce the ultimate project plan is possible, and may possibly work, doing so cuts at right angles across our approach of leading our project in a consultative way. Before we give in to that temptation to load up some software and start developing the first draft of the project plan in isolation, we should consider if there is a more effective and more collegiate approach that we can take, starting by focusing on the data needed and then moving out to the wider planning process.

Even if the outcome is eventually overridden, as we saw in our last field trip, the preparation of the initial cut of the project plan is an excellent opportunity to engage with the project team, stakeholders, and our wider "army of volunteers." This engagement will foster team commitment going forwards, because we are helping the team set *its own* project plan, taking *their views* and *their concerns* into account, and we are emphatically not dictating our view of what the plan should be and then imposing it upon them.[4] As in our field trip, engagement with the project team usually works better using a facilitated group approach rather than relying on a series of one-to-one consultations. This is because experience shows that facilitated group working tends to develop far greater creativity in problem solving by idea sharing amongst multiple participants.[5]

Whether in a facilitated meeting or otherwise, the first thing that we need to do is develop a short narrative description of each task and generate the basic set of task data upon which our planning will be based. This is the data that if we rush into using a software tool, we would be having to create later anyway. Irrespective of any improvement in the quality of what is produced by the team over that which we would have created alone, delegation to our facilitated team meeting is a great way to share the routine workload of creating task-based data and work package descriptions.

Table 5.1 shows a task data sheet containing the types of information that need to be detailed for each task, part 1 being the core information to which the team can contribute and part 2 being our information and assumptions upon which we can start to build the plan.

Table 5.1 Task Data Sheet

Task Data Sheet			
Part 1: Basic task data			
Task/WBS reference			
Task name			
Inputs			
Activities			
Outputs			
Quality standards			
Resources needed			
Predecessor tasks			
Successor tasks			
Part 2: Planning data			
Effort (days)		**Start date**	
Daily rate		**Duration (elapsed days)**	
Budget		**End date**	

In part 2 of the task data sheet are two terms which in project planning need to be used very precisely, and, before progressing, it is worth underlining the difference between task duration – i.e., the time period over which a task can be delivered – and the effort that it requires to deliver it. Both of these are key data items for our task data sheet and foundations for our project plan.

Effort. Usually denominated in days or hours, effort is a measure of the physical work required to complete the task – for example, Jan requires 160 hours (that is, 20 standard 8-hour days) to complete the architectural drawings.

Duration. The time period over which the required effort can be delivered, again usually denominated in days, weeks, or months depending on the size of the task. Taking our example again, Jan could complete this task starting on 1 June during the following four weeks – i.e., an elapsed period of 20 working days. If the project is delayed by more than a few days, however, she has one week's leave booked at the beginning of July, which will mean that the duration will increase to 25 days.

If it is decided to follow the approach to completion of the task data sheets of using facilitated meeting of interested parties, it is useful to split the agenda into the following parts:

- **Create task data sheets**. By starting with an introduction to the high-level plan and WBS, we can move the team on to creating the initial description, as shown in the task data sheet. If necessary, we can split the meeting into separate teams to each work on different project aspects or areas of expertise. It will be helpful if all the task data sheets are numbered – probably with the WBS reference.
- **Create a task-based timeline**. This stage of the meeting is moving our team firmly into step 4, creating the project plan. As in the last virtual field trip, we can create the horizontal project flow and timeline using the high-tech solution of a sticky note for each task, a few coloured pens, and the ever-present roll of brown paper stuck onto a suitable wall (other colours are available, and flip chart paper can be used instead). The brown paper is marked up into broad time periods and the sticky notes are referenced to their task data sheets and placed in their approximate position along the timeline. The timeline can also be marked up with the first thoughts on the period of delivery for each task and the linkages between tasks. Experience of working with such sheets to create a plan teaches that in order to maintain clarity, it is best to minimise the crossing of linkage lines, and it may be necessary to move sticky notes vertically up or down in order to achieve this.
- **Update task data sheets**. The teams holding the task data sheets are next asked to update them with details of the identified immediate predecessor and successor tasks.
- **Final review**. As a final stage, the meeting undertakes a sanity check on the results of their work and makes any adjustments that the team feels necessary.[6]

After the meeting, we need to undertake our own sanity review of the output as the final project plan will be "owned" by us as the project manager and will be submitted by us to the project board as our best assessment of the delivery path of the project. Consequently, when our task data sheets have been created and before we can be use them as input to the creation of the plan and the work package description, we need to evaluate and, where necessary, call on outside expert opinion to moderate each one, paying special attention to the consistency and validity of resource and cost estimates.

Step 4: Create the Project Plan

We looked at estimation and its perils in Chapter 2, and in the planning stage, we are about to face another round of detailed estimation, and in particular, we are going to be looking at both the duration and the effort required for each task. These are not necessarily areas which are conducive to resolving in a facilitated group meeting. The issue that we face is that we cannot be definitive too early in the planning process about a task's elapsed time as the resources available to deliver a task will influence the elapsed time it will require to complete. This conundrum means we have to base our initial elapsed time estimates on reasonable assumptions regarding resources – and monitor them to make sure they stay reasonable. Jumping ahead, it will also be possible to use the ranges of elapsed time calculated for each task to start looking at different scenarios for the size of the project team and the total time taken to deliver the project.

Displaying the Project Plan

In our task data sheets, and on the workshop's brown paper chart, we have captured all the information needed to start laying out our view of how the project will progress. But before we start to construct our project plan, we need to think about how the final version could be presented. Remember, though, that the project plan is not just a diagram, and our task descriptions will form an integral part of the final project plan.

Project plans can be shown graphically in a number of different ways, including:

Gantt chart. Chapter 3 introduced a simple Gantt chart as a way of laying out activities. This approach is great for showing how tasks progress along a timeline but less good for giving a clear picture of the relationships between different elements. There are many different software packages to help us prepare Gantt charts – including a Microsoft-provided Excel template which may be adequate for many projects.

Network diagram. This is a diagrammatic layout of how different tasks relate to each other. Each task or activity is referred to as a node and is joined to other tasks or activities (nodes) in a logical sequence by connectors, referred to as arrows.

There are two different ways of setting out a network diagram:

- **Activity on node.** This is seen as being the simpler type of network diagram to prepare. Our data about the task is attached to each node, as in the brown paper example shown in Figure 5.3. One major advantage of the activity on node approach is the ability to display further information about the task very clearly.
- **Activity on arrow**. This approach assigns our task data to the arrow between nodes and, whilst helpful in thinking through the project, may be less clear at communicating an overall picture of the project to others than the activity on node.

Whilst both versions of the network diagram method are strong on showing task relationships and dependencies, they are much weaker than a Gantt chart on showing overall project timing. The actual drawing of a manual network diagram is fairly straightforward and can be supported within most drawing packages or produced automatically from project planning software.[7]

Program evaluation and review technique (PERT).[8] This is a scheduling and planning technique that has been acknowledged as having been invented in the 1950s by the US Navy department responsible for building the Polaris nuclear submarines. PERT uses a highly specialised form of network diagram to display the results from a mathematical technique for calculating project timings. PERT may be overkill for simple projects, and complex ones will call for project management software which may use similar mathematical approaches in the background. Microsoft Visio includes a PERT template.

In our task data sheets, we have created a set of reasonable descriptions and estimates of the core data for each task, and in the output "brown paper" sheet, we have made a very rough draft of the flow of the tasks required to implement the project. We now need to use these as the foundations upon which we can start building our detailed project plan. In developing the project plan, we will be building an increasingly complex network diagram showing how tasks in our project will flow from one to another and use this to develop an achievable timeline. We can use the same process equally well whether we are preparing an overall project plan or a detailed implementation stage plan. Figure 5.3 shows what such output from a facilitated planning workshop could look like, including a first estimation of the time period over which the task may be delivered. Data about each task can be added, making it a modified "activity on node" network diagram.

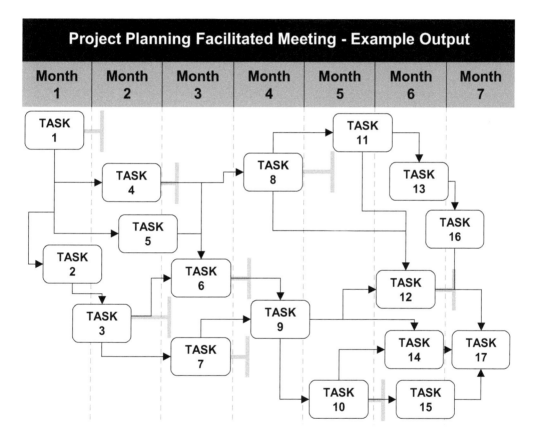

Figure 5.3 Project Planning Facilitated Meeting Example Output

This is probably the point at which we could choose to start using a software package to help, at least with the final presentation of the plan. However, using the software effectively depends on understanding what it is trying to do for us. Consequently, in the next section, we will step through the processes we would have to go through if we were to prepare the schedule of tasks manually, which will indicate what the software should be doing in the background. After that we will consider how a software package, in this case Microsoft Project, can help us with the planning process.

Make a Daisy Chain

Linking the tasks. Whether we are working manually or with software, one of our first tasks is to link every task to its immediate predecessors and successors. These links are called dependencies, and we have already captured this information on the task data sheets and the "brown paper" flow diagram, if we have one, but it will always be necessary to recheck it now. Using this data as a basis, we can start to refine a simple flow diagram or daisy chain of our project's tasks, as in Figure 5.4.[9]

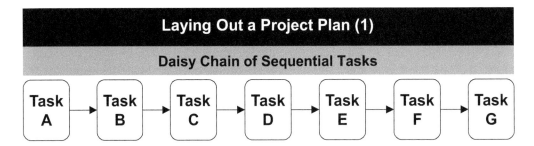

Figure 5.4 Laying Out a Project Plan (1) Daisy Chain of Sequential Tasks

Parallel tasks. Life, however, is seldom so simple for us as project managers that tasks can only be completed sequentially. In most projects this would expand the timetable for delivery far beyond the level the host organisation would accept. Consequently, as in the workshop output shown in Figure 5.3, we usually find ourselves planning to run the project with multiple tasks in parallel at the same time, as set out in Figure 5.5, effectively starting to build a simple network.

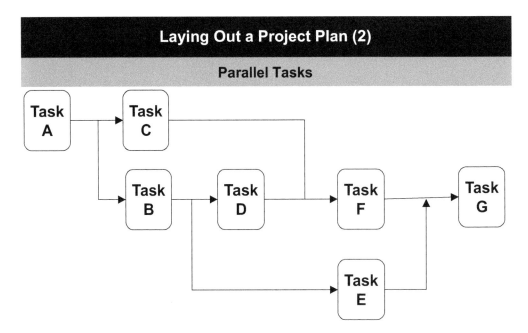

Figure 5.5 Laying Out a Project Plan (2) Parallel Tasks

Unlike in purely sequential tasks, Figure 5.5 shows that we get more choice on when tasks C and E are undertaken. We can start task C as soon as task A is complete, but we have only to complete it before starting task F. Similarly, we can start task E as soon as we have completed task B and only have to finish before starting task G. This flexibility to move when we deliver tasks forwards or backwards on our timeline gives us scope to manage the effective deployment of resources on the project. However, there is a matching set of additional dependencies for task F, which can now only start after both task C and task D have completed, and task G can only complete after task E and task F have completed.

Weave the Tasks into a Tapestry

Multiple workstreams. As linked dependencies become more complex, we can end up with multiple workstreams in our projects. For example, in the case of the accounts system replacement project, we would probably have to split the work first into two procurement streams, one for the accounting software itself and one for preferred cloud solution for hosting. Each of these would then lead into further separated workstreams for implementation. Later in the project, we may add communications, training, and rollout workstreams.

Consider the project shown in Table 5.2 with the following tasks and their elapsed time and linkages, all of which would have been extracted from the task data sheets. Note tasks 2, 6, and 13 are summaries of all tasks in the workstream.

Table 5.2 Task Schedule Data

Task Schedule Data			
Task Number	**Task Name**	**Elapsed Time**	**Predecessor**
1	Start Task A1	5 days	
2	**Workstream B**	**65 days**	
3	B1	10 days	1
4	B2	10 days	3, 7
5	B3	40 days	4, 9
6	**Workstream C**	**85 days**	
7	C1	10 days	1
8	C2	5 days	7
9	C3	5 days	8, 4
10	C4	5 days	9
11	C5	20 days	10
12	C6	10 days	11, 17
13	**Workstream D**	**60 days**	
14	D1	10 days	1
15	D2	5 days	14
16	D3	5 days	15, 9
17	D4	5 days	16
18	D5	10 days	17
19	End Task E1	5 days	5, 12, 18

Using this data gives us a true network diagram. Figure 5.6 shows the linkages in a multiple workstream project with three workstreams and crosscutting as follows.

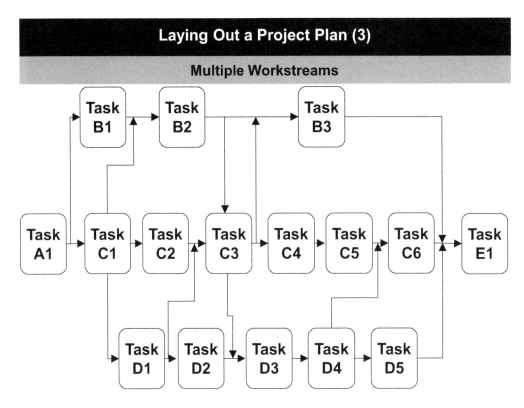

Figure 5.6 Laying Out a Project Plan (3) Multiple Workstreams

Although we may have multiple workstreams to deliver our project, as Figure 5.6 shows, there may be crosscutting dependencies between tasks in different workstreams. The linkages in this diagram are also important and indicate the timing of the relationships between tasks – for example, although there is a dependency between them, task B2 can start part-way through the delivery of task C1, and similarly task D3 can start before task C3 completes. To return to our accounts system example, although the software and hardware may be in different workstreams, we cannot specify the exact requirements of the hosting solution until we have chosen the accounting software and then understood its processing and storage needs. Increasingly the picture of the delivery of our projects normally resembles a network of interlinked tasks rather than strictly linear flows.

Create a Timeline

We can start to build in more sophistication to the network diagram for our project by amending the length of the task boxes to reflect the estimated elapsed time for each task. By aligning tasks more closely with the timeline, Figure 5.7 moves our understanding of the shape of our project forwards.

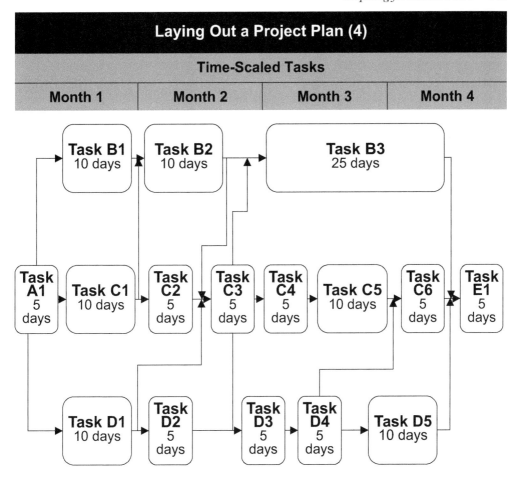

Figure 5.7 Laying Out a Project Plan (4) Time-Scaled Tasks

Building more complexity into our network diagram. The diagram now starts to reflect more closely how the project could operate, but there are increasing levels of complexity that we may wish to include. Our dependency relationships, for example, can have several different options:

- Only when we have finished task A can we start task B – a finish-to-start link (FS). This dependency is probably the most commonly used.
- When we start task A, we can also start task B – a start-to-start link (SS).
- When we finish task A, we can also finish task B – a finish-to-finish link (FF).
- In order to finish task B, we must start task A – a start-to-finish link (SF).[10]

We may also want to build in other rules or constraints into a link – for example, in a procurement workstream between selecting and announcing the winning bid in one task, there may have to be a mandatory delay to allow for objections before a contract can be awarded in the next task, or we may need to allow construction materials installed in one task to harden before we can move on to the successor task.

Figure 5.8 shows a network where some of the links have been moved to show that they are not simple finish-to-start tasks and have been annotated to show the number of days of additional delay on the link.

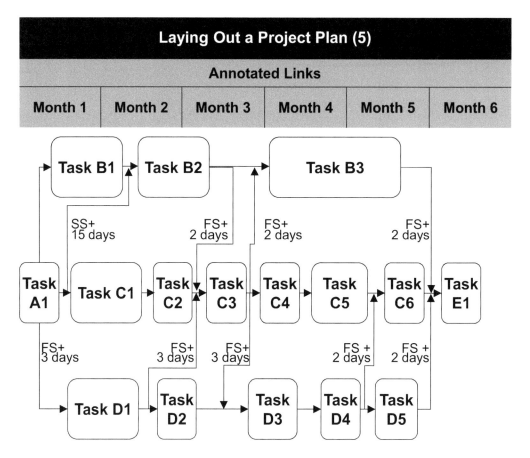

Figure 5.8 Laying Out a Project Plan (5) Annotated Links

Peopling Our Plan – Scheduling Our Resources

BE PREPARED, THERE IS ANOTHER SUM COMING UP – IT'S NOT HARD, BUT YOU MAY WANT TO LOOK AWAY AGAIN SHORTLY.

We now know how our project plan can flow if all the necessary resources are freely available when required for each task. Sadly, in the real world, our resources are often limited and have to work on multiple tasks. As well as our demands upon them, we need to understand during our planning that resources are not abstract interchangeable numbers of hours but are real people with disparate skills who also plan ahead to take days off for national holidays, vacations, childcare, and weekends, and some may only work part time.

LOOK AWAY NOW.

Consider the following totally fictious simple example, which shows how difficult it is to arrive at an estimate of elapsed time required to complete a task:[11]

One part of the design for our project calls for the building of a double skinned brick wall 20 bricks high and 50 bricks long – i.e., 1,000 bricks. Allowing for the double skin means that 2,000 bricks are to be laid.

BRICKS = 2,000

Whilst our buildings expert says that a really experienced bricklayer might lay 240 bricks an hour, it is unlikely that they could keep this up continuously, and so they have advised us that we should assume an average of between 600 and 700 bricks can be laid in an eight-hour day to an acceptable quality standard. So, dividing the 2,000 bricks by a target 700 bricks per day, at best we would require 22.9 labour hours, but if we can only achieve the lower figure of 600 bricks per day, then our estimated labour hours required rises to 26.7 labour hours.

MAXIMUM LABOUR HOURS = 2,000 / 600 = 26.7 hours
MINIMUM LABOUR HOURS = 2,000 / 700 = 22.9 hours

So far so good, but now we need to consider the elapsed time that the task will take, and this depends on number of bricklayers we can allocate to the task.

Using one bricklayer to build our simple wall, the required labour hours (22.9 or 26.7 hours) represents a duration for the task of between 2.8 and 3.3 eight-hour days. If instead we can allocate two bricklayers, the duration drops to between 1.4 and 1.65 days.

MAXIMUM ELAPSED TIME – ONE BRICKLAYER = 26.7 / 8 = 3.3 days
MINIMUM ELAPSED TIME – ONE BRICKLAYER = 22.9 / 8 = 2.8 days
MAXIMUM ELAPSED TIME – TWO BRICKLAYERS = 26.7 / (8 x 2) = 1.65 days
MINIMUM ELAPSED TIME – TWO BRICKLAYERS = 22.9 / (8 x 2) = 1.4 days

So even in this very simple example, our duration for bricklaying is now in the very wide range of 1.4 to 3.3 days depending on speed of the individuals in laying bricks to an acceptable standard and, crucially, on the number of bricklayers available to be assigned to the task.

With three bricklayers, of course, the elapsed time might be even shorter, but there is a very real danger that if we employ too many people, the working area will become overcrowded, productivity will fall, and total costs will go up.[12]

Now consider the increase in complexity of scheduling a project if we extend our example to consider a residential development of 150 houses with an average of, say, 8,000 bricks per house, making 1.2 million bricks to be laid instead of 2,000.

OK, IT'S SAFE TO LOOK BACK NOW.

Where there are no resource constraints or conflicting demands for different tasks to use the same people simultaneously, it is not difficult to manually assign people to tasks. Normally, however, we have to deal with situations in which we have to resolve competing demands for scarce people and attempt to minimise the impact of this scarcity on project delivery. Using a project planning software package to undertake the necessary calculations in this next stage of our project planning can make this task easier.

Now It's Time to Fire Up Software Support

Scheduling our example simple project to date has not been too difficult, as our network diagram shows. It is time to look more closely at the practical realities and the rapidly increasing levels of complexity that we will meet when planning a real-world project and how a software package can help us.

For now, we will assume that our project has tasks that can follow straight after each other with no delays between them. These are called finish-to-start links, and Microsoft Project would allow us, if we wished, to add delays between tasks if these are necessary. Taking our previous planning information set out in Table 5.2, we can extend the data by adding information on how we could allocate our project team of five people. These five are imaginatively called V, W, X, Y, and Z, and they will be working full time to deliver our project. As before, tasks 2, 6, and 13 are summaries of the tasks in the workstream; resources, however, are only assigned at the detailed task levels as follows (Table 5.3):

Table 5.3 Resourced Task Schedule

Resourced Task Data Schedule				
Task Number	**Task Name**	**Elapsed Time**	**Predecessor**	**Resources**
1	Start Task A1	5 days		W, X, Y, Z, V
2	**Workstream B**	**65 days**		
3	B1	10 days	1	W, X
4	B2	10 days	3, 7	W, X
5	B3	40 days	4, 9	W, X, Z
6	**Workstream C**	**85 days**		
7	C1	10 days	1	V, Y
8	C2	5 days	7	V, Y
9	C3	5 days	8, 4	V, Y
10	C4	5 days	9	V, Y, X
11	C5	20 days	10	V, X
12	C6	10 days	11, 17	X, Y, Z
13	**Workstream D**	**60 days**		
14	D1	10 days	1	W
15	D2	5 days	14	W
16	D3	5 days	15, 9	W, Y
17	D4	5 days	16	W
18	D5	10 days	17	W, Z
19	End Task E1	5 days	5, 12, 18	X, Z

We now have to schedule not only our original 16 tasks, but based on Table 5.3, we also have to schedule five staff to 35 competing task assignments and a total of 290 days of effort.

When life becomes this complicated, a software project planning tool is really going to help. But before we leap in and start assigning resources to tasks, we will step through the process of building our understanding of the project, this time using a software package to help us. The following examples use Microsoft Project as an example of available software.[13]

Creating Our Daisy Chain

Table 5.2 looks a lot like the data columns required in Microsoft Project to create a basic Gantt chart (purely by coincidence, of course), and the data are shown on the left of Figure 5.9.

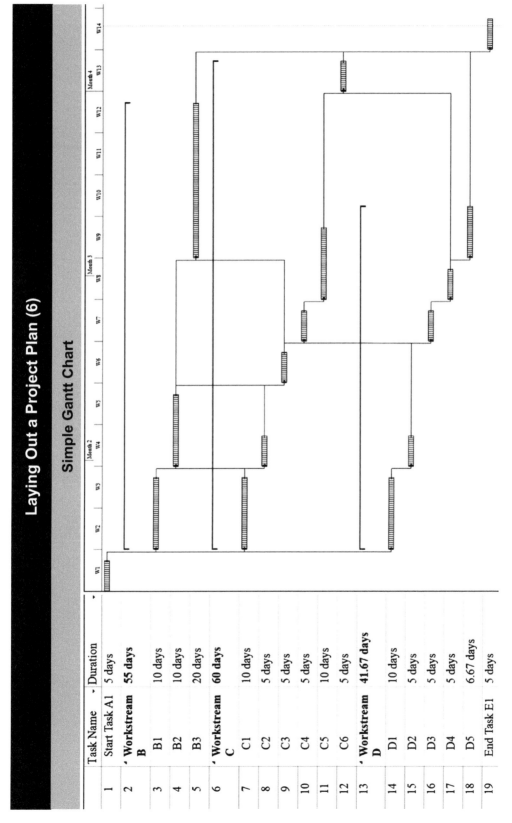

Figure 5.9 Laying Out a Project Plan (6) Simple Gantt Chart

There is one change here from our straight list of tasks; to each of the three workstreams a summary task has been added. This does not affect our scheduling, but, especially when developing large plans, Microsoft Project or similar software enables us to expose only the area of the project where we are working and show only the summary lines for the remaining areas.

Getting Critical

At this point we need to look at two further concepts in project planning:

Critical path. The critical path is defined as being:

- The sequence of tasks which when added together represent the longest elapsed time required to complete the project.

If you look back to Figure 5.4, with an elapsed period of 60 days, the critical path in our example works out to be the route A1, B1, B2, C3, B3, E1. As project managers, managing tasks on the critical path to avoid extending the length of the project is an area of our prime concern.

Float or slack time. These terms are commonly used interchangeably by many people, although there is a difference:

- Slack is how much later a task can start before impacting a subsequent task.
- Float is how much longer the task can take than planned – i.e., how much later the task can finish before impacting a subsequent task.

There is a further distinction between the concepts of float:

- Free float – the maximum period of lateness of completion before any subsequent task is impacted.
- Total float – the maximum period of lateness of completion before the completion of the project is impacted.

It should always be the total float that gets reported in the news when we hear that such and such a project is already two years late.

Our two definitions of critical path and float/slack have important implications for us as we monitor the progress on all tasks because:

- If a task already on the critical path has any delay in its completion, it will cause a delay in the project – i.e., it can have no free or total float time.
- If any other task – i.e., not on the critical path – experiences delays and will exceed its total float time, the critical path will change, and this task will now be on the changed critical path.

Figure 5.10 is another Gantt chart, produced using the same data, that reveals our project's critical path and free float time.

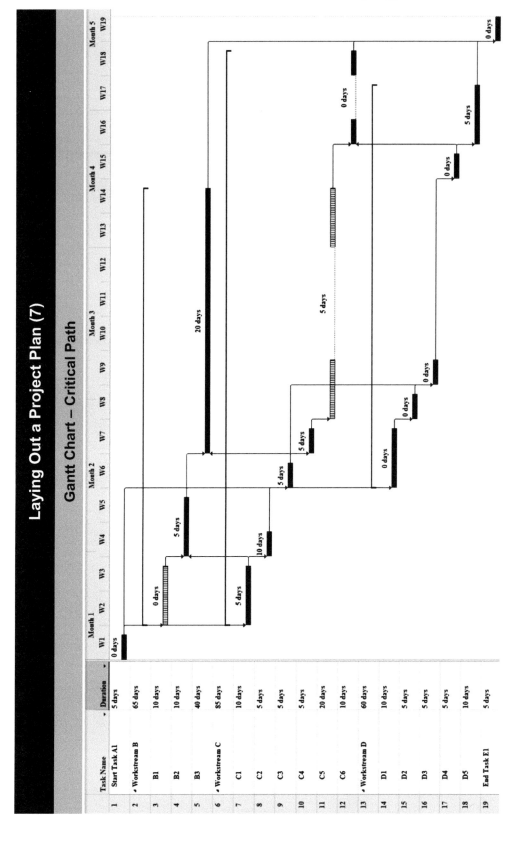

Figure 5.10 Laying Out a Project Plan (7) Gantt Chart – Critical Path

The heavy black bars in Figure 5.10 show those tasks on the critical path of the project. The lines with a number of days shown above tasks shows the free slack time.

Like in Comedy – Timing Is Everything

We now have a strong picture of the shape of how our project can be delivered; indeed, we may even have stuck a carefully drawn or software-prepared network diagram or Gantt chart up on the wall with a wonderfully clear timeline. But we still have to discover if the project is actually deliverable in accordance with this timeline because now is the time to address the people factor and answer the question:

• Do we have enough resources with the right skills to deliver the project plan?

There is, however, another equally pressing question:

• Will the right people be available when we need them?

As stated earlier, in real life project scheduling is much more complicated than just assignment of people to tasks and balancing workloads. This is because:

• Our dear colleagues V, W, X, Y, and Z all may have some annual holidays booked, including Y, who wants two weeks off to cover a trip to India to celebrate their 30th birthday with her wider family.
• W has five days annual leave carried over from last year as well, and company rules require that they use this up before 30 June.
• V, W, X, Y, and Z all want further time off for national and religious holidays.
• Z has childcare commitments and does not work on Tuesdays.
• We all know that V is waiting for a hospital appointment and may only be available for the first half of the project.

A good software package should let us record the availability of different staff members, essentially a staff calendar, and take this into account when calculating the project's delivery schedule. For the sake of the example, let's make an unrealistic assumption that we have access to all five people full time for the life of the project. Making the task assignments set out in Table 5.3 generates a new delivery time frame. In this case the detailed Gantt chart (Figure 5.11) shows a revised critical path and a lengthened timeline.

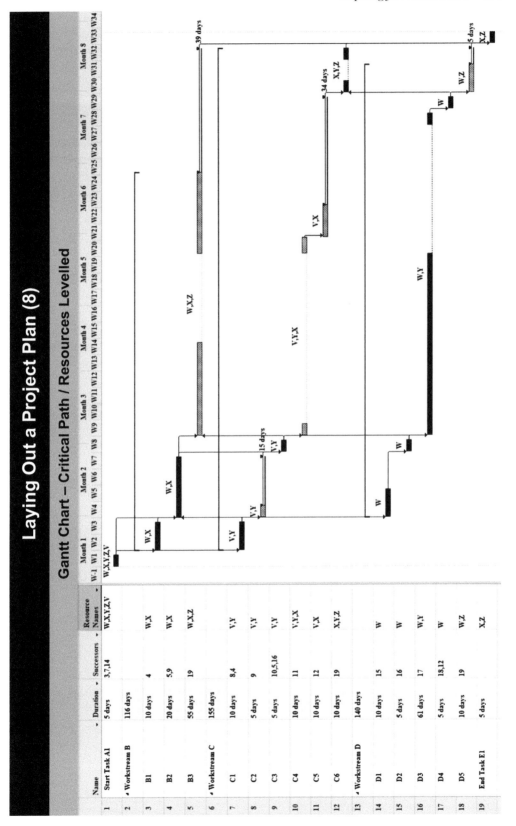

Figure 5.11 Laying Out a Project Plan (8) Gantt Chart – Critical Path/Resources Levelled

Scrutinising this carefully reveals that there may be issues arising from our allocations, but always carefully check the logic of the output. Recasting the information in our Gantt chart into a resource plan will show the causes of this delay more clearly. Such a resource plan for our example is redrawn as Figure 5.12.

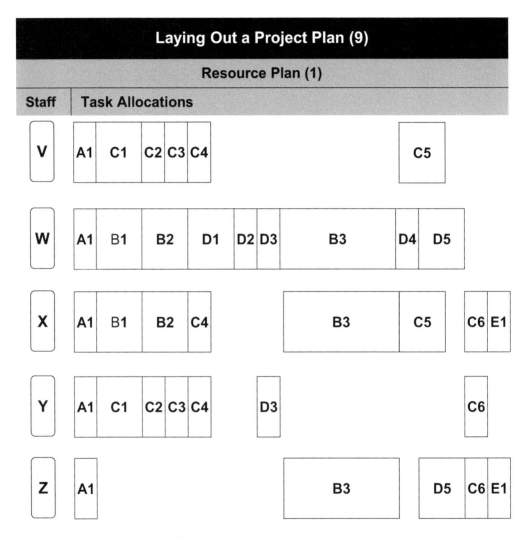

Figure 5.12 Laying Out a Project Plan (9) Resource Plan (1)

On closer inspection the major constraint seems to be our friend W, to whom we have allocated work in tasks in workstream B and also in all tasks in workstream D. There are also significant gaps in our use of resources – for example, our colleagues V and Z have not been allocated work in the middle section of the project. Oops!

This is our first cut of the resource plan, and at this stage, we may have to explain to our business sponsor why coming up with the final delivery schedule is going to take a little time. We can say something like, "Well, resource scheduling is always an iterative process, a bit like peeling back an onion really. In order to find the best solution, we have to work through one issue at a time until we arrive at a result which finds the optimal compromise between the effective use of resources and delivering as quickly as possible." This is highly plausible, should always sound convincing, and may even be partially true. Do you remember our friend Herbert Simon? What we are doing with resource planning is satisficing again. We do not have the luxury of infinite time to find the elusive "best" or "optimal" solution; what we have to do is to find one that fits our need for a compromise between available resources and not overextending the timeline.

Our first cut of the resource plan was problematic, so what do we do now? There are a number of areas that we can review and think through their implications in order to remove the constraints that our initial resource allocation placed upon our delivery schedule:

- Sometimes, in order to provide the best overall timing of the project, we may have to assign our second choice of person to tasks.
- Effort is only an estimate, and we may want to revisit the assumptions that were used in estimating effort at critical points in order to be able to squeeze down on our timeline.
- Available resources can often be varied within limits. We might be able to persuade some staff to adjust their holidays – not Y, of course, because they have bought non-refundable tickets, and anyway, Y's 30th birthday is a fixed date. We may have the option to increase resources on individual tasks, but only to a new reasonably achievable value – perhaps moving from one to two bricklayers to build our wall, or two software developers to interface the accounts package to our standard reporting software, or two crews to put up street lights instead of one. However, we probably would find it difficult to resource twenty-two bricklayers, developers, or street-lighting gangs, and in any case, diminishing marginal utility will kick in at some point as well.
- Elapsed time is often merely a calculation that equals effort divided by available resources, but if we have to allow time for a tenderer to respond or for concrete to cure, no additional resources will change the amount of time we need to allow.
- It may be possible for some team members to commence work on one task before other team members become available, thereby allowing the task to complete earlier. This has to be established on a task-by-task basis and is very much project specific.
- The task linkages that we established may not be quite as absolute as they first appeared, and a degree of task overlapping may be possible. Again, this has to be established on a task-by-task basis and is project specific.

So, as a first step, we can go back to the resource plan and identify how we might move task allocations around between staff. The results are shown in Figure 5.13.

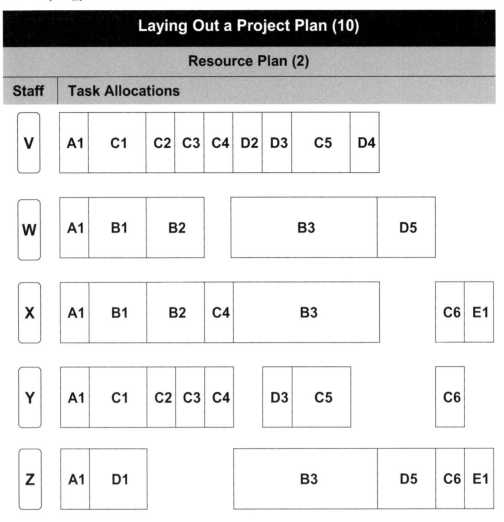

Figure 5.13 Laying Out a Project Plan (10) Resource Plan (2)

This is a better result, as we have brought the timeline back down a little and reduced some of the gapping. But changing task allocations is only one of the things we can do, and we may need to critically examine each of the other areas where changes can be made until we reach a satisfactory conclusion.

By being able to rapidly show the impact of such changes, software packages can allow us to experiment easily with different scenarios, which can be very time consuming if done manually. The different scenarios can help us understand what changes we need to make to our thinking in order to be able to deliver the project according to our desired, and often approved, timeline. If we cannot find a suitable compromise that allows us to work within the available resources, our different scenarios can also provide the evidence to support a discussion with the business about the need to increase resources to meet the effort required at certain pinch points in the project.

This does not necessarily mean increasing the budget, merely expanding the delivery capacity of the project at certain points and balancing this with offsetting reductions at other times. If, however, resources have to be found from other sources – e.g., outside contractors rather than internal staff – this may lead to cost increases.

Looking back at how increased flexibility might impact the next iteration of our resource plan, the new timeline is shown in Figure 5.14.

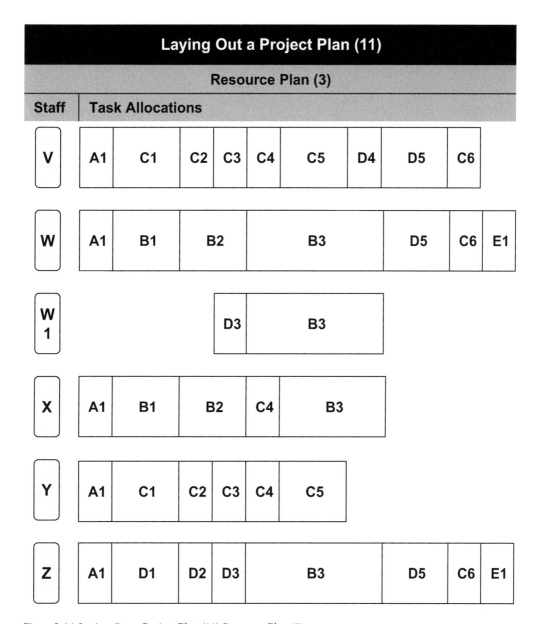

Figure 5.14 Laying Out a Project Plan (11) Resource Plan (3)

We have added a new pinch point resource, W1. This has allowed us to reallocate the 75 days of effort planned for task B3 across four rather than three staff and C6 across three staff rather than two, reducing its elapsed time by one third, and we have reduced X's commitment to task B3, allowing him to start a week later. Together with a further movement of D3 from W to W1, we have now produced a schedule in which no staff have gaps in their work programme, and we have reduced the timeline so we should complete at the end of month 4 instead of the end of month 6 in our first iteration and late in month 5 in the second one.

Follow the Money

Finally, we have a resource plan that underpins the project plan, and everyone is happy, and we can relax and get on with the delivery.

Not quite, because if it has not happened already, very shortly our phone will ring or an email will pop up in our inbox from the finance department. What is about to happen is that our nice cosy world of uncertainty using assumptions, plans, estimates, and projections is about to meet the harsh world of financial certainty based on cashflows, annual budgets, salaries, and invoices.

Our finance plan is also going to be a major tool for controlling our project. We will be monitoring our progress on our tasks against the costs that we are incurring and then projecting our outturn costs. This means that we, as well as the finance department, will want to make sure that our finance plan is as accurate as we can achieve in terms of both the level of expenditure and when we expect to incur it.

Time for a quick revisit back to the agency where we attended the project planning session. If you remember, it is a very major project that is going to transform the way the entire agency works.

It is now a few weeks after our first visit, and we are sitting in a corner in the project director's office when Doreen, the director of finance, comes in. She is a woman possessed of the steeliest of unflinching gazes – no doubt acquired during years as a government auditor in her earlier career where her job was disbelieving and breaking miscreants, from junior clerks to senior cabinet ministers, on a daily basis.

Ignoring us as unimportant, Doreen pins the project director with that stare.

"John, now the board have approved the innovation project, we need to talk about the funding."

"Well, Doreen, the board have agreed on a spend of up to £250 million over the next three years, although privately I think that we can bring it in for around £220 million, so I intend to hold back £30 million as a contingency reserve."

"John, the actual amount is not my worry, and I know that £250 million is a big sum. My concern is the impact that it is going to have on our cashflow over that time. Basically, and I know that you understand this, I need to know when and how much – what I am looking for is as much certainty as I can get."

"Well, you know, Doreen, there are still a number of imponderables, I cannot even be certain in which year expenditure will fall, let alone which month. What happens, for example, if the procurements take longer than we expected – or the suppliers have long lead times? I am doubtful if we even have the internal capacity to deliver everything, and that means engaging expensive consultants. You know that these guys are like wire coat hangers in a wardrobe – every time you open the door, they seem to have doubled in number. It is not straightforward, not straightforward at all!"

"John, on the wall behind you, I see that you have a huge printout showing your project plan with all the tasks and all the stages. I assume that this is your best guess at when things are going to happen.

All that you have to do is put two sets of numbers to each task – staff costs and supplier costs, and by supplier costs I mean absolutely anything anyone is going to invoice us for – and, most importantly, when I am going to have to find the funds. And I need it by Friday."

"How about Friday lunchtime your office. I'll bring up Jan, who actually owns the plan, and we'll go through it over sandwiches. Can you arrange for them? I'd prefer ham, but Jan is vegetarian and is usually OK with cheese or egg and cress."

Doreen, slightly mollified, agrees and leaves. John smiles broadly, picks up the phone, and says, "Jan, can you print out the project expenditure profile that we discussed? I want to go through it again before we take it to Doreen on Friday lunchtime – by the way, she is providing the sandwiches."

In terms of preparing a project finance plan, there is little more to do than apply a cost to each of our staff resources and project how these will look over time, add invoiceable costs to each task, and agree on the project totals to the budget in the business case. Whether we prepare this in a spreadsheet or use project management software is a matter of choice and the availability of a decent finance module in our chosen software. Then it can be straight on to the meeting and egg and cress sandwiches with the director of finance. Actually, we are not quite ready for that yet because, in moving from the resource plan to the finance plan, there are a few extra factors to consider:

- **Similar resources can have different costs, depending upon how we source them**. Our organisation may not even allocate the costs of existing staff to the project, so these costs for the project could be zero; other organisations may allocate salary costs, including pension and social security costs; and still others could add various overhead costs as well. Externally provided resources will normally be costed against the project.
- **Costs of individual staff go up over time**. A multiyear project may have to allow for increases both for general inflation and salary increases as staff progress through a grading structure. Similarly, externally contracted staff and material costs will usually have an uplift for inflation tucked away in the conditions.
- **Equipment purchasing costs present a whole series of major issues regarding timing.** The suppliers' sales people will push for us to take delivery as early as possible – their sales figures and commissions normally depend upon at least getting the goods out of the door, if not also invoiced and paid for. We, on the other hand, should be motivated to take delivery and pay as late as possible, especially in longer projects, because most equipment generates revenue costs. Usually this is at least insurance and maintenance. Even if we leave the equipment that we purchase in boxes until we need it, we will generate warehousing costs, and then we have to pay extra transportation costs when we need to move it to its final installation site instead of having had it delivered direct to site in the first place. Computerised equipment also generates costs in terms of operating software licences often based on the specification of the equipment, not the number of users – so we can pay the same costs on the day it is commissioned as the day it finally enters full use by thousands of users. Consequently, we may want to stagger computer equipment delivery and commissioning to match the project and any planned rollout. Also, as soon as any equipment has been commissioned, it usually will also start to generate running costs of some sort. Equipment also has a limited life span, and it depreciates. Somewhere in our business case, we should have allowed for replacing it. If we have bought everything on project day one and it is months – or perhaps years – before it is fully operational, we will have to fund any maintenance and other costs during the project. Also, our NPV calculations will be

impacted because we will have a shorter opportunity to offset equipment costs against benefits before it needs replacing. In extreme cases, especially if the operation of the project has been delayed, some equipment may even have reached "end of life" and need replacing before it is operational.

In general, the rule is to have equipment supplied as late as possible and to purchase in phases, with everything required being ordered, delivered, and commissioned on a just-in-time basis. For example, if our project requires training multiple staff to use new machinery, we should consider having the equipment delivered and commissioned for the training centre first and then aligning subsequent deliveries with our rollout programme according to the numbers and base locations of staff to be trained and transitioned to live operation.

In our example project, we will need to meet the finance department's requirements for a schedule detailing when they may need to provide the funding. In addition to our staff resources, we have added some fixed costs, including paying for as much equipment as possible in the last week, although on such a short project, this really makes little difference – but if it were 14 months instead of 14 weeks, it would have a large impact.

Our costs for each task are as shown in Table 5.4.

Table 5.4 Costs by Task

Costs by Tasks			
Task Name	**Total Cost**	**Resource Cost**	**Fixed Cost**
Total Costs	**£366,469**	**£79,469**	**£287,000**
Start Task A1	£18,000	£8,000	£10,000
B1	£11,400	£6,400	£5,000
B2	£11,400	£6,400	£5,000
B3	£37,600	£17,600	£20,000
C1	£116,400	£6,400	£110,000
C2	£5,200	£3,200	£2,000
C3	£5,200	£3,200	£2,000
C4	£6,800	£4,800	£2,000
C5	£23,200	£3,200	£20,000
C6	£4,800	£4,800	
D1	£5,200	£3,200	£2,000
D2	£3,600	£1,600	£2,000
D3	£2,600	£1,600	£1,000
D4	£2,600	£1,600	£1,000
D5	£24,269	£4,269	£20,000
End Task E1	£88,200	£3,200	£85,000

Based on our project plan, this gives a finance plan to both share with the organisation's finance department and use as an input to our financial control of the project, as in Table 5.5.

Table 5.5 Weekly Costs by Task

	Total Cost	Wk 1	Wk 2	Wk 3	Wk 4	Wk 5	Wk 6	Wk 7	Wk 8	Wk 9	Wk 10	Wk 11	Wk 12	Wk 13	Wk 14
Total	£366,469	£18,000	£8,000	£125,000	£12,000	£8,200	£5,200	£9,400	£4,200	£29,049	£25,620	£4,400	£24,400	£4,800	£88,200
Start Task A1	£18,000	£18,000														
B1	£11,400		£3,200	£8,200												
B2	£11,400				£3,200	£8,200										
B3	£37,600									£4,400	£4,400	£4,400	£24,400		
C1	£116,400		£3,200	£113,200												
C2	£5,200				£5,200											
C3	£5,200						£5,200									
C4	£6,800							£6,800								
C5	£23,200								£1,600	£21,600						
C6	£4,800													£4,800	
D1	£5,200		£1,600	£3,600												
D2	£3,600				£3,600											
D3	£2,600							£2,600								
D4	£2,600								£2,600							
D5	£24,269									£3,049	£21,220				
End Task E1	£88,200														£88,200

Weekly Costs by Task

Milestones Marking Out the Path

We have now developed a workable project plan, and maybe we have been through a number of iterations to make sure the resources and finances mesh with what the project has to deliver. Now we have to take a step back and look critically at the project plan and answer this simple question: "What are the important events in the life of the project that show we are making progress?"[14]

These events are very specific to each project and could include, for example:

- The project board signing off on the business case.
- Completing and formally closing stage 2 of the project plan.
- Completing the first prototype of an IT solution or new product.
- Pouring the concrete for the foundations of a building or completing the roof.
- Passing user acceptance testing.
- Opening a new facility.

These significant events are called milestones, and like milestones along a road, they are waypoints on our journey. Unlike our tasks, they have no resources attached, and they are only points in time, not extended periods.

Reflections

If possible, using the same project as in your previous reflections, spend some time considering the following exercises:

- Put together a project plan showing the principal stages of your chosen project, together with the main events in each stage. As a form of practice, ask if you can use your organisation's chosen project planning or management software for this, but if not, you can work through the process manually or choose a software package that can be accessed for free on the internet.
- How would you avoid the situation in the field trip where the business sponsor imposes an irrational deadline after the team have completed a detailed project schedule?
- Take one stage of the project and develop a sample resourced stage plan, including resolving any resourcing conflicts. Ideally for this exercise, this stage plan should contain around ten tasks; certainly limit yourself to ten if you are not using a software package.
- Assign costs to each of the resources in your stage plan, add any material costs, and use these to prepare a finance plan for the stage showing when you expect the different costs to be incurred. You will need to decide when in each task material costs will be incurred. Note: When doing this for real, you may need to ask your finance department how they handle the timing of purchases. As project managers, we always need to know what we have left in our budget to commit, but you may find that, whilst this is of interest to them, their main driver is knowing when they can expect to pay an invoice.
- Create a task data sheet for each of the tasks identified in your stage plan. Check back to see that the stage plan is still rational and make any required adjustments.
- Design a template for a work package description and populate it for two sample tasks.
- List the key milestones in the life of your project and plan how you will celebrate reaching them.

Notes

1 OK, I hoped you spotted that we are down to six; at this point Why can normally be dropped for a while as we should have answered that in the business case.
2 Setting an unrealistic expectation for delivery of a major project in this way was hopelessly optimistic and finally rebounded on the business sponsor, who was reported to have retired early, and the project was also scaled back massively.
3 There is the so-called Hofstadter's law, which is sometimes discussed in relation to software development projects. This law extends our understanding of time estimation by building in a recursive loop and says that it always takes longer than you think, even when you take into account Hofstadter's law, which, of course says that it always takes longer than you think. Hofstadter is genuinely useless for real estimation, as the recursive loop means the answer of how long it will take must always tend towards infinity. Whilst this may be very accurate for some projects, we should not try using Hofstadter as the basis of planning, unless we are really not intending to deliver any time before our universe fades into oblivion.
4 Of course, some of us may be trying to use team engagement as a smokescreen to both impose our views on the team and generate their commitment. Be warned, inconceivable as it seems, we might just be wrong, and our team might just come up with better solutions than we do.
5 See Appendix 1 for suggestions on how such a facilitated meeting can be managed.
6 It is always worth taking a photograph of the sheet before removing it from the meeting room – sticky notes have been known to fall off . . .
7 If we can't even draw a straight line using a ruler, a package is going to make our output look far more professional.
8 Another contrived acronym, sorry again, but PERT has become a commonly used term.
9 If we are creating this diagram manually, it can be useful to go back to the sticky note and brown paper method to give us complete flexibility to move things around and capture our final result for drawing up later.
10 This last one is here for completeness, it's not often used.
11 I really do know even less about professional bricklaying than I do about sewage treatment – but please just bear with me on this.
12 A classic example of the economic law of diminishing marginal utility. Just like what happens when we have our fourth large slice of chocolate cake . . .
13 Plenty of other project planning and project management software packages are available,
14 Alternatively, you could ask, "What are the points in the project that we will celebrate with a [insert name of favoured beverage here] when we reach them successfully?"

Bibliography

UK National Audit Office, 12 January 2013. *Over-Optimism in Government Projects Report.* Available at: www.nao.org.uk: www.nao.org.uk/wp-content/uploads/2013/12/10320-001-Over-optimism-in-government-projects.pdf.

6 Project Team Creation and Management

Round Up the Usual Suspects

Aim of this chapter: To set out the issues around identifying and leading a successful team, including drawing on internal resources, multi-organisational collaboration, and remote team management, as well as identifying special issues around remote teams and remote working.

Learning outcomes: To be able to identify the most appropriate sources and key attributes of the staffing required to deliver the project. To understand different management styles and how they may be appropriate to different circumstances, including the management of staff and teams working remotely.

They say that it takes a village to raise a child. If this is true, we can extrapolate and posit that it may take at least a medium-sized town to deliver a major project. For example, it was reported that at peak times over 3,500 people were working on the Empire State Building in New York (Jackson, 2010) and 13,000 people were working on the Channel Tunnel between Britain and France (Eurotunnel, 2022).

From now until the handover and closure of our project, we need to create roles and recruit, focus, motivate, and monitor the contribution of each member of a variety of disparate groups of people in order to deliver our project plan. We need to do this continuously, every day. We, as the project manager, are responsible for fostering and managing the relationships, not only between the organisation hosting the project and these members of the project's wider social organisation (our team) but also often between these team members as well.

To be fair, most of us will not be managing a project of the size and complexity of the Empire State Building or the Channel Tunnel. However, there are principles and activities of which we should be aware that will apply whatever the size of the team that we have to build for our project. Even if we are given a free hand to bring on board whomever we want, there are still a number of key areas that we have to think through carefully in order to develop the team that is going to deliver our carefully crafted plan. If we are not given a free hand, then we will still have to go through the same thought processes about our requirements and then use the results to justify our demands. These processes are laid out in Figure 6.1.

DOI: 10.4324/9781003405344-7

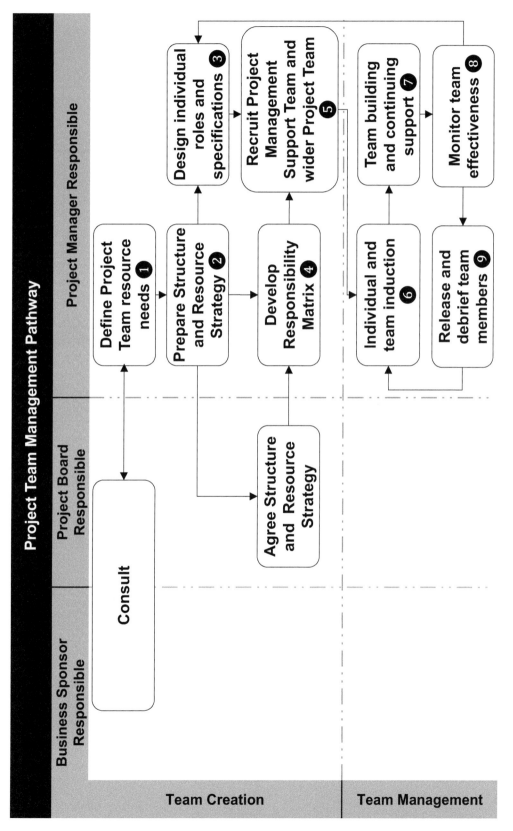

Figure 6.1 Project Team Management Pathway

But, before we go further into this, it's time to visit our continuing case study again.

An email has arrived from the director of HR to the project manager of bio-energy from effluent.

Re: Resourcing the bio-energy from effluent project
Dated 5 January
Hi,

I've just seen the management summary of the business case for the bio-energy from effluent project, which I believe you're heading up. I noticed there's no mention of our new company-wide management apprenticeship/trainee programme.

This scheme has been mandated by the board, and all departments must now adhere to it. We're aiming to grow our own top technical and managerial talent by taking in well-qualified school leavers and university graduates and putting them through a series of managed work placements, each lasting six months. This work experience is coupled with a series of two-week study periods spent on intensive courses led by a series of different universities. It's the company's objective that each person successfully completing the full scheme will obtain a degree, either a good honours degree or a masters, and also move on to obtaining full chartered membership of a relevant professional body whilst in a substantive post.

The CEO and I see your project as an excellent opportunity to introduce the management apprentices on the programme to the full depth and reach of our business. The first intake is due to start at the beginning of April, and I'd like you to plan to take a minimum of 10 people into your project. In October we'll have a second intake, and you'll also need to take an additional 10 from amongst those. In addition, we'll also be rotating the first intake around other departments. So, to be absolutely clear, this means that from 1 October, you will be required to provide 20 apprenticeship places. Of course, I'll be happy to accommodate you if you can take more than this.

Please get in touch with my apprenticeship team as soon as possible to put this in place.
J. Stevens
Director of Human Resources (Corporate)

It's your decision, as project manager, how you react to this one. But how will being the nursery class for up to 20 school and college leavers impact on the deliverables of the project and the other resources that need to be brought in? What sorts of things can be negotiated here to comply with company policy and defend the project, or is another approach required?

The email from the HR director prompts two tricky questions that we must consider before going through the steps to create our team:

- Who will make up our project team?
- How should we choose to manage the individuals that will be part of our team?

Who Are Our Project Team?

As we agreed at the outset, a project is always delivered by a temporary social organisation of individuals created for the specific purpose of delivering the project's objectives. It is this temporary social organisation that we now need to create, focus, and motivate in order to deliver the project successfully. But before going off and recruiting people to different roles, we first have to understand what constitutes a project team. This is a deceptively easy question, and some approaches to project management will define the project team as being just "those individuals engaged to help the project manager to deliver the project's objectives." The truth is, however, that this definition is less than adequate. Firstly, it starts out by excluding one of the most important members of the team – the project manager – and fails to recognise that the truth of who constitutes our project team is much more subtle. In reality, our project will be delivered by a far wider spectrum of people than just those "engaged to help the project manager," and this wider spectrum will have major implications for our approach to managing the team.

We need to start by recognising that the individuals who will form the team may belong to different formal organisations, or at least may come from different departments, with differing management styles, working methods, employment practices, and procedures. They are also all individuals with their own motivations, interests, strengths, prejudices, jealousies, rivalries, personal goals, demands, and limitations. Remember from our project scheduling examples that Y wanted two weeks' holiday in India for a 30th birthday celebration and Z had childcare commitments on a Tuesday? In reality, we might also want to avoid scheduling V and X to work together if their interpersonal alchemy isn't beneficial. In all cases, we, as the project manager, are the person responsible for managing not only the relationships between the project and the members of the temporary social organisation but often also the interpersonal relationships between them.

Even the term *management*, in its normal business sense, is probably not the most useful one when it comes to projects. Often even those individuals in the parts of our team we can control directly are only loaned to us and remain under their original management structures. Other individuals are only partially engaged with us and are completely outside our control. For example, we can ask for a member of our organisation's or a supplier's staff to be removed from our project, but we usually have no power to implement management sanctions against them such as disciplinary action or dismissal.

What does our "temporary social organisation" really look like? Starting from the lazy definition of "individuals engaged to help the project manager," we can split the "individuals engaged" fundamentally among:

- Individuals working to support us as the project manager in shaping and controlling the project – often called the project management team. This is really better seen as the project support team because it usually has a wider role than management alone. In some organisations with an enterprise project management office, the way the project operates this support will be tightly controlled centrally. Some of the functions of this part of the team may be performed centrally as a service to the project.
- Individuals engaged in creating the project's deliverables. These may be direct employees, temporary staff, or provided by our suppliers. These team members are very specific to the nature of the project and could range from general labourers to skilled tradesmen, software developers, architects, engineers, and even rocket scientists if the project calls for them. These we can label as functional specialists.
- On some projects, especially those involving business change, we will be engaging end users and other consultees who are going to be affected by the project.
- Our wider social organisation will also consist of multiple different elements not under our direct control. These different elements, both groups and individuals, include suppliers, contractors,

professional advisers, and staff in other departments on whose services we depend – e.g., human resources, procurement, and finance, gradually fading out through to our stakeholders, the industry, and the wider public.

Figure 6.2 shows not only the different layers that form our project team but also indicates how our direct control over individuals will diminish as we move outwards from the centre.

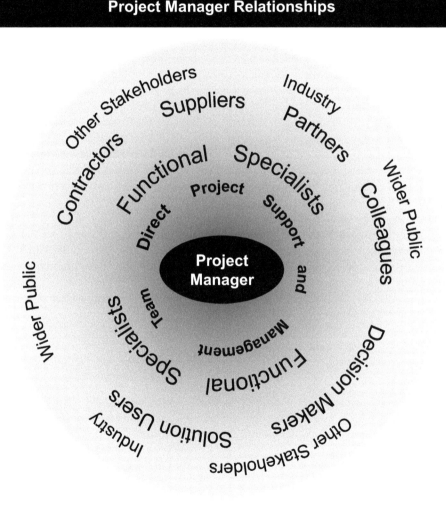

Figure 6.2 Project Manager Relationships

Choosing a Management and Leadership Style

At the start of our journey, we considered what a project manager actually does every day. Now, we need to consider how a project manager does it. The answer, of course, is that what we do is mostly through leading, directing, and organising other people. Consequently, a large part of our role as the project manager is about managing the relationships which are essential in order to deliver our project's objectives. In practice, some of our relationships are not just outwards to team members and stakeholders, as shown in Figure 6.2, but upwards as well with the business sponsor, the project board, and through them to the wider management of the business and its stakeholders, who may be different from the project's stakeholders. Managing other people's relationships can be challenging. Imagine bounding in one morning through the door of the large open office in which the client's and supplier's teams were co-located, only to see the respective managers of each with arms folded, backs to each other, staring angrily at a different part of the ceiling. In this case, it took days to get them even facing each other again, let alone talking.

A complete discussion of organisational psychology as applied to management and leadership practice is not in our scope, but a brief introduction to some elements of thinking on leadership behaviours will help inform the choices we have. It has been perceived that there are two types of managers. The first type believes that workers are naturally lazy, work only sufficiently to support their chosen lifestyle, and therefore need an authoritarian manager giving highly directional management instructions, backed up by a "firm hand" to ensure they perform effectively. The other type of manager considers that most people find satisfaction and motivation in doing their work well and see themselves as offering help, support, and guidance to help their staff to perform effectively. As an example, Steve Jobs of Apple fame is reputed to have said that he did not recruit intelligent people to tell them what to do; instead, he recruited intelligent people to tell him what to do. This strongly suggests that his instincts were towards the more social, supportive style of management.

In reality, there is a continuum of styles between these two, and at different times and in different situations, managers may have to veer closer to one or the other in order to achieve their objectives. As project managers we are task focused, aiming to deliver the project within our agreed-upon boundaries, but this does not mean that we always need to adopt an authoritarian approach. The ability to select the appropriate positive style of leadership or management for any given context is an essential element of any successful manager's toolbox. The special nature of a project, however, with its broad spread of relationships, requires the project manager to understand and be able to deploy an unusually wide range of positive management styles to different individuals and groups, often simultaneously. It is probably true in line management as well, but in the case of the project manager, it is more appropriate to refer not to management style but to relationship leadership, or simply relationship styles.

Irrespective of whether we naturally tend towards being an authoritarian or a supportive and coaching leader, there is another dimension to add to our relationships which may decide how we should manage them. These three crucial, but very simple, concepts help define our freedom to act – power, authority, and influence.

Power. This is the degree of power that the organisation has granted to the project manager in order to be able to control the delivery of the project. In practice, as project managers we may have very little direct personal power granted to us. Our project staff are often loaned by other departments and remain the responsibility of their line managers, and decisions on our contracts for both suppliers and contractors are usually made by the business and often administered by permanent departments. We may not be granted power to place orders or sign off on invoices for payment.[1]

Authority. Authority is much softer than power and is based upon the personal knowledge, skills, and confidence developed and projected outwards by the project manager. This authority may be over a wide area of knowledge broadly relevant to the project but equally may be limited just to the knowledge of the project itself. A good indication that a project manager is gaining authority is when the project manager's views are genuinely sought by a specialist on resolving a technical issue. Authority is only built over time and is the reciprocal of respect for the project manager by the team, partners, and colleagues.

Influence. Influence is the softest factor of all. The level to which we achieve this can only be seen in results. Influence, for the project manager, is the ability to take forwards the ideas and views usually generated by the project and present them in such a way as to cause decision makers and stakeholders to take them into account when arriving at a decision or view.

Each of our relationships with others draws differentially on these three fundamental concepts depending on our relative standing to each other. The way that we relate to a direct subordinate will be naturally different to how we relate to a contractor, a decision maker, or a stakeholder, because the three elements exist in different proportions in each relationship (or may be absent completely). We would, for example, expect to have no power in relation to a stakeholder, but we would seek to establish influence with them.

Personal integrity is our core starting point for each relationship. Although we all have a leaning towards a preferred style, we can choose how we present any transaction within that relationship to the other party or how we react to any transaction presented by them. Ideally, our choices regarding the style of each relationship transaction should:

- Respect all parties.
- Reflect the outcome that we wish to achieve in the short term.
- Foster a positive relationship in the longer term.

Achieving a balance between these three is not always easy. You may have seen the press stories of a former British prime minister hurling at least three mobile phones a week and the occasional office stapler at the wall, and again of a senior UK cabinet minister who was publicly outed for shouting and even swearing at civil servants. Whilst staff may comply with requests from people showing such behaviours in the short term, would such a relationship style suit us as project managers trying to foster positive long-term co-operation and participation in achieving the goals of our projects?

It sounds cold and calculating, but, in general, we should select our management style in accordance with the context, the setting, and the result that we wish to achieve for each of the different parties.

In writings on management styles, some terms seem to be used interchangeably, and sometimes with varying meanings. Table 6.1 contains a broad, but by no means exhaustive, description of management styles that we could chose to adopt as project managers in different settings.

Table 6.1 Project Management Behavioural Styles

Project Management Behavioural Styles	
Behaviour	**Features**
Authoritarian	Controlling all activity at a detailed level. Staff are given precise instructions and monitored to ensure they carry them out exactly. They are discouraged from offering thoughts or suggestions for improvements. Failure to obey instructions may lead to sanctions and dismissal. This approach allows for quick decision making and clarifies roles and responsibilities. It may be useful for junior or inexperienced teams but can be demotivating and fail if the manager is not present.
Autocratic	Less dictatorial than the authoritarian behaviour. The manager will attempt to convince staff that the decisions being imposed on them are correct and will be beneficial to the staff.
Directive	Although the manager is still controlling all activity, they listen to discussions on the best ways of implementing decisions already taken and adjusts accordingly
Bureaucratic	The manager relies heavily on predefined processes and actively discourages any activity outside them. Contributions and ideas from the team are only considered where they are within the boundaries allowed by the processes. There is a clearly defined hierarchy and specified roles and responsibilities for all team members. Staff may be forced into working in separate silos with communications between them being only through the defined formal channels.
Supportive	Also called paternalistic. Management still make decisions and may explain that this is because they hold the experience and wisdom to do so but will always take staff interests into account. This style also includes focusing on staff development and training; however, the knowledge and skills gained may not be properly exploited.
Coaching	Sometimes also called the servant leadership style, in which the manager is the servant. It is also a supportive style and casts the manager primarily as the coach and mentor of the team, with this role taking up much of the manager's time. Rather than disciplinary measures, the focus is on helping team members learn and grow from their mistakes. The focus is therefore on longer-term development of the team, which may cut across the short- and intermediate-term needs of the project.
Transforming	Efforts are concentrated on encouraging those engaged on the project to move forwards beyond their areas of comfort and motivating them to raise the level of their own achievements. This usually requires the project manager working alongside others, supporting and inspiring them and demonstrating a high personal work ethic.
Pace setting	Leading from the front, delivering instructions, and setting the pace of the project. This could be seen as part of our core role as project managers, but as a management style, however, it implies setting very high or challenging targets in an attempt to drive forwards.
Visionary	The visionary manager is one who is seen as being a source of creative solutions and tends to centre on the "bigger picture" and explaining goals rationales whilst leaving the team to focus on the details and giving them the freedom to deliver the necessary tasks with minimal interference. Consequently, it can deliver benefits of fully engaging team members and others to deliver creative solutions, but it can also present risks of failing to meet project delivery requirements.
Persuasive	This is the outward-facing counterpart to the autocratic style but relies on the manager's influence to convince others that the unilateral decisions taken are for the benefit of the wider project. A decision may be justified in terms of the process used and the underlying reasons for it.

Project Management Behavioural Styles	
Behaviour	**Features**
Consensual	Also called consultative. The opinions and thoughts of the whole project are sought from team members and stakeholders. The final decision will be taken in accordance with the roles and responsibilities for the project but will consider all the information received. This approach underlines much of the information gathering approach recommended in Appendix 2.
Engaging	This is the core approach for self-empowered teams. Management may give the team guidance, but all team members, and others, are actively encouraged to participate in the decision-making process. This can be time consuming and can result in decisions that reflect the lowest level of agreement rather than that which would produce the best outcome.

None of these terms should be seen as negative; they are all merely appropriate or inappropriate in each context. For example, sometimes it is necessary to be authoritarian, perhaps if a supplier or staff member is not delivering as required and our "softer" approaches have not worked. In these cases we may have to state very firmly what the consequences will be if they do not change the way they are working and deliver on their contract.[2] Our chosen styles are not fixed, and we can often migrate from one style to another in response to the same event – a situation with a new member of staff or team that starts out as seeming to need to be a directive or even autocratic transaction could move swiftly into coaching, pace setting, and transforming if we discover that the individuals involved are genuinely struggling and needing support.

Figure 6.3 shows how our definitions of power, authority, and influence underlie the list of terms used to describe leadership styles and indicates for which types of our relationships as project manager these may be appropriate. It also highlights that our change focus, lack of permanent staffing, and defined timetable may limit our potential choices of style compared to a line manager and indicates that other styles are always more applicable to other relationships. As in all personal interactions, however, none of the boundaries are fixed.

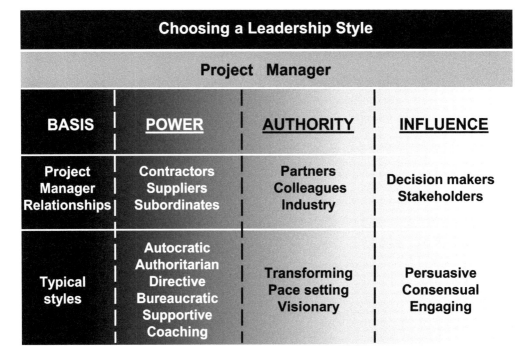

Figure 6.3 Choosing a Leadership Style

A Final Word on Management Styles

Although there are no negative styles, only ones that are not suitable for a particular transaction, there is, however, one very dangerous trap into which we can all fall far too easily – overindulgence in adopting the bureaucratic style. As should be apparent by now, by its very nature, project management is process driven and can be documentation heavy. Consequently, there is a real danger that we can become obsessed by the project management processes and their required documentation at the expense of our real leadership and management of relationships across the project. In short, we can become the prisoners of project bureaucracy. The bureaucratic style, therefore, needs to be kept in balance with other approaches, as it is just one of the useful management styles that we require. Our most bureaucratic of processes, however – compiling our weekly project manager's report on a Friday – might happen only after every single person needing us to be transforming, visionary, coaching, pace setting, persuasive, autocratic, consensual, consulting, directive, and so on has left the building for the weekend!

Creating and Managing the Team

To recap on the process for creating and managing project teams set out in Figure 6.1, we can break creating our team down into a series of steps:

- Define resource needs over the life of the project.
- Prepare and agree on a project organisation structure and resourcing strategy.
- Specify individual and team roles and objectives.
- Develop a matrix of leadership and other responsibilities for the different roles.
- Recruit the project support team and functional specialists.
- Induct the team into the project, its objectives, and its processes using a carefully thought-out onboarding process.

Ongoing processes will then include:

- Team building and support to individual team members.
- Monitoring team effectiveness.
- Team member release and debriefing.

Step 1: What Resources Do We Need?

The good news is that we have already completed the definition of our resource requirements. During the planning stage, we identified the resource requirements for the project from the bottom up by estimating the resources of different types required for each task, and then, after resolving any resource conflicts, we calculated the minimum numbers of staff of each type required for each task. We can now easily prepare the required picture of our project's resource needs and show how they will change over time.

Step 2: How Should We Structure and Resource Our Team?

Before we decide on our team structure, it's time for a site visit. This time, we're meeting a business sponsor who has an unusual suggestion to make.

We're visiting an initial meeting between a team representing a professional services company and Daniel. Daniel is a main board director from a government agency and a former client of the company but not someone either of the team have met previously. He has asked them for guidance on a particular issue regarding creating a project team.

This time we are invited guests, and we've been promised tea and biscuits. This will be a plate of standard public sector organisation biscuits, and it is guaranteed that they will be mostly plain but with one lonely chocolate biscuit. Therefore, we have a dilemma as guests. Do we take the chocolate biscuit or not?

We are shown in by his personal assistant.

"Thank you very much for coming in," Daniel says as we shake hands. "Please sit down. Did you find the place OK? We've got some tea or coffee, and please help yourself to the biscuits."

You do so, politely taking a plain one.

The personal assistant shows in the visitors. Both are dressed nearly identically and carrying identical company issued bags.

Daniel turns to them. "I have a couple of observers in the meeting, I hope that you don't mind. Although your company worked for us on some tricky projects, we haven't met before. So, please let me introduce myself and explain what we're trying to achieve. I'm an executive board director of this agency, and our main business activity centres around the independent review of case decisions by other bodies. For each case being reviewed, we build up a body of evidence in a case file, conduct whatever research our professional staff deem proper, and then reach and publish a decision.

"I've been made the business sponsor for modernisation of our operations. This will mean both computerising internal processes and creating a linked citizen/industry portal which allows for electronic submission of documents and makes all the documentation in the case file available to the public. This will be a radical change for what has up to now been a paper-based organisation.

"The two elements, internal systems and the portal, will have a different balance of issues. Some require driving change with external stakeholders, others require internal innovation, and yet others require dealing with the development of systems to support new ways of interacting with citizens and with industry. I've decided to create a managed programme consisting of two projects, one focused on the internal and external workflows and the other on the portal development. Our initial assessment is that this is going to take about three years to deliver into full operation.

"The agency has had little experience of initiating such a major innovation or delivering the necessary new and complex systems. We also need to reduce any risk of potential failure on our existing services and to safeguard our position of respect with stakeholders, including politicians, the industry, and the public. As you are aware, your company was selected in a tendering exercise a couple of years ago to provide us external professional support. Our contract still has three years to go, and I've decided to ask you to provide us with two experienced project managers to help us deliver on time, to cost, and within government guidelines for IT-related projects."

At the word experienced, *both of the visiting team members nod ambiguously.*[3]

"My problem is that although I know that your company has the expertise and experience to deliver projects of this nature, you lack experience in our particular area of business, which is governed by very tight legislation. We can cover over a hundred different types of cases, each of which has different rules, which are contained in different primary and secondary legislation. If we put a foot wrong anywhere, I assure you that it always results in parliamentary questions to ministers and a judicial review of our decisions.

"I've come up with a solution to staffing the project teams, which I'd like to ask your opinion about. Then we can move forwards with drafting the terms of reference for the two project manager roles. It's quite simple. I thought we could establish each project with one project manager paired on an equal

basis with one experienced manager from inside the agency as a full time project leader, with both reporting to me as the responsible board director. My challenge will be creating a matched pair that will pull in the same direction."

You ask if you can address a question to Daniel. *"Isn't that an extra cost overhead for the project?"*

"Yes and no," Daniel replies. *"In reality, whoever fills the role will have to be back-filled, but we wouldn't charge any costs of existing staff against the project. It's not our way of doing things."*

The team leader thinks for a moment.

"I can see several major advantages to adopting this approach, but it will depend on the people that you, and we, choose to lead each of the projects. Our experienced project managers are all used to working in a situation of uncertainty where information is often late, not of the best quality, or sometimes not there at all. We start work based on assumptions that we refine as we go on as better information becomes available. When we arrive at a new organisation, we open up our metaphorical toolbox containing everything a project manager needs to get a project running, and we then handle whatever the project throws at us. Your project leader will have to be happy working with uncertainty and with a completely new set of tools, processes, and techniques."

The speaker pauses and sips their drink before continuing.

"From what I understand about the agency, your staff are either administrative process workers gathering information and setting up the review processes or professional staff reviewing the case and arriving at a balanced decision. Neither of these seems suited to the world of uncertainty, flexibility, and responsiveness of two high-profile projects."

"Actually," Daniel says, *"I have a couple of staff in mind, both of whom have a different approach to life and whom I think would respond well to the stimulating experience of delivering projects. I also thought that to give them ownership, I might appoint them not to the project but to a role of service manager with long-term responsibilities for managing the two solutions and ensuring that we get the projected cash and non-cash benefits from them."*

The team leader considers Daniels views and replies, *"In principle it seems to be a great idea. I'd focus the project leaders' roles on areas facing the business internally and its clients externally, including organising stakeholders, managing communications internally and externally, leading requirements definitions, testing – and when we get there, benefits management. This would leave our project managers to lead on creating the project infrastructure, managing finances, monitoring and reporting progress, running procurements, and managing suppliers. One last thing. I feel that we need to make it clear to both the project leaders and the project managers that they're a team. While they may lead in different areas, they should both cover the full range of each project and be able to deputise for each other."*

Daniel is nodding.

"Thank you, that's all very helpful. I'll prepare role specifications for both project leaders and project managers. I've already spoken to your director about availability, and I don't want to lose any more time. Can you two start on Monday?"[4]

For both of two projects in our site visit, the concept of the project leader being an expert in the business appointed to lead jointly with the specialist project manager worked extremely well. Both the project leaders came to their long-term role at the end of the project with a full knowledge of how and why the solution was developed and an established set of relationships with suppliers and stakeholders across the organisation and its clients. The concept also allowed for much greater resilience in each project team, as two people had a complete picture of the project, its progress, and the issues being currently progressed.[5] This resilience meant that there was continuity, and when one of the project leaders moved on, there was no adverse impact on the project.

In order to manage our project effectively, we will need to identify how our resources are to be organised and to prepare an organisation structure. Sometimes, as on our site visit, a business leader working with the project manager in a business change project is helpful, and in other cases, they may not be. In any case, the larger our project, the more we will need staff to be organised into coherent teams. Each team will usually require a leadership structure, with the team leader accountable to us for the team's performance. Without such an approach, we can end up trying to manage disparate professional resources across a wide span of individuals, and this will increase the complexity of the project manager's role.

The basis for the structure will depend on the nature of the project, and in some cases it may be appropriate to focus on professional skill areas. For example, we could separate out analysis and design from physical delivery. In other cases we may want to create teams responsible for end-to-end delivery of a part of the project, or we may mix the two. An example of a mixed structure would be an IT system built and installed centrally but rolled out by a series of parallel regional implementation teams.

We also need to identify where we will draw the different types of staff needed to populate our project. These issues may have been addressed previously, but we will need to set out and gain confirmation from the project board on where and how we will:

- Recruit staff to the project on a temporary basis.
- Obtain staff on loan from elsewhere in the client's organisation. This also requires considering who will fund any necessary temporary replacements for the staff released.
- Use external staff on an agency or labour-only contract basis.
- Deliver tasks under contract from suppliers.

As ever, we need to document our conclusions in a strategy document, and our resource strategy must address our seven key areas by answering questions such as:

- Why is a separate resource strategy required?
- What are the main resource requirements for our project, and how will we structure them into a manageable project team?
- When will the resource management activities take place?
- Who will be responsible for ensuring that the resource needs are monitored and adapted to the evolving needs of the project?
- How will team and individual reviews and other events be triggered?
- Where will resources be located?
- How much is it anticipated that resourcing the project will cost during each time period?

Step 3: What Do We Want People to Do? – Specifying the Roles

As the starting point in our recruitment process, for each of the task areas appropriate to our project, we need to:

- Design the job roles and required performance standards that we will expect from our support team.
- Estimate the time and numbers of staff that will be required.
- Build a skills profile for each role.

Our starting point, however, is the design of the job roles, and we need to prepare a specification for each role within the team with the only exception being where tasks are delivered wholly by suppliers under contractual delivery terms. Ideally, when designing the different job roles, we would work with the designated lead professionals in each of the teams that we have identified in our project structure. Often, however, these individuals may not be in place at this stage, so it may be appropriate to make a first cut at the role specifications and refine them later as the team leaders join the project.

Whilst the role specification compares directly to the line management process of preparing a job specification, including tasks, reporting lines, qualification and skill requirements, and standards to be achieved, there are also some significant areas in relation to the project that we should include:

* The project's major objectives.
* The project's critical success factors and the expected contribution from this role to meeting these factors.[6]
* Personal responsibility of all project team members to collaborate openly and honestly in the delivery of all tasks across the project – i.e., requirement to adopt a collegiate rather than a siloed approach to working.
* Personal responsibility of all project team members to identify and notify their team leader or the project manager of any areas where they feel that changes to the project may be required or desirable.
* Personal responsibility of all project team members to identify and notify any new risks or issues arising using whatever mechanisms that we put in place for this.[7]

Role of the Project Support Team

Whilst the roles required to deliver the major outputs from the project will be varied and highly dependent on the type of project, there is a large amount of commonality in the requirements for the functions to support the project manager. In Chapter 1 we identified key areas of the project manager's role in the oversight and monitoring of the project delivery process. In many projects the level of activity necessary to discharge these areas of responsibility requires the project manager to create a support team. Depending on the size and complexity of the project, this team can become a very large.

These activities are typically supported by a core group, which may often be called the project management office but would more descriptively be called the project support team.[8] Whatever it's called, this team's responsibilities are focused on two areas:

* Ensuring that the administration and governance processes we have designed for the project run smoothly.
* Acting as the central point to support the wide variety of parties which are working to deliver or are interested in our project.

Project Support – Technical Skills, Competencies, and Attitudes

Whether we have the ability to hire new staff or, as in many projects, have members of the project support team loaned temporarily from other parts of the organisation, when deciding to accept them into the project team we have to consider three equally important elements:

* Match with the technical skills competencies required for project team roles.
* Personal attitudes to work generally and to the project's outcomes.
* Ability to fit within a new and dynamic team.

TECHNICAL SKILLS

We have already covered project planning in detail (see Chapter 5) and identified circumstances where a specialist project planner was a very useful addition to our project support team (see Chapter 4). In some projects the support team may have additional responsibilities directly attached to it – for example, a large project may have its own human resources team, or one with a heavy contractual workload could have its own legal resources. Set out here are the usual remaining major activity areas for the project support team. For each activity area is a brief summary of the activity and the type of main skills we should look for when searching for staff to support us.[9]

Task progress monitoring. Overseeing of the delivery tasks designed to achieve the project's goals. On a regular basis, this activity requires checking up on how all staff and teams charged with delivering a task within the project plan are progressing and then updating the plan and associated forecasts. The staff engaged will require:

- A thorough knowledge of the project.
- An understanding of each of the task areas and their potential delivery issues.
- Ability to interact with different functional specialists to obtain and critically evaluate progress information.
- Expertise in any project management software packages we plan to use, including project planning tools, or the ability to acquire it.
- Ability to present complex information in summarised and tabular formats.

Financial management. All projects burn money, some at an alarming rate. Authorising expenditure and keeping track of budgets, committed sums, invoiced amounts, and reported expenditure of every item at the lowest level of the work breakdown structure is a major task – and often a major headache. But without it we cannot report on the true state of our project and forecast financial outcomes. The project's finance team will need to:

- Include a qualified accountant or at least a very experienced finance department staff member, who should be highly analytical with the ability to analyse expenditure patterns and identify and report variances from expectation.
- Have strong work scheduling skills to ensure that the project finance team are able to meet multiple reporting deadlines, as the project reporting cycles will usually be based on the project's stages and will differ from the organisation's time-based financial cycles.
- Have a thorough grasp of the general financial management policies and processes within the organisation.
- Possess excellent communication and analysis skills, the ability to build strong relationships, and the capacity to work under the pressure of tight deadlines.

Performance reporting. Collecting and disseminating project information regarding current progress and forecasting future progress and status. Reporting clearly and truthfully to business sponsors and stakeholders throughout the project life cycle, including closure of the project. Members of staff preparing performance reports should:

- Have advanced spreadsheet skills and be able to collate information from disparate sources and prepare useful, often graphical, analyses of data.
- Be skilled in the use of any business intelligence software preferred by the host organisation.
- Be able to use the software packages available to the project to visualise and prepare the necessary project management dashboards attuned to the needs of each group of users – e.g., project management, project board, stakeholders.

Document management. Projects usually generate reams of documents, some still on paper but now more usually in electronic forms. In some cases, this is mainly core project management documentation. In other cases, it can include drawings and other designs, specifications, programme listings, test results, photographs, videos, and, of course, correspondence and emails. Whilst systems can help, keeping track of all project documentation, maintaining version control, and ensuring that anyone with the need to access documents is verified and given access to the most appropriate version needs a nominated person (or a whole team on a large project). The required skills are:

- Technical ability to establish and manage a document repository and the professional ability to advise on the structure of the repository.
- Very close attention to detail to ensure that all documentation is fully accessible, cross-referenced as necessary, and version controlled.
- Process oriented to ensure that documentation is properly reviewed, authorised, and verified, coupled with the ability to enforce timely submission of documentation to the repository.

Quality monitoring. The early establishment (and subsequent enforcement) of criteria and processes to ensure that all project deliverables meet required quality levels is essential for meeting any project's goals. The quality team is responsible for ensuring that internal control processes are followed by all parts of the project and for liaising with all other quality reviewers. There are two basic types of quality monitoring with which the team are concerned:

- Evaluation of the central processes and documentation being used to organise and manage the project delivery – i.e., those areas under the control of the project support team, for example answering the question, "Is the business case still fit for purpose?"
- Co-ordination and review of the independent assessment of the processes followed by the project, including any testing of the output produced by the functional specialists. Examples of this may be in an engineering project, where the team will ensure that designs, test results, and final deliverables from a supplier are passed to a specialist quality advisor for assessment or, as in UK government sector projects, where the team will liaise with an independent gateway review team.

The skills required by the project support staff responsible for overseeing quality monitoring are:

- A thorough knowledge of the project.
- An understanding of each of the task areas and their potential quality issues.
- Ability to interact with different functional specialists to obtain and critically evaluate quality information.
- Strong organisational and planning skills to ensure that all aspects of quality monitoring, whether internal to the project or external, can be co-ordinated and progressed without detriment to planned project activity.

Procurement and contractual management. This is a continuous activity from designing the procurement and managing the tendering stage through to monitoring contract delivery and contract closure. There are often two levels of procurement and contract management. The administrative and clerical work may be carried out within the project but operating under the advice, supervision, and review of a specialist procurement team (within larger organisations) or a contracted adviser (in smaller ones). Where difficulties are experienced, the specialist procurement team may take over the lead in contract management and, when necessary, escalate issues for legal advice. The project support staff with responsibilities for procurement need:

- An understanding of the organisational and legal requirements of the procurement process.
- Very close attention to detail and accuracy. Where documents such as specifications or invitations to tender become part of a contract, accuracy can be critical. Consider the difference

and potential contractual problems arising further on in the project if our list of landscaping requirements was meant to include "fruit trees, and flowers" and instead an extra comma was inserted, leading to "fruit, trees, and flowers" being supplied.

- Strong communications and interpersonal skills, for dealing both internally with project staff and externally with potential suppliers.
- An ability to work under pressure to meet tight deadlines, to ensure that procurements are operated with any legally imposed timescales as well as meeting the demands of the project plan.
- Well-developed skills in the use of office-based software products, including any specialist procurement and contract management software.

Change management. Managing changes to the project is an area of major concern in some projects. The project support team needs to be able to control and track all requests for change. Controlling and monitoring the change can be complex and covers:

- Internally generated change – for example, "We would get much faster take-up if we changed A around and had two Bs as well; there's only marginal extra cost and no additional time will be needed."
- Externally generated change, either such as, "The regulations have changed and now instead of being left as plain metal everything needs to be painted with two-centimetre-diameter blue spots, which will cost £Y extra and take an additional four days," or the totally unexpected event, "We found it wasn't a solid wall but one filled with rubble, and when we cut through the doorway, seven tonnes just came out. Choking dust everywhere. It took us fourteen hours of effort to get it out of the building and clean up. You now have to arrange removal, which is again extra cost."

The team needs to be able to categorise changes, respond rapidly, and employ appropriate change authorisation and tracking processes to cover those changes which can be planned as well as those that require immediate action and post-change authorisation. Their required skills include:

- A thorough knowledge of the project.
- A detailed understanding of each of the task areas and their potential change issues.
- Very close attention to detail and accuracy.
- Strong communications and interpersonal skills to ensure that any conflicts arising from the change process are minimised.
- An ability to work under pressure to meet tight deadlines to ensure that, where necessary, change requests are properly authorised to enable work to continue.
- Well-developed skills in the use of office-based software products.

Risk and issue management. All projects face risks and issues, and their continuing assessment, management, and control is a central role for us as the project manager and for our support team. Identification, quantification, and the management of risks faced by the project are considered in Chapter 8. In some projects a dedicated risk manager or risk management team will be required. In others the risk management process will be operated by the project manager, with or without administrative support. If the project is of a scale that dedicated staff are required, their key skills and attributes include:

- Understanding of the risk and issue management process and the evaluation of risks and issues.
- Attention to detail in the recording and tracking of risks and issues and their mitigation.
- Strong communications and interpersonal skills to ensure that any conflicts arising from the risk management process are minimised.
- Well-developed skills in the use of office-based software products.

In addition, the completion of a recognised project risk management accreditation by at least some members of staff working in this area is desirable.

Communications. Well-planned and managed communications with stakeholders, and frequently with wider members of the public as well, are often essential, uniting team members, stakeholders, and others behind the project's goals. Poorly managed communications can lead to unnecessary and ill-informed opposition to the project's objectives, adverse publicity, and public comment when a project runs into difficulties. In turn, this can lead to poor overall performance by the project. It is therefore worth taking extra care to select support team members with the appropriate communications skills. The types of skills required are:

- A thorough knowledge of the project and its aim and objectives.
- An understanding of the information needs of the stakeholder community and, where necessary, the wider public.
- An understanding of the importance of the project to the wider organisation.
- Very close attention to detail and accuracy.
- Very strong written and verbal communications and interpersonal skills.
- Ability to write and present persuasive communications.
- An understanding of the strengths and weaknesses of different communications channels.
- An ability to work under pressure to meet tight deadlines.
- Well-developed skills in the use of office-based software products, including presentation packages and web and document publishing tools.

Benefits management. Where it is decided that the project will have initial responsibility for benefits management, the most appropriate strategy is usually to agree that the post-project benefits manager will commence *during* the project. The benefits manager should then work either as part of the project support team reporting to the project manager or in very close liaison with it. In practice, as project manager, we will often have less responsibility for appointing a benefits manager who will have a longer-term role in the organisation, but we can offer advice on the balance of skills needed. These skills include:

- Exceptional understanding of the area of the organisation's business to be impacted by the project's outcomes.
- Very strong interpersonal skills, including the ability to negotiate with business managers to ensure that the anticipated benefits of the project are achieved.
- Numerical analysis skills, including the ability to use any business intelligence software preferred by the host organisation.
- Ability to use those software packages available during the period of the project's operations, and afterwards during "normal" operations, to visualise and prepare the necessary benefits dashboards attuned to the needs of each group of users, managers, and stakeholders.
- Ability to present to different groups, including users, managers, and stakeholders, on the progress of the benefits management plan.

ATTITUDES TO WORK

It is clear that project work is unlike working in a line operational role, and this applies especially to the project support team. The second element for us to assess when considering people to fill the roles in our project support team is our need for team members to demonstrate the following personal attitudes and attributes:

- Ability to work to the many deadlines and pressures which abound in the uncertain world of projects.
- Adaptability and flexibility to support other team members under pressure at different times.
- Creativity in problem solving and responsiveness to changing circumstances.

As project managers, we need to ensure that we recruit the team members with attitudes to work that at least do not cause future management difficulties. Consequently, whilst assessing attitudes is more subjective and challenging than mapping technical skills and competencies to a matrix of requirements, it is equally important.

TEAM DYNAMICS

The final element that we should assess is the candidate's match with the overall team of the wider project team. Our task is to create a team with a positive dynamic to deliver the project, and we need to consider how candidates might be able to contribute to that dynamic or at least not disrupt it. One business sponsor explained it this way:

> You don't want me to loan you my best process workers for your project team. These people all have their heads down and just get on with the job, smoothly and efficiently. What you need is people that can do this, certainly, but who can also look up from time to time, look around, and think: are we doing this in the best way?

There is widespread guidance available on the types of people that we need within teams.[10] But we also need to consider our own preferred leadership style. A team of mainly very senior and experienced staff might require a more hands-off style from us, and one consisting of more junior, inexperienced staff may require a more directive approach, probably backed up by coaching and support.

It is worth considering using probing scenario-based interview questions and personality assessment questionnaires to help establish whether a candidate would make a positive contribution to our new team. In some critical roles, it may also be worth engaging outside expert assessments, and these should cover wider attitudes to work as well as team dynamics.

Step 4: Who Will Be Responsible? – Loading the Responsibility Matrix

There is a further document we should prepare that is especially useful on larger projects or those which have deliverables being provided by different teams. Put simply, we cross-reference the tasks in the work breakdown structure to each of the teams identified in the resource plan and then downwards into individual roles within the team. The resulting responsibility matrix will be a main component in our control of the project as we move into the delivery stage. The responsibility matrix is a key document for us when managing larger projects by:

- Providing an overview to teams or individuals showing all of the tasks in which they will be engaged.
- Demonstrating how their work will mesh with that of other teams in delivering the project. This is especially useful during the onboarding sessions for staff new to the project team and for resolving crosscutting issues arising between teams.
- Providing the baseline for control and reporting mechanisms during the active implementation stages of the project.

A simple example of a visual responsibility matrix mapping the responsibilities of different teams against WBS elements is set out in Table 6.2, but this could be equally applied to individuals within the project or to tasks at a lower level.

Table 6.2 Team Responsibility Matrix

IT Accounting System Replacement Project Team Responsibility Matrix									
WBS Element				Project Management	Design	Procurement	Development	Quality	Communications
1	1		Project initiation	■					
	2		Business case	■					
	3		Project plan	■					
	4		Project process development	■					
	5		Stage quality review	■				■	
	6	1	Next stage plan	■					
		2	Current stage closure	■					
2	1		Project team creation	■					
	2		Detailed requirements		■				
	3		Solution design		■				
	4		Procurement strategy	■					
	5		Procurement processes			■			
	6		Contract agreement			■			
	7		Stage quality review	■				■	
	8	1	Next stage plan	■					
		2	Current stage closure	■					
3	1		Phase 1 development				■		
	2		Phase 1 testing					■	
	3		Phase 1 acceptance	■					
	4		Phase 1 public launch						■
	5		Stage quality review	■				■	
	6	1	Next stage plan	■					
		2	Current stage closure	■					
4	1		Phase 2 development				■		
	2		Phase 2 testing					■	
	3		Phase 2 acceptance	■					
	4		Phase 2 public launch						■
	5		Stage quality review	■				■	
	6	1	Next stage plan	■					
		2	Current stage closure	■					
5	1	1	Project closure report	■					
	2	2	Project quality report					■	
	3	3	Lessons learned report	■					

Step 5: Recruiting Our Team

As most client organisations will have defined recruitment processes already in place, our role in recruitment as project manager is to engage with those processes and brief the staff who normally manage and operate them. We also need to stress to them the importance of timeliness in the recruitment process and their role in helping the project deliver on time.

Not all team members will come to us through a formal recruitment process; some may be through temporary internal transfers. Staff currently working elsewhere in the client organisation can bring a whole range of specialist knowledge, local experience, and contacts which can be very useful to us in establishing and delivering the project.

There is, however, a health warning that should come with all staff being "offered" to us as internal loans. An old saying refers to the unfortunate habit of forcibly conscripting (pressing) men into the armies and navies of many countries during the 17th and 18th centuries: "A willing volunteer is worth ten pressed men."[11] The same is true of our project team, and we want to attract willing volunteers rather than having staff forcibly transferred to us. Any team member who is "pressed" upon us is likely to require more time in formal onboarding and may be less willing to be supportive at times of peak pressure. So, if managers offer to loan internal staff to our project, we need to explain to them that a key part of our recruitment process is attracting potential team members who are already motivated to help us deliver. We should always ask any managers offering staff transfers for the opportunity to stimulate interest in target staff groups and seek individual requests to join the project. We must then subject the potential transferees to the same assessment processes that we will apply to any other applicants to join the project.

Many project teams will have a membership which varies according to the project tasks being completed at any particular time. At any one time, we need to have in place only those people required to complete the current tasks and to be recruiting and onboarding staff with the skills and experience required to complete our next set of tasks. Recruiting the team is, therefore, a continuing process throughout the project. Taking a building project as an example, our dynamic "delivery" team membership might begin with architects, surveyors, and engineers and then progress through a whole range of building trades from groundworks staff all the way through to painters and decorators. Each of these groups of staff are released from the team as their tasks are completed, but remember we will may need support to handle any changes to project outputs or resolve any quality issues that emerge later.

Initially, we will normally be looking to populate the major roles in our project support team. Later, as project momentum builds, we will want to fill the remainder of the roles that we have identified in that team and in parallel be recruiting the staff who will be delivering the first outputs of the project.

When considering the skills required for the project support team, we identified that attitudes as well as technical skills and competencies were important, and this can be applied to the recruitment of any member of the wider project team. In summary, there are only three questions we have to answer when deciding whether to recruit an individual to work on our project. These are the same questions on which we need to satisfy ourselves when recruiting for any roles:

- **Technical skills**. Can the person do the job?
- **Attitudes to project work**. Will they do the job?
- **Team dynamics.** Will they fit in with ourselves and with the rest of the team?

Step 6: Bringing Our Team on Board

Especially in larger organisations, there may be standard induction processes and training requirements that every new employee will be required to follow. In addition to these, our objectives for a project induction and orientation process for all new project team members (whether they are employees of the host organisation or not) are to:

- Enable every member of the team to understand the business context of the project and their personal contribution to achieving the project's goals.
- Transfer our knowledge of the project and what it is trying to achieve to new starters.
- Ensure the team members can use the various tools required to deliver their part of the project and that they understand the processes, quality, and safety standards the project will be using.
- Empower all team members to contribute to wider discussions and activities such as those in our processes for defining requirements and managing risks, issues, and changes.

There are many ways in which we can engage with new team members to achieve these objectives, and these may range from a one-to-one briefing session in a very small team to a fully structured programme for large teams. Such a structured programme would justify investment in areas such as online learning courses, preferably to be taken before or as soon as a team member starts, and regular scheduled awaydays for new team members. There are, however, some simple but successful practices that we can use in any project, large or small. We should, of course, use common sense in selecting which actions to employ and when, as not all actions are fully applicable to all team members. For example, if we are adding a contract bricklayer to cover a holiday period, we may bother less about orientation and focus instead on health and safety. The types of simple good practices that we can draw on include:

- The welcome pack.
- The first day.
- Workplace partner.
- Feedback.

The welcome pack. Some larger organisations will have a new employee welcome pack that will contain much of the information needed for new team members, such as general policies and company-wide practices and processes. Given that some of our team members may already be employees while some may be working for other organisations, we need to prepare our own "welcome to the project" pack, some of which may be directly "borrowed" from the employee welcome pack. Whilst our welcome to the project pack should contain both the broadest overview of the project and details of key personnel, it should primarily focus on the information required to enable the new entrant to function as part of the team. The pack should cover important areas such as access to buildings, IT facilities, how time spent on the project should be recorded, and travel and safety policies. It is always useful to send the welcome pack as part of engaging with the team member prior to their first day. When sending the pack to a team member, it is useful to add a personal schedule compiled by their team leader showing where they need to be and when and what their induction and other activities will be for the first few days.

The first day. New starters need to be welcomed with enthusiasm on their arrival. This may be our job as project manager, or it may fall to their team leader, in which case we ought to meet them later as part of their initial orientation. Again, this is an even more essential step if the team is working virtually. Do be wary of throwing too much at a new starter on the first day, and if there is an option to spread out the orientation and induction over a short period, this can be helpful in allowing them to absorb and reflect on the information we are giving them. A critical part of this phase of the induction is to ensure that new starters fully understand the project's objectives and their expected contribution to meeting them.

Workplace partner. Assigning each new starter an existing team member as a formal workplace partner or "buddy" can be extremely useful in guiding them through the quirks of the organisation and showing them how things work in reality as well as in theory. The ideal person will be working in the same project area but should not be the starter's line manager if that can be avoided. Whilst this is a recognised strategy in conventional teams, in distributed projects where team members may have less opportunity to form direct relationships, it becomes crucial. It also has the advantage of offering continuing support after our initial team induction process and any formal welcome events have finished.

Feedback. A short time after starting, it is useful to ask the new team member for feedback on the induction process. Depending on the size of the team, this could be undertaken by the project manager or by the team leader and reviewed by the project manager. As ever, we need to ask two key questions, always finishing on the positive one:

- What aspects of our induction do you feel we should strengthen or change?
- What aspects of the induction went well and do not need to be changed?

Step 7: Team Building

Now we have at least the core of our able and competent project support team, and in due course, we will also be bringing on board the first of our functional specialists. Our task is to build this group of disparate individuals, probably from different organisations and certainly at least from different departments, into a high-performing team focused on delivering our project. Particularly where we are engaging external suppliers and contractors, our relationships are focused through the lens of the contract, and their rights, responsibilities, rewards, and penalties will impact upon the ways we can organise and lead our team. Team building is a continuing activity arising from understanding and motivating the individuals that we have recruited; it is not a sporadic series of awaydays where people are challenged to get their team across a river using only two pencils and a rubber band (fun though some may find them to be).

Appendix 2 provides an overview of how partnership working has developed in delivering complex projects. It also has some important conclusions which highlight potential conflicts in team member motivation in all project teams that we should be addressing in team building. We have to recognise that, important as our project is, project team membership is always temporary and forms only part of our team members' careers. The effects of this are as follows, with the exception of contractors engaged through agencies:

- Many team members still owe allegiance to the employing organisation, to whom they remain managerially responsible.
- Team members have relationships with different parts of their employing organisation which may exert pressures of different sorts upon them that run counter to the project team's requirements. Different organisations may exert pressures on their staff in different areas.
- Team members may identify solely with their employer and their employer's objectives and not with the project and its objectives.

Figure 6.4 shows how these effects may emerge to cause team members to have motivations not aligned with delivering the project.

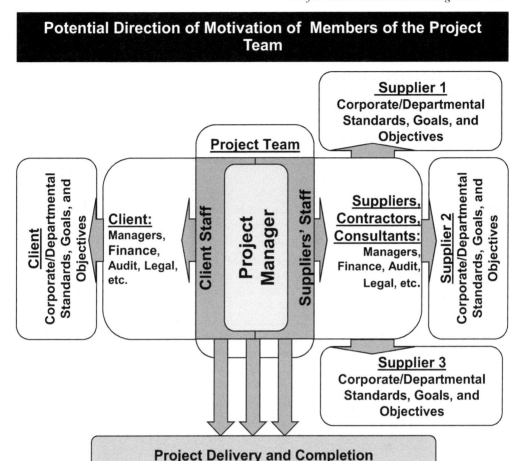

Figure 6.4 Motivation of Project Team Members

Our aim as project manager is to create a cohesive team that identifies with the objectives of the project and to provide support to team members so that they can be open with us when they feel they are being subjected to conflicting pressures from ourselves and from their "home" organisations. Sometimes we can help by recognising those pressures ourselves and changing our demands. At other times we may need to intercede with our suppliers and client organisation departments to find acceptable solutions that will ensure that staff feel able to deliver for us. To illustrate the point, trying to organise a site visit for team members employed by multiple organisations can be a nightmare, especially if overnight stays and international travel are involved. Every organisation involved will have different travel rules regarding flights, trains, and taxis, and some will insist on central booking for their employees. Subsistence rates or permitted meal costs will vary between companies and can seemingly be cut on a whim.[12] Trying to get the whole set of team members on a site visit and be able to eat together, let alone stay in the same hotel, can be virtually impossible.

The major question that we face is how to build a high-performing team. There are a number of issues to address and some basic techniques and ideas that we can adopt. One particular issue we need to always bear in mind when looking at these suggested techniques is the wider local culture within which our team, or different parts of it, will operate. This is even more vital where our project team is drawn from different organisations or different regions, countries, or continents. We must always recognise that actions which may be seen as acceptable in one country or culture may be regarded as either ineffectual or excessive in others.

Before we start those team-building exercises, awaydays, and the other sorts of activity beloved by team-building experts, we need to consider this simple three-part process for building an effective team.

- Selection.
- Understanding.
- Copying success.

Selection. Good news, the first step to creating a high-performing team is to select the right people, and this we have already done. We have rigorously applied our selection criteria, and we pushed back when sneaky attempts were made to transfer other teams' poor performers onto us. Any project will fail to deliver if we fail at this important step of getting the right people with the right attitudes and skills at the core of the project, so we do need to have been very firm with the client organisation and suppliers over the people we can accept.[13]

Understanding. We need to build an understanding of the motivation of *each* team member. During the selection process, we have gained *some* understanding of the attitudes of the people that we have appointed, and now we need to ask a question about each of them – "What will it take to build *this perso*n into a high-performing team member?" The answer may cover the type of relationship they need from us, the pressures they are under from their own parent department or organisation, as well as any personal orientation or team onboarding activities.

Copying success. Our third step is to recognise those factors that high-performing teams have in common and work hard to replicate them within our project. There are many different lists of common factors of high-performing teams, but they centre around three main areas:

- **Trust**. We need to actively build trust and confidence between team members, as well as between team members, the team leader, and ourselves. Building and maintaining trust is a continuous activity, given that team members:

  - Change over time.
  - Come from different organisations which will also change over time and put different pressures upon them.
  - Are all individuals with their own perspectives on life, work, and our project.

- **Safety**. If we are to engage the full abilities of our team, they must be free to challenge and free to put forwards different ideas. This can only be achieved by creating an intellectually safe working environment. For example, a facilitated discovery exercise will fail if everyone feels the need to agree with the most senior manager present.
- **Purpose.** Spending time with team members to ensure that they understand the purpose and direction of both the team and their role within it is critical to fostering their self-motivation and their commitment to delivering to the quality standards and in the time frame that the project demands.

Figure 6.5 shows how values and behaviours that we need to create for our team contribute to these three foundations.

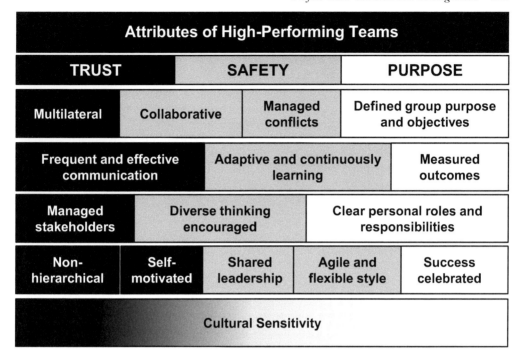

Figure 6.5 High-Performing Teams

Step 8: Monitoring the Effectiveness of Our Team

There are usually only two motivations for monitoring team effectiveness. The first is to confirm the current level of achievement, and the second is to be able to improve it. At one level we have an advantage over line managers in that the effectiveness of our team in delivering against our plans is being monitored and reported each time we compile any form of progress report. Team performance, however, reflects the performance of each of its members against the three key attributes which we assessed during the selection process:

• Technical skills.
• Attitudes to project work.
• Team dynamics.

Some of an individual's performance can be monitored via assessing and providing feedback on the quality of their output and their record of meeting time and costs targets. However, in order to maintain an effective delivery process over the whole life of the project, we need to be not just measuring current overall progress on tasks but assessing how well we match the identified attributes of a high-performing team as in Figure 6.5. This means that we need to allow time in one-to-one sessions with each team member to explore how well they perceive that the entire team, including ourselves as project manager, match the 16 attributes of a high-performing team set out in Figure 6.5 and where they feel that improvements need to be made. This can be an experience which is challenging and uncomfortable for ourselves, the team leaders, and the team members. Its success will reflect the levels of mutual trust and respect that we have established.

Step 9: Release and Debrief Team Members

It is almost too straightforward to say that the most cost-effective time to release a member of the project team is when their allocated tasks are completed. Quite often this is true, but the problem we can encounter on some projects is that even if all of a team member's planned allocated tasks are completed, it may be that the project will subsequently require the team member's input. This is especially true where we are relying on the team member's knowledge of the work that they have carried out. A simple case might be a software development. We employ a specialist to develop a web front end so that customers can enquire online about the status of their accounts. The specialist completes the work, which unit tests very well, and with no problems being found, we release them from the project. Fast forward a few weeks and a test failure in another system element means that we need to make alterations to the web front end. Ideally, we wish to recall the original developer as the most effective way to make the required changes, but if we cannot negotiate such a recall, we would have to adopt other, less desirable strategies. Sometimes it is possible to mitigate such risks if, for example, instead of releasing key team members, we can assign them to other suitable tasks for a period or negotiate a recall agreement with their home department or employer.

Using a formal debriefing interview with team members at the end of their time with the project is an undervalued way of obtaining further feedback on performance – both the team's and our own. Unlike staff exit interviews by line managers, most of our exit interviews will be with team members recruited only for a temporary period, and so we are not seeking information about staff retention issues. Instead, each exit interview provides us with the opportunity for personally thanking the team member, or quite often a supplier, for the contribution they have made to the project. This is also an opportunity to learn from team members:

• How they perceived the project, the team, and the work itself.
• Their feelings about our management style and the culture we created.
• Any wider views on how we could improve the delivery of the project and the performance of the project team.

It is possible, if we wish, to close an exit interview by "borrowing" from our approach to eliciting information from stakeholders and asking the team members two questions:

• What three things do you think that we should change to improve the delivery of the project?
• What three things do you feel that we do well and should continue to do?[14]

Project Management at a Distance

Dispersed or remote projects add a dimension of difficulty in choosing our style for managing particular relationships. This is because some types of projects are increasingly being delivered by whole teams that are partially or completely remotely located. We may never be able to physically meet some members, or even leaders, of our team. Consequently, we will need to develop effective remote communication and control processes to underpin our relationship with them. Frequent examples of remote and dispersed working include:

• Software development projects being carried out by staff located in a supplier's own facility rather than at a client site, and thus physically distanced from the user community. On one project the supplier even had two teams working, one in the UK and one in Australia. Whilst this permitted two-shift working and led to a quick turnaround of urgent issues, it also led to communications problems. On another European project, one part of the software development was handled by a sub-contractor in Hong Kong, and the prime contractor actually prevented contact between them and the client organisation!

- The preparation of building and construction plans being "offshored." This has the result that designers have no first-hand knowledge of the site and are less able to resolve design issues arising during construction.

Overlaying this dispersal of teams has been the increasing trend towards individual remote working, where individuals are able to leverage low costs for technology and communications and move away from full-time location in a central office.

Seemingly, this movement towards dispersed working is in direct contradiction to one of the lessons of effective partnership working between organisations – i.e., that there is a need to co-locate teams from the different organisations to create a single project team.[15] Therefore, there are three interlinked questions facing every project manager where parts or all of the team's members are physically dislocated:

- How should our behaviours as project managers be modified to make distributed working effective?
- How should we set our expectations of others to foster integrated working, irrespective of the location of team members?
- How can we achieve a balance between the two poles of integration and isolation and obtain the advantages available from each?

The models of management behaviours from directive to supportive can help us answer the first of these questions. Remote working means that we have far less visibility of our team members and therefore less control at any moment over how the work is being done (or not done) by our individual team members. Authoritarian- or coercive-type management behaviours will, therefore, inevitably work less well. Coupled with this, our team members will have fewer informal mutual support mechanisms from colleagues when working remotely, also pushing us to adopt more supportive behaviours.

As project managers, when managing remotely we have a major advantage over line managers. This is because our project should be working to clear expectations of time, cost, and quality for every task, and we would naturally cascade these expectations down to team members in our role definitions, the responsibility matrix, and our inductions and team management activities. We can therefore expect all members of the team to have a clear understanding of their personal objectives and their contribution to the successful delivery of the project. Adopting a supportive management approach means that wherever they are located, we would, in any case, expect each team member to be self-directed towards achieving these objectives and for us to be monitoring their work outputs and providing support and guidance to them when necessary. This approach works very well as a basis for remote management but may need us to undertake specific additional activities and make adaptations to ensure that it is effective. These additional activities are covered in the following sections.

Building Additional Trust

Trust is key. It is possible to use technology to monitor team members' activities moment by moment, but we need to face down our inner authoritarian manager and ask ourselves, "Do we really care if Anita, Bob, or Charlie nips out to the dentist/walks round the garden/picks the kids up from school/takes a post-lunch power nap?"

What we really want is the work done to the specified quality and within the specified time frame. We may set fixed times for contact, but otherwise adopting a flexible approach to how the team members organise themselves around their natural energy cycle and other responsibilities should increase rather than decrease their productivity.[16] Time zones can be important factors in dispersed project teams, and, if so, we need to be flexible in determining contact periods and expected working hours. If Anita in New York needs to talk to Charlie in Singapore, someone is starting early or staying late. That could mean an earlier finish or a later start for that person. Always remember that we want our team to keep their energy for crises and big pushes, not working regular extra-long days.

If problems subsequently arise with a particular team member, then we may have to work on changing individual attitudes. This might require a team member to personally have to work under more authoritarian control until trust is rebuilt.

Embracing Communications Technology

The increasing availability of high-quality multiparticipant video calls and conference calling, without the need for dedicated video suites and communications links, means we can now assemble a project team meeting whenever required with all team members.[17]

Every organisation will find different solutions, but we need to consider how we replace both formal and informal physical meetings with planned video-enabled communications. There are certain technical prerequisites to ensure high levels of user acceptance:

- A video-conferencing solution that is easy to connect to, whether from a fixed location or using a mobile.
- Ability to share screens easily.
- High-quality connections.
- Anytime calls initiated by any team member – it shouldn't be necessary to fight for a virtual meeting space.

With the communications technology now in place and facilitating our team communications, our next step is to design how best to engage with the technology to integrate our team whether they are working remotely by themselves or as part of a remote-based team. Solutions that are being used include:

- Daily stand-up video meetings.
- Regular formal team video meetings.
- Virtual water cooler/coffee point conversations.
- Video one on one.

Daily stand-up video meetings, as per the agile approaches, are a fixed 15 minutes each day to report in on task progress, set priorities, identify potential obstacles, and keep building the team dynamic. There is a catch here – just be aware that the reasoning behind a stand-up meeting is that the mere act of standing tends to keep people focused and attentive and the meeting short. All this can be lost on a video call if half the team is lounging in purpose-designed gaming chairs, so the video stand-up meeting may need more active time management.

Regular formal team video meetings. This is where we can cascade corporate messages, deal with wider team management, and discuss project issues in more depth. With a remote team, it is even more important to allow anyone to add to the agenda, ideally in advance rather than springing a surprise under "any other business." It is also useful to think through how to avoid the situation of a video meeting where the project manager talks and others listen. One way to encourage a two-way conversation is for the calls to be chaired by different members of the team and for various team members to lead on different items. These meetings need to be more frequent than with a physically located team – possibly every two weeks if the norm for a physical team would be monthly.

Virtual water cooler/coffee point conversations. In a physical team, these conversations may happen anywhere, be that in a lift, by the coffee machine, or at the printer. These are usually by chance. They may be just social chats, but they can be "under the hood" conversations where small issues are resolved and emerging worries discussed before they become problems that need to be escalated. One solution to facilitating such informal discussion is a regular short and very open drop-in meeting. If any conversation gets too long or detailed, it can be hived off into an ad hoc call or added to a team video meeting agenda.

Video one-on-one. Increasingly, one-on-one video calls can be used in all circumstances, business and private, to replace telephone calls. Under remote project management, the project manager needs to be making regular direct contact with each team leader and key team members to ascertain how they are feeling, how their work is progressing, and whether the project manager needs to take any supportive actions. The team leaders should be holding similar conversations with the rest of their team. Whether it is strictly necessary to be a cameras-on video rather than an audio-only call is a matter of preference and etiquette.

Building Relationships between Team Members

The virtual water cooler conversations can help this. If the team is dispersed, the adoption of the buddying strategy for new colleagues discussed earlier is even more useful as a way of combating perceived isolation.[18]

Being Aware of the Time Zone and Holiday Trap

Earlier we identified a communications issue if the project has remote team members based in different countries and time zones, but awareness of team working hours becomes extremely important for us more generally. Setting a regular weekly video conference for 9 a.m. (Berlin time) is unlikely to be popular with UK-based staff, and even if it does suit those in Delhi, it is completely unmanageable for anyone in the US. There are many online meeting planners that can help with this problem and also take into account daylight saving time, helping to avoid twice-yearly confusion.

Equally, we need to be highly sensitive to local holidays when thinking about deadlines, schedules, and meetings. For example, if the weekly team video call is set for a Monday, UK attendance will be spotty as half of the UK's public holidays fall on a Monday. Our New Yorkers will hate us if we give them a big deadline of 9 a.m. on 5 July, there will be no *amour* from team France if we call a crunch team meeting on 14 July, and none of our Irish team members will welcome us if we set our virtual awayday for 17 March.

Fostering Additional Sensitivity

In all meetings we may need to take positive action to ensure that all views are heard. When using video calls, cultural differences come into play as well. We may need to adopt some of the techniques (see Appendix 1) to ensure that views are heard from team members who have a cultural reluctance to speak, as well as to manage those from more assertive backgrounds whose willingness to "shout out" could otherwise drown out other voices. This is especially important if some of the attendees are also not native speakers of the language of the meeting.

Feeding Back

As a routine part of our communications with the project team, certainly in one-on-one meetings and others where we feel comfortable, we need to ask for feedback. This is doubly important if all or parts of our team are working remotely. We need to ask what works for the team, where we need to adapt, and how we can improve the ways we are working remotely. If this feels a little daunting, try asking the following question directly to each participant at the end of every meeting: "In order to deliver our project better, what behaviours do I as project manager, or we as a team, need to stop doing, start doing, or start doing much better?"

Reflections

Before reflecting on the content of the project, reflect instead on yourself and your management style:

- On a scale of 1 (directive/authoritarian manager) to 5 (supportive manager), where would you place yourself and why?
- In what circumstances would you adopt a directive approach?
- What changes would be required to move you in the direction of a supportive manager?

If possible, using the same project as in your previous reflections, spend some time considering the following exercises:

- What roles would you require in your core project support team?
- Take one role and prepare a bullet point list of responsibilities.
- For the same role, prepare a specification of the ideal person and two interview questions designed to allow interviewees to explain how they meet two of the key requirements.
- Are there any lessons from how remote teams can be managed that you would build into the ways you would manage any project team?
- Using a work breakdown structure for your project, how would you allocate responsibility for each of the tasks?
- Map out the timetable an "onboarding" day for a new team. How would you use part of the day to address creating a high-performing team?

Notes

1 On many occasions, I have had responsibility for multimillion-pound projects but have had to get a departmental manager to sign the order for refreshments, even if we only wanted to be hospitable and offer coffee and that essential plate of biscuits to visitors. The biscuits provided were, of course, only plain, as chocolate, sandwich biscuits, and custard creams were definitely outside the organisation's policy!

2 Although, of course, invoking those consequences may involve many other consultations with different parties.

3 Nodding is a great listening technique. Maybe you are indicating to the speaker that you agree, maybe that you understood, or even, perhaps, that they should keep talking in the hope that you eventually can work out what they mean!

4 This approach should come with a warning. Experience of having now adopted this solution several times teaches that there can be one particular downside. Part way through the project, the client may decide that as it now has effective project governance, a great plan, a project leader trained in our approach to project management, and enough knowledge, they want to save their money and complete the project without us. They may even forget to invite us to the launch party. No, I am not still annoyed. Why do you ask?

5 It also made it much easier for the project manager to go on holiday knowing that someone else would take over the weekly reports, field the day-to-day panics, and soothe the business sponsor and the supplier as necessary!

6 Our "golden thread" again.

7 These are discussed in Chapter 8.

8 Another personal terminology gripe – it is not an office, but a team, and it is not responsible for project management, which is the project manager's responsibility. I might accept project manager's support team.

9 Chapters 7 to 13 include more detailed examination of the areas of activity covered by the project support team.

10 If you are new to this area, the concepts in the Myers Briggs approach are a good place to start.

11 This process often involved being knocked unconscious and waking up the next day several miles out to sea!

12 One company even refused to pay for a tea or coffee for staff whilst travelling unless the journey time exceeded four hours, when it would pay – but only for one.

13 Try practising this sentence in front of a mirror: "As you are aware, this is a business-critical project, and I need to do everything I can to ensure that we deliver. Above all, this means getting the best people. I am very sorry, but S and T just don't match the criteria for the people we need."

14 Always ending, as I said earlier, on a positive note.

15 See Appendix 2.

16 Just in case anyone was paying attention, I occasionally used the delay function to send emails at about 2:30 a.m. I also had a colleague who preferred to work from late in the evening through midnight and into the small hours of the morning.

17 Also, let us not mourn the passing of the crackly audio conference call, and let us do everything we can to ensure that such events become a very distant and fading memory. These meetings all seemed to have similar characteristics:

- A dozen people in a room were treating it as a normal meeting – i.e., squabbling, chatting, talking over each other, rustling papers, spilling coffee, and eating all the biscuits.
- An inadequate microphone, which had been placed in the centre of an oblong table, meaning some people were much further away from it and barely audible, especially when other people were squabbling, chatting, etc.
- Other people dialled in on phone lines of various qualities and often from unsuitable locations – "We don't need to know that the 10:30 to Bristol/Baltimore/Beijing is delayed; please, could whoever is in a railway station mute themselves."

18 Even if it just gives the team members a safe space to complain about the project manager.

Bibliography

Eurotunnel, 2022. *The Channel Tunnel Interesting Facts* [Online]. Available at: www.eurotunnel.com/uk/build/.

Jackson, K.T., ed., 2010. *The Encyclopedia of New York City* (2nd ed.). New Haven: Yale University Press.

7 Communications and Engagement

Who Is Holding the Stake?

Aim of this chapter: To introduce the key role of the project manager as communicator and to set out ways of identifying, classifying, and managing stakeholders.

Learning outcome: To be able to use key tools to help identify, evaluate, communicate with, and engage stakeholders.

Most of the processes in the initial phase of our project have been linear. We complete the work, get it authorised, and move on. Updates may be required periodically – for example, to the business case when the information or assumptions upon which it was based change. Considering stakeholder communications and engagement moves us forwards into actual delivery of planned actions.

Why Worry about Communicating or Engaging?

According to a Project Management Institute survey, two in five projects fail. Half of all those failures are related directly to ineffective communications (Project Management Institute, 2013). The implication is that 20% of projects fail because of poor communications. The survey also found that the two biggest problems were:

- Not communicating the benefits of the project clearly.
- The wrong type of language being used for the specific recipient of the communication, including using jargon and project management–related terms.

In addition to these systemic problems, the project manager may encounter and have to overcome obstacles to effective communication in other areas. These may be:

- Cultural, as in, "We don't share anything with our finance department/the regulator/our suppliers until we absolutely have to. We certainly wouldn't engage them in a discussion about what we intend to do before the main board has made a final decision."
- Political/personal, as in, "I don't tell them anything as they always get things wrong and cause trouble" or "You can't trust anything they say because they're always pushing a personal agenda."

Successful projects therefore require an effective strategy for managing communications and engagement with each of our stakeholders. This strategy must address both the generic issues identified by the Project Management Institute and those unique to the project. Those unique issues include the current and historic relationships between the host organisation and the stakeholders and their attitudes towards the project and its objectives. Our strategy and the associated plan of events and activities will need to cover the whole life

DOI: 10.4324/9781003405344-8

of the project, and we will need to monitor the degree to which each planned event or activity achieves its objectives. Our stakeholder communications and engagement processes therefore need to contain feedback loops in which we can monitor the effectiveness of each event to ensure that:

- Our project learns how well any communication or engagement event meets its objectives.
- We can analyse shortfalls and be able to identify and implement changes to improve the performance of the next communication or event.
- We can update our understanding of their needs with new information we have learned about our stakeholders.

The main steps in developing and delivering the stakeholder communications and engagement strategy are set out in Figure 7.1.

Before we consider the separate activities in stakeholder management, you have received another email in your role as project manager of the bio-energy from effluent project. This time it has been relayed to you by the investor relations team. It originated from an investment analyst in the insurance company which is your largest shareholder, but you have been asked to comment before a reply is drafted.

Re: Bio-Energy from Effluent
Dated 3 March
I understand that you are proposing to develop and deploy a process that generates energy and fertiliser from raw sewage. As part of our ethical and climate change reduction investment review processes, we are required to establish:

1. That all investments we hold in certain of our funds are carbon neutral or negative.
2. That companies in which we invest have robust procedures in place to demonstrate their contribution to carbon reduction.
3. To ensure that we are informed of progress on a regular basis.

I would be grateful if you could let me know how you intend for this to be progressed.
    Kind regards,
    Chris James

How are you going to react? Again, it's your choice, but this is someone who will need to be satisfied. Are there any other investors, pressure groups, or others who will make the same demands? Does the company even have a policy on greenhouse gas reduction?

What Exactly Is a Stakeholder?

The easiest definition of a stakeholder is anyone, individual or group, interested in or affected by our project and its outcomes. This definition of stakeholder covers many different types of interest and often a wide range of different individuals and organisations (as shown earlier in Figure 6.2). We should, however, clearly differentiate between communication with the project team itself and with the wider community of stakeholders, as in Figure 7.2.

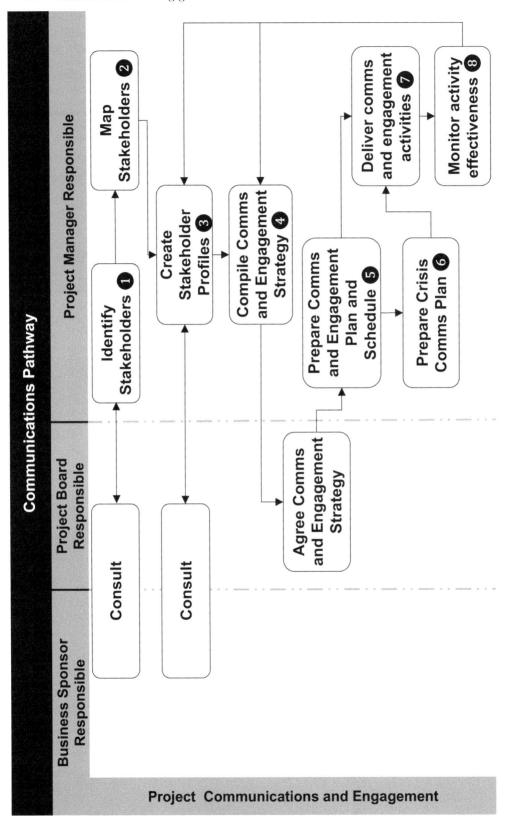

Figure 7.1 Communications Pathway

Project Manager Relationships

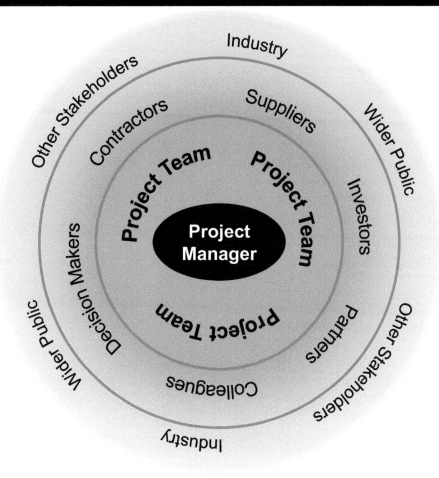

Figure 7.2 Project Manager, Team, and Stakeholder Relationships

Real-life relationships can be more complex than shown in Figure 7.2 in some projects. For example, some individuals provided by suppliers or agencies will be members of our project team, but their employers will be stakeholders. This complexity can lead to conflicting pressures on individuals, as discussed in Chapter 6.

As the lead on the project, it usually falls on us as project managers to be responsible for overseeing the management of the stakeholders as well as being the prime source of information for them. This is not to say that the project manager will have the prime relationship with each stakeholder. Each of our stakeholders will initially often bring to our project a different set of existing formal and informal relationships. They are also likely to bring preconceived ideas about what we are setting out to achieve, or should be setting out to achieve, and about organisational standards of assessment, as well as objectives and emotional responses. Each stakeholder may also see that they will be advantaged by the project or suffer a disadvantage – even where this is an unfounded fear. Consequently, their overall initial stance on our project will be one of only three states:

- Support – Positive.
- Indifference – Balanced.
- Opposition – Negative.[1]

These three states define our objective in working with each stakeholder, which is to:

- Understand and overcome any negative aspects in their attitude (remember, even positive supporters may have some reservations).
- Increase, or at least maintain, the levels of positive support from those stakeholders in a position to influence the delivery or outcome of the project.

When defining our strategy, we may have to prioritise our resources (which are almost always scarce) in favour of convincing doubters with the potential to block or undermine our progress, at the expense of those of our supporters or the indifferent who have less influence.

Understanding Communications versus Engagement

Once again, we are faced with terms that are often used interchangeably, or at least with no clear differentiation between them. One differentiation between communication and engagement is based on distance, with engagement being likened to a contact sport and communication being something more distant.[2] With instant communication technologies, this is increasingly less useful as a guide. Probably a better way is to consider the direction of interaction:

- Communication is the outward dissemination of information from the project to the stakeholder, with the caveat that we will seek feedback on its effectiveness.
- Engagement is a two-way information exchange between the project and the stakeholder which may be initiated by either party, with the expectation that the project will evaluate information received and that this may lead to variations in the project.

To highlight the differences between the two, consider the case of a planned new railway line. Following the initial ground surveys and design work, the project publishes information on several alternative possible routes. This publication takes the form of a press release, letters to affected landowners and local authorities, website updates including a series of computer-generated videos of 3D fly-pasts along each route, and a glossy 30-page publication. All these are communication activities. After all that initial communications activity, in order to understand the issues relating to each of the alternative routes, the project holds a series of one-to-one meetings with landowners, conferences with local authorities, public meetings in village halls, and web- and interview-based opinion surveys. These are all engagement activities and events.

Accepting the Project Management Institute's research as backing up the empirical evidence that every successful project will need to communicate or engage effectively with a wide range of stakeholders, we have to put in place stakeholder management processes for the following, at the very least:

- Identifying the stakeholders.
- Determining the communications media to be used overall and for each stakeholder.
- Designing and approving the messages we wish to send.
- Monitoring the response to the messages.

In addition, we will need to engage very closely with a narrower range of our stakeholders, either because they are directly involved in the project's chain of authority or because we genuinely need input from them to improve the outcomes of the project. For this engagement to be effective, we will need to design a series of events to facilitate our desired two-way information exchange.

From the outset we also must gain support from the stakeholders for them to *want* to invest their time in engaging with the project. Examples of different techniques for fostering stakeholder engagement include:

- For a government business change project, creating a two-day event in an exhibition hall for approximately five hundred staff to participate in the choice of preferred solution, as well as including over one hundred external stakeholders. Five suppliers gave a series of presentations each day, and the staff members' evaluations of what they saw formed part of the selection criteria. A much wider take-up was gained because the project manager took advantage of the relative locations of the office and the exhibition centre and was able to move the staff between them by boat, making participating in the event a fun experience. It was also turned out to be much cheaper than using buses, which made it easier to get the expenditure on the boat trips approved.
- In another case, this time for a new IT system for use by several hundred professional staff, a representative user group of 20 was created, but each member of the group was tasked with building a local group of another 10 to 20 staff for whom they would act as a rapporteur. The take-up was very strong because the staff genuinely felt that they were involved in the project from the outset.
- Finally, on a project for a health IT system modernisation, user engagement was fostered by a simple competition to name the project. The winner, who was chosen by the project board, received two bottles of champagne (paid for by the project manager – who else!), a trivial actual cost but a tremendous gain in staff engagement with the project.

How Do Stakeholders Behave?

Before developing our stakeholder communications and engagement strategy, instead of taking a virtual field trip, we are going to consider the following contribution from an immensely experienced project manager, mentor, and tutor on issues around managing stakeholders and business sponsors.[3] This project manager, who wishes to remain anonymous,[4] shares two salutary examples of how important these issues can be to our project's outcomes and, because we are dealing with individuals, how personal factors external to our project can both shape their responses initially and cause alterations in them over time.

If I think about why some of my projects have failed recently, it comes back time and time again to sponsorship and stakeholder alignment at the top. Let me illustrate with two examples.

A Major International Telco

Several years ago, I was hired to reverse a trend of failing projects in a South African operating company. The organisation had once been a leading light in the project management (PM) field, but a chief executive officer (CEO) had disbanded the enterprise project management office (EPMO) five years earlier and sacked all the PMs as a cost-cutting exercise (reputedly he'd said, "My line managers can do project management alongside their day jobs – it's not that hard").

So, I started a quick business case to re-establish the EPMO and some very basic processes and controls and to grow lots more competent PMs – with short-term experienced hires to tide us over. I got management committee (MANCO) approval and set off. Within a year, the going got tough – lots of pushback from line executives who'd been used to firing from the hip and resented this new Centre of Excellence setting rules for them – even though the same people had sat round the MANCO table and promised the CEO their full backing for my new model. I became a hot potato and got passed from one executive to another over the next two years (I had seven different executive sponsors in three years!). Eventually, the one with the most to lose (the CFO – who'd taken advantage of the lack of an EPMO to build a finance empire to control the other executives) could stand it no longer and engineered my termination. This was despite overt support for me from the CEO, whom you would think ought to have been able to overrule him.

The lesson here is twofold:

- *Even if your initial sponsor is very strong (and mine was chief information officer, CIO), your success in organisation-wide transformation will be dictated by the extent to which his/her colleagues share his/her views. If he or she leaves (and mine did after the first year), don't just assume continuing support (even when everyone is giving you the right noises, as they did with me), but pause to do an engagement health check, be honest with yourself, and be prepared to challenge the CEO if you're not happy with the results. I believed what everyone was telling me, even when their actions didn't always tally, and that was my downfall.*
- *Never assume the obvious. In the end, I came to realise that the chief finance officer (CFO) had more power in the organisation than the CEO. I was a direct threat to the CFO, and when I looked for support from the CEO, he ran for cover. My only saving grace is that the CFO only lasted 12 months after me!*

A Leading African Financial Institution

I was hired with the explicit remit to build an EPMO by the guy who was my initial sponsor in the example prior. He told me that he'd got executive committee (ExCo) approval and that all I had to do was to design a model, get it approved, and build it. In just six months.

It started out brilliantly – I met each of the executives individually and got ringing endorsement for my model after only two months. I took it to ExCo, and that's where the trouble started. I got approval, and a big thank you from the CEO, and began to implement. But I had not taken the time to understand the culture properly and made a fundamental assumption that turned out to be false.

This particular organisation had originally been set up by the founder, who was a national icon, and the organisation still retained many of his personal values. So, no-one ever dissented decisions openly that they thought had the CEO's backing, but merely worked subsequently to undermine them if they didn't like them. And the executives, as with the first example, had got used to running their own shows and didn't like the idea of some central authority like an EPMO telling them what to do. Also, as I came to realise eventually, the CEO did not have the guts to apply consequence management – he was shooting for a major promotion to be CEO of the parent company and could not afford for there to be any messy fallouts on his watch. So as soon as the going got tough, the consensus started to disappear.

Eventually my new sponsor (I'd moved from CIO to CFO by this time) paid me off – he said it wasn't my fault but that my work had shown that the organisation wasn't, after all, ready for a revolutionary idea like an EPMO! My successor left after three months, and her successor lasted only another two – so I guess the organisation still wasn't ready.

Both these examples illustrate my point that sponsorship and senior stakeholder alignment are (to me) the most important factors in eventual project success. The books will tell you that the PM should ensure these two factors are strong at all times. But they don't tell you HOW to do this in the incredibly complex and superheated, fast-changing environments that are today's ExCos/MANCOs. Even my 47 years of project management experience wasn't enough to stop me falling into the "obvious" elephant traps. Twice. Successively!

So maybe the overriding lesson is "prepare to fail sometimes – just dust yourself off and try to do better next time." At the end of the day, as I always tell my students, project management is an art, not a science.

Step 1: Who Are Our Stakeholders?

The first step in developing our communications and engagement strategy, as in Figure 7.1, is to determine who our stakeholders are. The list can be quite extensive. On one government business change project, in addition to multiple internal stakeholders, the formal external stakeholder management strategy covered, one very senior government minister, two government departments, ten government regional offices, two professional bodies, one devolved administration, and over three hundred local authorities. In all, 104 named external stakeholders were consulted throughout the life of the project. This was in addition to two major groups of involved staff which totalled over five hundred people, the project board, the business sponsor, and the board of directors. Subsequently, selected suppliers, contractors, and a continuing operations team were added to the mix. Whilst this was probably an extreme for a relatively small project, a diverse community of stakeholders is not unusual.

The project board and the business sponsor are a useful starting point, and as sources of information, they can be invaluable in not only identifying who the stakeholders should be but also helping us understand their potential views on the project. Holding a series of one-to-one or small group conversations with the members of the project board about stakeholders is also an excellent way of ensuring that they, as stakeholders and decision makers themselves, are engaged in the development of the communications and engagement strategy from the outset.

Step 2: Mapping Our Stakeholders

In order to manage the diversity in our community of interested organisations and individuals, we need to be able to analyse and group them in some way. There are many suggested ways of grouping stakeholders, but probably the most often encountered in project management is one that focuses on the relationship between the stakeholder and the project. The method is known by an acronym drawn from its initial letters of RACI:

- **Responsible.** Stakeholders who must complete a task or make decisions about the project.
- **Accountable.** Stakeholders who must approve the work of the project. In our project structure, this is the business sponsor and possibly other decision makers in the host organisation.
- **Consulted.** Stakeholders from whom we require input before completing a particular task and obtaining approval from the business sponsor. These stakeholders may be internal staff or departments. For example, representative users whose input is sought at the design and testing stages of an IT system or internal auditors. Alternatively, they could be external, and in some industries, they may include a government regulator or a professional body.
- **Informed**. Stakeholders who should be kept in the loop. They require progress updates and notifications of decisions from us about the project, but we do not need to formally consult them and would not expect them to contribute directly to any task or decision. In commercial projects, this could include major customers or shareholders, and in the public sector, this could include local and national politicians and the wider public.

Whilst the RACI analysis is a useful first step in thinking about who our stakeholders are, it is not always definitive, and there is another tool, stakeholder mapping, that is frequently used in formulating a comprehensive communications and engagement strategy. This is because stakeholder mapping gives us a more in-depth understanding of our stakeholders, the impact that their actions and decisions can have on our project, and the information that they desire from us. These factors can be critical in ensuring that we develop and maintain strong and positive stakeholder relationships throughout the project life cycle. Conducting a thorough analysis at this stage, using a tool such as stakeholder mapping, provides a solid foundation for our communications strategy development and planning and will identify the risks that could arise from stakeholders who are negative, uninterested, or feel excluded. The completed analysis will also help us prioritise our investment of time and resources and focus on the information needs of key stakeholders.

The stakeholder mapping approach analyses all stakeholders on two axes: the stakeholder's level of perceived influence (or power) over the project and their amount of interest in it. The result is a simple two-by-two matrix, as shown in Figure 7.3. This matrix can be used to group stakeholders into one of four types, each of which will require different communications and engagement goals and activities:

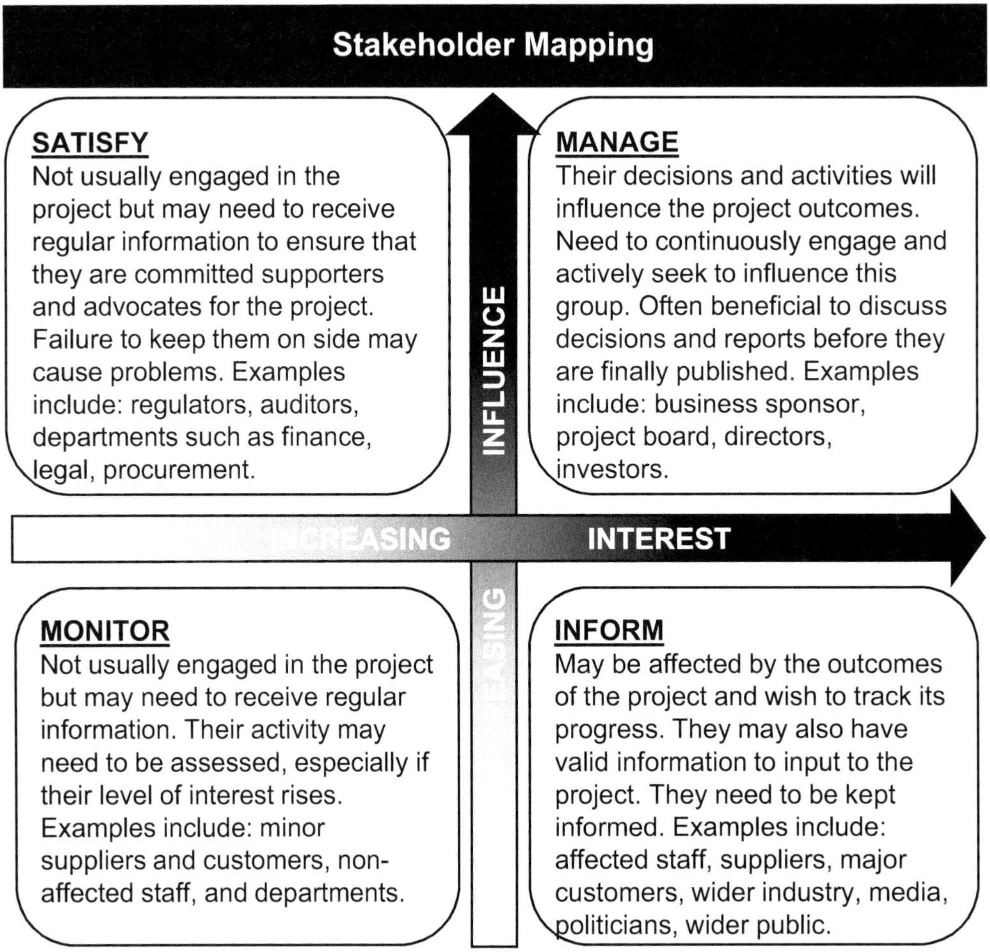

Stakeholder Mapping

SATISFY
Not usually engaged in the project but may need to receive regular information to ensure that they are committed supporters and advocates for the project. Failure to keep them on side may cause problems. Examples include: regulators, auditors, departments such as finance, legal, procurement.

MANAGE
Their decisions and activities will influence the project outcomes. Need to continuously engage and actively seek to influence this group. Often beneficial to discuss decisions and reports before they are finally published. Examples include: business sponsor, project board, directors, investors.

INFLUENCE

INCREASING   INTEREST

MONITOR
Not usually engaged in the project but may need to receive regular information. Their activity may need to be assessed, especially if their level of interest rises. Examples include: minor suppliers and customers, non-affected staff, and departments.

INFORM
May be affected by the outcomes of the project and wish to track its progress. They may also have valid information to input to the project. They need to be kept informed. Examples include: affected staff, suppliers, major customers, wider industry, media, politicians, wider public.

Figure 7.3 Example Stakeholder Map

We can refine this concept further by assigning scores to the interest and influence of the different stakeholders. Table 7.1 considers how such an analysis might work for a specimen project, in this case a government-to-citizen service modernisation project. In this example, scores for interest and influence are assessed on a scale of 1 (minimum) to 8 (maximum).

Table 7.1 Stakeholder Assessment

Government-to-Citizen Service Modernisation Stakeholder Assessment		
Stakeholder	**Interest**	**Influence**
Board of directors	8	8
Affected employees	7	6
Internal auditors	3	6
Sponsoring ministry	6	5
Other gov. depts.	4	5
Professional bodies	2	5
External auditors	1	5
Principal service users	4	4
Devolved administration	6	3
Local authorities	5	3
Members of Parliament	5	2
Other employees	4	2

Figure 7.4 shows how the data in Table 7.1 might look when displayed on a simple stakeholder map.

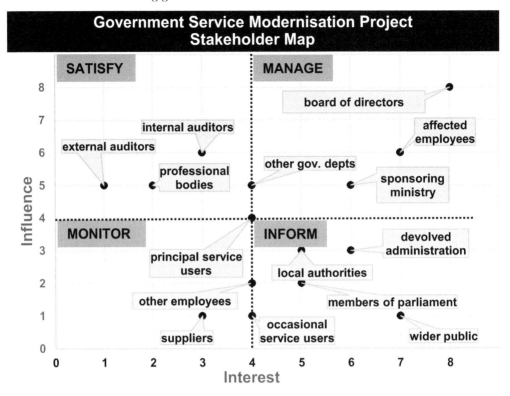

Figure 7.4 Example Stakeholder Map

Step 3: Stakeholder Profiling

In addition to preparing our stakeholder map, we need to build a separate profile for each stakeholder. This stakeholder profile should:

- Include the administrative part of the data needed for our stakeholder records – i.e., core contact data.
- Consider the contributions that we expect from the stakeholder.
- Note the types of engagement and communications we think the stakeholder will prefer.
- Seek to establish the position that each stakeholder is likely to take in relation to the project.

Having already sought their initial views on who should be regarded as a stakeholder, it can be a very productive exercise to work through these areas for each of the stakeholders with the business sponsor and the members of the project board.

Stakeholder mapping and profiling is not a one-time operation. As the project progresses, stakeholder interests will naturally change. For example, those with a policy interest will be much more concerned at the business case stage and less engaged with detailed progress reports. However, those affected operationally in the completed project will at least maintain and will usually increase interest during the detailed delivery, testing, and handover phases. We also need to be prepared for the views of stakeholders to change for other reasons, perhaps in response to changes in personnel, political, financial, or other external factors, or sometimes for no immediately discernible reason at all.

There are several areas we need to think through in relation to each stakeholder and then include in their profile:

- Desired stakeholder reactions. We should have captured any desired movement in views on the project in each of the stakeholder profiles, and we need to tailor our planned activities to foster that change. This means that, for example, a sceptical but influential stakeholder may need to receive additional information or a higher level of contact than a strong supporter of the project.
- Selection of the most appropriate channels and media to use to communicate and engage with them. Although each stakeholder's preferred communication channels may have been captured in their profile, we also have to consider how stakeholders can be grouped – for example, to receive targeted mailings and notifications (postal, electronic, or personal). This will make the communications process much more manageable. We may also have to add several individuals as contact points for a single stakeholder organisation and consider if each will need a dedicated point of contact in the project. For example, the most senior contact in a funding body might be assigned to the business sponsor as the project contact point, but contact with their operational lead could be made the responsibility of the project manager. We also need to think through who within the stakeholder should be invited to engagement events, such as a requirements discovery workshop, or for testing or inspection visits. This can be a difficult area and requires tactful negotiations to ensure that the most appropriate stakeholder representatives are engaged.

A word of caution. The profiling data that we collect can be personal to an individual and can represent a judgement on them by the project. This data may be covered by data protection legislation, and the project manager should ensure that all data is collected and maintained within any legal obligations. Even if legal rules do not apply, it is professionally responsible to ensure that data is kept secure and that any judgements are soundly based and use only temperate language. It goes without saying that it would be counterproductive and maybe even career limiting if a negative stakeholder profile were leaked suggesting, for example, that the CEO had the attention span of a fly and was more interested in golf than the company.

Table 7.2 captures the minimum information needed for each stakeholder. Whilst this works on a manual basis for smaller numbers of stakeholders, if larger numbers are involved, it is useful to retain this data in a contact management system. Such systems may also be used to track which communications each stakeholder has received as well as their responses or absence of responses.

Table 7.2 Template Stakeholder Profile Record

Stakeholder Profile			
Stakeholder			
Primary contact(s)			
Primary contact details			
Address	**Email**	**Office tel. number**	**Mobile tel. number**
Relationship with project		**Preferred contact methods**	
Influence over project decisions/outcomes		**Assessed influence score (0–8 highest)**	
Interest in project decisions/outcomes		**Assessed interest score (0–8 highest)**	
Stakeholder Category	**MANAGE/SATISFY/MONITOR/INFORM** (delete as appropriate)		
Summary of known initial opinions/views on the project			
Desirable changes of views on project			
Proposed communications/ engagement	1 2 3 4		
Created (date)	**Created by**	**Reviewed (date)**	**Reviewed by**

Step 4: Compiling Our Communications and Engagement Strategy

As identified in Chapter 4, when creating our communications and engagement strategy, we will have to answer our seven basic questions again:[5]

- **What** do we want to communicate/engage about, and what quality standards will be applied to our communications?
- **Why** do we need to communicate or engage on any particular aspect of the project?
- **When** are the appropriate times in the life of the project to communicate or engage?
- **How** are we going to communicate or engage – i.e., what communications channels will we use, how do these vary by stakeholder or over time, and how will we know if they have been effective?
- **Where** will our communications be prepared?
- **Who** will prepare, authorise, and action our communications with different stakeholders?
- **How much** should we invest in ensuring effective communications?

If the pause before project plan approval was used to draft the communications and engagement strategy, we now need to review it to take account of the data contained in our completed stakeholder profiles. Our completed strategy needs to contain sufficient information for the project board to understand how we propose to communicate and why. Therefore, in answering our seven questions in the strategy, we need to think through and then present at least the following information:

- The criteria that we have used for selecting and mapping our stakeholders.
- The process that will be followed to review the initial strategy and any updates to it.
- The process to be used to develop and agree on the communications and engagement plan which will schedule each of our activities in this area.
- Key responsibilities for defining, compiling, approving, and actioning communications and engagement events.
- Key responsibilities for receiving, actioning, and escalating positive and negative feedback from stakeholders.
- Policy regarding the use of different communications channels and technologies for different groups of stakeholders.
- Communications standards and their enforcement including, for example, use of Web Content Accessibility Guidelines, plain language,[6] punctuation and standard terminology, and multilingual standards, and quality control where required.
- Details of how our effectiveness monitoring process will operate, including how different types of communications and engagement events will be evaluated.
- Legal and other requirements for data confidentiality, usage, and disposal.

The following items could also usefully be added as appendices to the strategy:

- An extract from the stakeholder map showing how different stakeholders have been classified or grouped.
- An outline of the communications and engagement plan showing how the needs of different stakeholders will be met during the life of the project.
- The emergency/crisis communications plan.

Step 5: Developing the Communications and Engagement Plan

Most of the information needed in our plan is highly project specific and depends on our assessment of the needs of the different stakeholders, as contained in our stakeholder profiles. Set out in the following section is an example of how information can be extracted from a stakeholder profile and included in a communications and engagement plan.

The example is based on a project to construct a major road scheme, which will be delivered by a consortium arrangement consisting of a UK county council, an engineering consultancy, and several supply chain partners. The sample completed stakeholder profile in Table 7.3 is for the company that will be acting as the designer for the project.

Table 7.3 Example Completed Stakeholder Profile

ZZZZ Major Road Construction Partnership Stakeholder Profile			
Stakeholder	Acme Construction and Design Company		
Primary contact(s)	Jane Smith, CEO; Fred Bloggs, Director of Finance		
Primary contact details			
Address	**Email**	**Office tel. number**	**Mobile tel. number**
AAAAA BBBB CCCC	janesmith@xxyyzz.com fbloggs@xxyyzz.com	01234 567890	N/A
Relationship with project	Supply chain design consortium partner	**Preferred contact methods**	email telephone
Influence over project decisions/outcomes	Jane Smith sits on Strategic Partnership Board. At a tactical level, the supplier's local contract manager will need to agree on and resource their delivery plan. The supplier will provide the design for acceptance by the Strategic Partnership Board.	**Assessed influence score (0–8 highest)** 6	
Interest in project decisions/outcomes	This is the company's most major road construction design project to date, and they are looking to use it to enhance their reputation for innovative use of materials and rapid build designs.	**Assessed interest score (0–8 highest)** 8	
Stakeholder category	**MANAGE/~~SATISFY/MONITOR/INFORM~~** (delete as appropriate)		
Summary of known initial views on the project[7]	Jane Smith is very enthusiastic, but her director of finance, Fred Bloggs, is known to be worried about the exposure of a relatively small company to financial and reputational risk. Jane is also a local resident and is an advocate of the project for reducing the congestion in the town centre.		
Desirable changes of views on project	Reassurance of finance director on financial and reputational risk, especially risk to be taken by lead supplier.		
Proposed communications/ engagement	1 Initial supply chain conference to highlight contract requirements, provide customer service training. 2 Indicative work plan for supply chain to understand scope of contract. 3 Launch of contract amongst supply chain, publicising to trade press. 4 Briefing note to explain the contract/relationship between lead supplier and council, including how work to be undertaken and who to contact. 5 Portal on lead supplier website to allow questions to be submitted. Jane Smith also receives comms as part of the Strategic Partnership Board.		
Created (date)	**Created by**	**Reviewed (date)**	**Reviewed by**
01/11/XXXX	Jeff Jones PM		

Building from the bottom up, we can take our basic data from the stakeholder profiles, analyse each stakeholder's requirements, group the stakeholders, and summarise all the activities required for each grouping, as shown in Table 7.4.

Table 7.4 Comms and Engagement Activities

ZZZZ Major Road Construction Partnership Communications and Engagement Activity Schedule			
Stakeholder	**Action**	**By Whom**	**By When**
Contractor/ council supplier partnership	Senior personnel confirmed in all organisations to sit on the Project Strategic Partnership Board	Lead supplier/council	October XXXX
	Terms of reference for Project Strategic Partnership Board agreed	Lead supplier/council	October XXXX
	Detailed breakdown of roles of Project Strategic Partnership Board and project management team agreed	Lead supplier/council	October XXXX
	Monthly progress reports on the procurement process, mobilisation stage, and delivery	Partnership project manager	Monthly from October XXXX
Council business sponsor	Weekly progress reports and update meeting	Partnership project manager	Weekly from preferred bidder announcement January XXXY
Chair Project Strategic Partnership Board	Weekly summary progress reports	Partnership project manager/council business sponsor	Weekly from preferred bidder announcement January XXXY
Supply chain partners	Supply chain conference to highlight contract requirements, provide customer service training	Partnership project manager/lead supplier commercial team	February XXXY
	Indicative work plan produced to supply chain to understand scope of contract	Partnership project manager/lead supplier commercial team	March XXXY
	Launch of contract amongst supply chain, publicising to trade press	Partnership project manager/lead supplier commercial team	Contract start date
	Issue briefing note to explain the contract and relationship between lead supplier and council, including how work to be undertaken and who to contact	Partnership project manager/lead supplier commercial team	April XXXY
	Portal on lead supplier website to allow questions to be submitted	Lead supplier	March XXXY
Council members	Issue council members' briefings once a month during mobilisation to update on progress in lead up to start date	Partnership communications manager with approval from council	Monthly from preferred bidder announcement January XXXY

ZZZZ Major Road Construction Partnership Communications and Engagement Activity Schedule			
Stakeholder	**Action**	**By Whom**	**By When**
	Launch event for council members to explain how the contract will work, the benefits expected including pain/gain mechanism, and delivery time frames	Partnership/ council corporate communication teams	Contract award date March XXXY
	During contract delivery, issue quarterly members' briefings	Partnership communications manager with approval from council	Quarterly from contract start April XXXY
	Arrange for site visits at fixed points in the delivery cycle, including opening event	Partnership/ council corporate communication teams	Dates to be agreed
Residents, local media and businesses, MPs	Issue news release to announce contract and explain the benefits	Council corporate communication team	Contract award date March XXXY
	News release on launch of new contract with comment from council and lead supplier	Partnership/ council corporate communication teams	Contract start date April XXXY
District and parish councils	District council member briefing to explain the contract and relationship between lead supplier and council, including how/when work to be undertaken in the affected areas	Partnership project manager and council service manager	Contract award date March XXXY
	Issue briefing note to all members and parish councillors	Partnership corporate communication team prepare, council corporate communication team approve and issue	April XXXY
Partnership delivery staff, including trades unions	Briefing by council at preferred bidder stage	Partnership project manager and council service manager	January XXXY
	Initial team – onboarding staff drop-in day to explain contract and structures and orientation, including mandatory health and safety training	Partnership project manager and council service manager	March XXXY
	Subsequent joiners – introductory video and online orientation e-learning, including mandatory health and safety training.	Partnership/ council corporate communication teams	April XXXY
Other interested bodies	e-Newsletter, monthly throughout contract	Partnership/ council corporate communication teams	January XXXY
Consortium and council staff	Staff focus group set up with representatives from across the joint organisation	Partnership project manager and council service manager	March XXXY

The next stage is to recast this plan into a timed schedule ready for input into the project plan. Table 7.5 shows how the first page of the timed schedule might look. When considering repeating events and activities, it may be useful to break out the initial preparation and agreement of standard templates for reports and presentations as separate tasks with their own required resources and to schedule the repeating activities separately.

Table 7.5 Communications and Engagement Plan

ZZZZ Major Road Construction Partnership Communications and Engagement Plan					
Action	**By Whom**	**Target Stakeholder**	**Frequency**	**Start**	**End**
Senior personnel confirmed in all organisations to sit on the Project Strategic Partnership Board	Lead supplier, council	Contractor/ council supplier partnership	–	October XXXX	October XXXX
Terms of reference for Project Strategic Partnership Board agreed	Lead supplier, council	Contractor/ council supplier partnership	–	October XXXX	October XXXX
Detailed breakdown of roles of Project Strategic Partnership Board and project management team agreed	Lead supplier, council	Contractor/ council supplier partnership	–	October XXXX	October XXXX
Monthly progress reports on the procurement process, mobilisation stage and delivery	Partnership project manager	Contractor/ council supplier partnership	Monthly	October XXXX	October XXXZ
Weekly progress reports and update meeting	Partnership project manager	Council business sponsor	Weekly	January XXXY	October XXXZ
Weekly summary progress reports	Partnership project manager, council business sponsor	Chair Project Strategic Partnership Board	Weekly	January XXXY	October XXXZ
Issue council members' briefings once a month during mobilisation to update on progress in lead-up to start date	Partnership communications manager with approval from council	Council members	Monthly	January XXXY	October XXXZ
Briefing by council at preferred bidder stage	Partnership project manager, council service manager	Partnership delivery staff, including trade unions	–	January XX XY	January XXXY
Staff e-newsletter, monthly throughout contract	Partnership and council communications teams	Other interested bodies	Monthly	January XXXY	January XXXY

Step 6: The Emergency or Crisis Communications Plan

Ours is a well-managed project. Our plans have been drawn up, the right team assembled, risks and issues both identified and controlled. What can possibly go wrong? The answer is just about everything, from a supplier going into receivership to a global pandemic shutting down our supply of materials and our operating sites. When a completely unexpected and unplanned event hits the project, the viability of the project may be completely undermined or the very stability or reputation of the client organisation may be threatened by the project's sudden failure. Although many organisations will have documented plans for emergency situations covering their normal operations, we need to consider how the project, which is, after all, only a temporary adjunct to the organisation, will respond. An extremely key consideration is how and what the project will communicate to its stakeholders and by whom. Preparing such a document is insurance that we hope will never be needed.

The following are important points to bear in mind:

• Whilst the unforeseen event will determine the nature of the communications needed, it is extremely helpful to have a skeleton crisis communications plan. Our plan should be short and succinct. Ideally it would form an appendix to the overall authorised communications and engagement strategy, although it does need to be a completely standalone document.
• The plan will need to contain contact details of the agreed-upon crisis communications team, emergency authorisation levels, escalation routes to business management, and contact details for key stakeholders and the media.
• As the untoward event may occur at any time, key people in the plan will need to be able to access it from off site and out of hours.
• As with any crisis management plan, regular scenario-based testing is always better as a learning exercise than having to learn during the crisis.[8]

Step 7: Deliver Communications and Engagement Activities

Our strategy and our plan for each stakeholder and an associated chronological activity schedule are all in place. We now have to start to plan each of the activities in detail. Some will be one-off events, and others – for example, sending out newsletters, press releases, or briefing notes – will be recurrent activities. All these activities must be treated as separate tasks and broken down into a series of timed, costed, and resourced sub-tasks, just as we have to do for everything else that we plan for the project. There are two specific areas of quality in our communications with stakeholders to which we will need to play close attention:

• Addressing stakeholders in the most appropriate terms.
• Saying what we need to say consistently in all channels and to all stakeholders.

Communication Is in the Eye of the Beholder

Whenever we intend to communicate with anyone, we need to think very carefully about our words – not only about what we have to say, but how we are going to say it and to whom we are saying it.

Are you ready for another virtual field trip? In fact, this one is going to involve more than one journey to London. We are going to be indoors, so no hat, coat, gloves, or scarf needed on these trips.

We find ourselves in what was probably built as a grand London townhouse for a wealthy aristocrat in the mid-19th century. All white Portland limestone and some grand public rooms with floor-to-ceiling windows, but converted long ago into offices. It is currently occupied by the governing body of a major and very traditional academic institution. They are looking to replace their ageing management systems with an integrated one that will enable them to modernise and streamline their core operations and disciplinary processes.

During weeks of research and preparation of the business case by a team of expert analysts and consultants, Steve, the director of finance, who is the business sponsor, and Ian, his chief information officer, have been kept in close touch with and, where necessary, have guided the emerging analysis. We are shown into the office of the director of finance, who is meeting with Alan, the project manager, and the project team to discuss the final document, which is going to go directly to the management board for approval, and the accompanying presentation.

Alan: "Just to sum up where we are. We all agreed with the management board at the last meeting that we can discount doing nothing as by this time next year the existing system will be effectively dead. So, as you can see what we have come up with is five different options. Starting with the last, and most expensive, this is what the highly comparable, but smaller, ABCDE Agency have just installed. The software industry calls this the Rolls Royce solution, and like a Rolls Royce, it is far more expensive and has a much wider range of features than most of the others – although probably the agency will not need some of them. Anyway, we are not suggesting this, as it would be far too expensive, but following the same theme, at my suggestion you agreed that we call the other options the BMW, the Volvo, the Ford, and so on to reflect different levels of sophistication and costs."

Steve: "Ian and I are still fine with that, just a little tidying up and we are ready to go to the board with the business case. I have spoken with all the board members, outlined the options, and they are all in agreement with our conclusions. The meeting should just be a formal rubber-stamping exercise and we can get on with the procurement. Please prepare the final version by tomorrow midday so that it can go into the next board meeting document pack. Please, can you finalise the PowerPoint presentation as well. Use the agency's standard templates – my PA will ensure that you have the latest versions as the branding has changed recently. The meeting will be on Tuesday next week at 11 a.m. You can use the time after the paper and presentation are compete to get on with starting to prepare the project plan and engaging with the procurement manager."

The project has got the green light, and the director of finance is happy . . . but let's skip forwards to next Tuesday, when they meet with the main board, which is the most powerful group of stakeholders, and see what happens.

This time we are back in London on an already warm day and are going to observe part of a management board meeting. Probably best to leave your jacket off as well on this one, as it is going to be in yet another overheated, cramped, very old-fashioned meeting room.

The selected option, the Volvo, is clearly out in front by a considerable margin, and Alan, the lead presenter, puts his argument for it very strongly and clearly. But as each of the board members has already agreed individually, the team are not expecting to be in the meeting room for very long. At least they are right in that.

Having been kept waiting for an interminable period whilst the board had deep discussions on some topic or other, the project team are brought in to present the business case. We follow on in and manage to find a space to stand at the back. Once again, it is a very stuffy office with too many people in it, sat round an oblong table. Somehow a projector and screen have been squeezed in. Empty and half-full coffee cups are in evidence, as is an empty plate which had held the biscuits – the board has obviously had their refreshments whilst we were kept waiting.

Alan squeezes around the seated board members and is now standing by the screen ready to start. Taking a deep breath, he launches into the details of the different options and why the recommended one is preferred. Ten minutes later, his presentation ends with absolute silence from the board.

The chair puts down her coffee cup and says, "Alan, thank you and the team for coming today, but the board have decided to reject your report. Will you all please leave us. Now."

The momentary silence seems to last for hours, and then, without a word, the team turns and leaves. We follow them out.

Anxious to find out what was going on, the next day we call Alan to ask for an explanation.

'Well, I was as staggered as you were. Everyone was speechless. It was so unexpected and, I might add, I thought abrupt to the point of being downright rude. We waited for the meeting to finish and had a debrief with the director of finance yesterday evening, and he really wasn't happy with the way board members behaved either.

"You really will not believe it when I tell you what the core problem was, and it was all my fault for not understanding and meeting the needs of my audience. The difficulty that they had was that in present-ing the options, I named each one after a vehicle – from Mini up to Rolls Royce (with apologies to the manufacturers). I just used it as a shorthand way of communicating costs, features, and complexities. Steve, the director of finance, had discussed the content of the approach and each option with key board members before he agreed on the final board paper and the presentation. But obviously he had not raised the option names, why would he? Before we got into the room, the board had already rejected the paper because they claimed that they did not understand the option names chosen and what they meant.

"Steve, Ian, and especially myself were guilty of not using language relevant to them. With the excep-tion of Steve, who has come from a city accountancy firm and commutes in from the suburbs, the board are highly metropolitan academics. Most of them don't drive, and none of them would ever see the need to own a car. Two of the professors live close enough to cycle to work, some 'live in' their college, and the rest only ever use public transport. Steve explained to me that for them cars come in three types; black London cabs that you can hail on the street, minicabs that have to be booked, and other cars and vans that get in the way of the first two. When talking to them, our using car brands as analogy was, to say the least, pointless. They were just meaningless words to them. To make it worse, Emma, a junior consultant who had just started with us, did say to me that she thought my choice of option names was not very inclusive, but I said we had used it before and it would be OK. A clear example of me being insensitive to the client, and I will be scoring Emma very high on emotional intelligence when I do the end of assignment assessments. So, we have agreed that I am going to replace the car brand names with option 1, 2, 3, etc., and I will resubmit the paper to the next board meeting. I will let you know how it goes."

True to his word, exactly a month later, Alan calls back. "They loved the paper, accepted our recom-mendation, and have given approval to proceed. The only trouble is that not only has the project lost a month but the all-important impetus has gone and probably cannot be recovered. My consultants in the team have had to be moved onto work for other clients, and I am not sure if I can get them back in time for the next stage. This is a tremendous loss of knowledge to the project.

"If I can summarise, my lesson from this is don't assume that you and your audience have any-thing in common. It is essential that you communicate using language in a way that makes sense to your audience and to research this in advance. Just imagine trying to convey a complex message using

examples of British tea brands or American supermarkets to an international audience. Please, always put your audience first, because failure to do so can have a major impact on the acceptability of what is being said. In this case I obviously didn't pay attention to this, or really listen to Emma's warning, and the project has suffered. I will not make the same mistake again."

No one on the presentation team will ever forget that day; it was too painful and too deeply etched into memory. Still, no lesson is ever wasted.

Consistency in Our Communications

Every message and event will be project specific, but our greatest difficulty is to ensure consistency across a network of multiple channels, operated by multiple contact points within the project, to multiple stakeholders. To achieve this consistency requires a very strictly managed process, and it is necessary to ensure that all messages and engagements, wherever and by whomever they are originated, are co-ordinated by a role within the project support team.

Planning and controlling the messages emanating from even a medium-sized project can be a complex task, and even more so when delivered under time constraints. In our government-to-citizen example, particular pressure could arise from adverse comment in the press or from formal parliamentary questions being asked to ministers. Similar pressures can arise unexpectedly even when a project is running well, and our communications arrangements need to be able to with-stand sudden peaks of interest in what we are doing.

It is essential that all drafts of project communication letters, emails, website content, videos, press briefings, training courses, etc. should be subject to a documented review and release process with an instruction that the message or event can only be actioned after authorisation. Each communication, once approved, should be issued with an explanatory briefing note and instruction to action to all project contact points and persons with a channel responsibility. This can be a very wide circulation list and on a large project could include the website content manager, graphic designer, engagement event leader, e-newsletter team, and every individual with personal contact responsibilities.

Remember, members of our wider project team also act as informal channels of communication, and consideration should be given to preparing a separate "for information" briefing to them when major announcements are about to be made.

Step 8: Monitoring Activity Effectiveness

As continuing successful targeted communication and engagement is an important factor in the success of a project, it is vital to keep checking back with the stakeholders to assess how effective we are being and whether we need to amend what we are doing. In order for our monitoring to work, whenever we design an event or communication, two questions need to be answered:

- What would constitute success for this event or communication?
- How are we going to assess whether we achieve it?

Take, as an example communication, a press release about the project. Let's say we're announcing meeting a critical milestone. We can stipulate in our plan for the communication that success would be positive coverage in specified target media and that we will measure this by subscribing to a clipping service. Identifying the target media is important rather than just counting all press coverage. For one press release aimed to increase interest by the UK call centre industry, full marks were obtained for getting several column centimetres of coverage in the targeted *Times* and *Financial Times* in the UK and no marks at all for the coverage Irish *Examiner* and the Australian *Sydney Morning Herald*.

We can be very specific about events and set a policy covering the whole project, such as:

- Every attendee at an event, whether in person or remote, will be asked to complete a short survey.
- The event survey will be designed to establish the event's effectiveness and establish any areas for improvement.
- Advantage will be taken of the opportunities provided by the survey to ask generic open questions about how the participant perceives the project's wider communications and engagement.

Responses to our survey can normally be stimulated by asking versions of our standard open questions such as "What elements of the event went well for you?" and "What do we need to stop doing/start doing/need to do better? As in an interview, it is always best to finish on a positive note by asking "what we do well." If, however, we are looking to do a numerical analysis of our interactions, then asking the participant to score aspects of the event is very useful, and this can be supported by opportunities to add comments on why a certain score was given.

Reflections

If possible, using the same project as in your previous reflections, spend some time considering the following exercises:

- List and map the major stakeholders in your project.
- Where do you expect to encounter the main barriers from stakeholders, and what strategies might you use to overcome them?
- Take two stakeholders and prepare stakeholder profiles for them.
- Selecting one type of stakeholder, prepare a short communication regarding the launch of the project which:

  - Communicates the overall objectives of the project.
  - Outlines the benefits for them.
  - Uses language to which they can relate.
  - Identifies what they should expect next and where they can get further information.

- Prepare a schedule of possible communications and engagement events covering the initial stages of your chosen project up to and including a contract award (or comparable event). The schedule should include details of target stakeholders for each communication, communication method, approval process for release, and how you intend to monitor the effectiveness of each event or communication.

Notes

1 On one industry reform project, there was a very poor history of communications and engagement by the government regulator in trying to introduce change. By the time I arrived, strict security protocols and guards had been put in place to prevent aggrieved members of the industry from storming in and taking over the regulator's offices – again.
2 Postal chess, perhaps?
3 It is this person whom I must thank for the warm and comforting phrase that should be on the front cover of every project management handbook: "This project will take as long as it takes and cost as much as it costs."
4 Just as we might, if we knew where the bodies were buried – metaphorically speaking, of course.
5 No apologies for repetition, but it really should not be necessary by now.
6 For example, the Crystal Mark standard for plain English.
7 Remember to be mindful of how judgements about people are recorded.
8 Running through a crisis scenario or two might be a more useful team-building task for the communications team than the exercise where they have to bridge an imaginary river with two planks, a rubber duck, and a knotted handkerchief!

Bibliography

Project Management Institute, Inc., 2013. *The Essential Role of Communications*, Newtown Square, PA: Project Management Institute, Inc.

8 Managing Risks and Issues

What Could Possibly Go Wrong?

Aim of this chapter: To introduce the concepts of risks and issues and to consider how they may be managed.

Learning outcome: To be able to differentiate between risks and issues; apply the principles of identifying, controlling, and mitigating risks and issues throughout the life of the project; and use basic risk and issue management tools.

Inevitability of Risks and Issues

In 1785 the Scots poet Robert Burns wrote:

> The best laid schemes o' Mice an' Men Gang aft agley,
> An' lea'e us nought but grief an' pain, For promis'd joy!

> (Burns, 1785)

A leading Prussian general[1] during the 19th century, Graf von Moltke the Elder, is normally summarised (or misquoted) as making the same point about not being able to plan everything with certainty; he wrote that no plan survives first contact with the enemy's main force (Moltke, 1995).

Whilst a project is not a battle or a war, both Graf von Moltke's and Robert Burns's words should be wake-up calls for the novice project manager and a reminder that our project plan, so carefully crafted, was only an opinion based on assumptions and drawn on the information that we had at the time. Risk and issue management is about how we can reduce the impact of our inability to see beyond that first contact with the enemy and the risks and issues that will arise when the project starts to engage with reality.

Assume that at this point we have now been through the growing pains of the project and have completed the planning, estimating, writing of business cases, and even building of the team. We are now about to transfer our beautifully crafted ideas, estimates, and plans into project deliverables. It is now time to reveal a truth which must be understood by all project managers: real life will try everything to stop us achieving the project's goals. It is down to us to predict, and where necessary to manage, all the threats and obstacles that project delivery will encounter. We call this risk and issue management.

DOI: 10.4324/9781003405344-9

All projects are about delivering a change, and with change comes risk. For us, risk always starts the moment the project is conceived and continues until the project is complete and the outcomes are delivered and signed off. During that period, different risks or issues will have been identified. Some of these will have been treated, some will have been tolerated, and some risks will have matured into issues. Other risks will have been left behind as being no longer relevant because the project has progressed past the event that was threatened by that particular risk. Risk management is not therefore a task that we can undertake at the start of the project and move on; managing risks and issues is an inescapable part of our responsibility during the whole of the project's life cycle.

Risk management itself is an area that has spawned a whole new profession with alternative ways of assessing and valuing risk. There is even an ISO standard on risk management (International Standards Organisation, 2022) and a whole series of guidance papers, academic reports, and separate qualifications in the management of risks.

Understanding the Management of Risks and Issues

Managing risks and issues is always an integrated part of our responsibility for managing the delivery of our project. It is not an add-on set of processes or duties that can be performed outside of our control, even if the client organisation has a risk management team. Risk and issue management is a critical area. As project managers we have to review the tools, skills, and competencies available to us and question whether we are able to manage them effectively with the resources that we have or whether we need to engage additional tools, train up existing project staff, or bring in additional expertise to support us – including from a corporate risk management team. Failing to recognise our need for effective resourcing for the management of risks and issues means we will simply be trusting the successful delivery of our project to good luck. This is not recommended.

Even if we do have to bring in outside expertise to support us, we cannot delegate responsibility for risk and issue management and will still need to understand the basic concepts involved. Our continuing goal for our project risk management activity is to ensure that all risks to the project, and the issues arising, are identified, captured, monitored, and controlled. There is a simple golden rule in controlling risk that once a risk or issue has been identified, it must be managed and cannot be ignored. This holds true even if the risk or issue is felt to be trivial or is within a situation where a decision has been made to accept it without taking any action.

As with our other areas of responsibility, we need to plan our approach to project risk and issue management. We need to agree on the principles upon which risks and issues will be managed, and we need to follow an iterative management process that we will operate throughout the life of the project, as set out in Figure 8.1

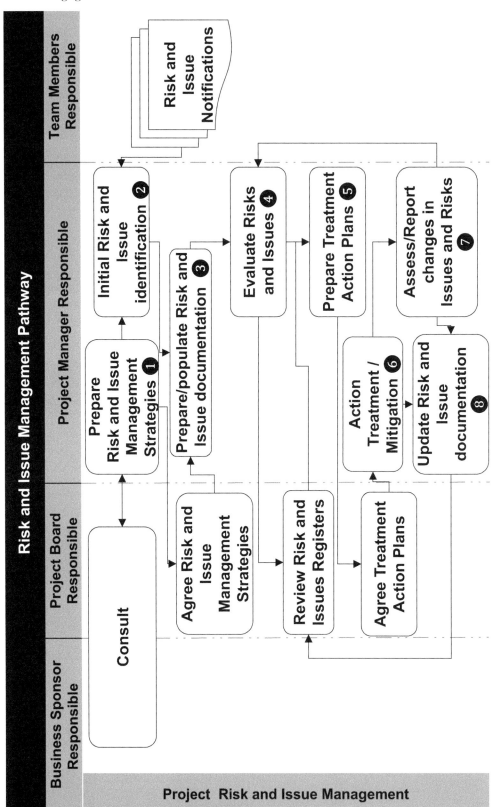

Figure 8.1 Risk Management Pathway

What Are Risks and Issues?

The words *risk* and *issue* are often used very loosely and frequently interchangeably, but because the management of risks has significant additional process steps and control opportunities over management of issues, we need to be precise. Let's expand on our definitions from Chapter 1:

- **An issue** is a current problem, concern, opportunity, or particular situation where something has happened and where we must make a decision about whether we need to respond now to either reduce a negative impact or exploit the opportunity for a positive one. It often seems that our major concerns are with constraining the negative impact of issues, but we should always consider the potential upsides. An example of this would be if we develop and launch a new product or service which has a far greater uptake than we expected. For example, there are, of course, downsides arising from unexpected success. How many times do we hear "Within minutes of the launch of the new online booking service, the website crashed" or "I'm sorry, but there's now a waiting list of three months for the delivery of our new model"?
- **A risk** is where something may happen in the future that would impact our project and its outcomes – either positively or negatively. As with issues, most often our concern seems to focus on constraining any negative effects.

Our risk countermeasures include:

- Regularly reanalysing the likelihood of a risk occurring and its potential impact on the project if it does occur.
- Actioning countermeasures to reduce the *probability* of the risk occurring.
- Actioning additional countermeasures now to minimise the *impact* of the risk *if* it becomes an issue.
- Planning countermeasures to take *when* the risk matures into an issue.

Just to underline the point, in our discussions throughout this chapter, please bear in mind that an *issue* is simply a *risk* that has happened. Reversing the distinction, a *risk* is a potential *issue*, but one we hope we can deal with before it bites us! In truth, we always hope that we have spotted a risk and taken steps:

- To reduce the probability of it occurring – i.e., to make it more unlikely to mature into an issue.
- To reduce its impact if it does.

Types of Risks and Issues

In addition to likelihood and impact, there is another broad dimension to risks and issues that affects how we are able to manage them. This is externality. Some risks and issues that arise are completely external to the project. These may be due to many things, including weather and/or natural disasters, political and legislative changes, changes in society, evolution of technology, and organisational policies.

An example of externally generated risks and issues arose during the development and pilot deployment stages of a city's unified transport ticketing project. The planned solution was overtaken by external factors many times, including:

- Telecommunications technology developments during the project. Before the rollout commenced, new generations of mobile telecoms technology twice required changes to the mobile communications equipment that had to be installed on buses and trams.

- Chip and pin bank cards were launched in the country, adding a different possibility for traveller authorisation processes.
- The government introduced a policy requiring the adoption of a single national identity card for accessing all government and municipal electronic systems. This meant the use of the project's planned personalised travel cards would be banned. This external change nullified a large amount of systems development work on designing the systems and printing the personalised travel cards, which had included an engraved image. The change also required scrapping the pre-ordered hardware planned to prepare and issue the personalised travel cards. As the national identity card project was in its early stages, the ticketing project would also have to be considerably delayed.
- Unforeseen requirements to protect heritage sites meant that card-operated entry gates could not be installed at all stations and had to be replaced with card touch-and-go readers as on the buses.

We can often prepare our project (to some extent) to meet these types of external adverse events, and we may in some cases be able to have some influence over them. On one government project, it was determined that there was a risk of legal challenge to working electronically with customers, which could potentially have been judged to be outside the scope of ageing legislation. The mitigation was to get the law changed very promptly. It may have helped, of course, having the country's deputy prime minister as the project's political sponsor. However, there are many areas such as weather-related disruption, wars, pandemics, and natural disasters where we can have no influence and may not be able to prepare effectively.

Other risks and issues that arise are internal to the project, and these we should be able to manage and control more easily. They range over many different areas – for example, supplier failure, design and production problems, inadequate staffing, poorly understood requirements, and inadequate costing. A classic example of how to manage an internal risk comes from COVID-19 vaccination programmes. Early in the pandemic, many governments identified a probability of at least some of the vaccines under development failing to work sufficiently effectively or to deliver on time. They mitigated these risks by arranging contracts with multiple vaccine suppliers to supply, in aggregate, far more vaccine doses than would be required. Governments had, therefore, built in at least some resilience to supply failures stemming from internal project risks.

Let's look at what can happen when unexpected real-life issues and our project plans come into conflict by taking another of our virtual field trips.

This time, you should wear a warm, waterproof coat and a hat. We're in a small European country, standing by a bus stop on a main road just outside the busy city centre. There's a chill north wind blowing, and it's raining. You pull your coat more tightly around you.

It's almost 8 a.m. on a November morning. The traffic on the main carriageway is jammed solid, but the bus lane is clear. In front of us, on the edge of the kerb, stands a stainless-steel pole, probably a bit over two and a half metres tall. Fixed on top of the pole and projecting back over the pavement is a large, illuminated sign showing when the next few buses are due to arrive.

As far as such street furniture goes, the sign is pretty attractive – and much larger than many similar signs in other cities. It has been carefully designed and positioned so that it can be read by both the bus passengers standing at the stop and the car drivers in the traffic jam. The idea is to drop the hint gently and repeatedly to these drivers that there are plenty of buses and that we would all have a much better life if they stopped sitting there in the traffic, jamming up the city, warming the planet, and chewing their steering wheel in frustration. After all, instead of being cramped in a tiny polluting tin box, they could be sat relaxed on the bus instead. Perhaps they could be doing a sudoku, reading a novel, silently cursing the deviousness of the compiler of a cryptic crossword, drinking a coffee, eating a second breakfast, or even catching a few extra minutes of sleep.

At the bus stop, we spend a few minutes watching the predictions update, the buses arrive almost exactly as predicted on the sign, and these fulfilled predictions being deleted just as the bus pulls away. I explain that you can see the same information on your phone, although we would have missed the main point of standing here, which is the sign itself, rather than the wizardry behind getting, processing, and displaying the predicted arrival times.

If you look more closely, you can see that our sign has two main elements; the head, which receives information via text messages and displays them on a double-sided display, and a shiny stainless-steel post. Hidden inside the post is a third element, the electricity supply.

Back in the welcome warmth of the project office and nursing a much-appreciated coffee and some delicious biscuits provided by a very hospitable project manager, we ask if we can look back at the documentation recording the risks and issues associated with each of the three elements of the sign. It turns out that there were some issues that could not have been foreseen and so never arose as risks at all. All issues were resolved at the expense of changes to materials, extra costs, and especially delays to the project. The project manager, a tall man with prematurely greying hair, provides us with a copy of the issues register in the form of a very large spreadsheet and talks us through the main problems that arose with the sign:

__The display.__ The country is bilingual. All public signs have to respect this, using text in both Language A and Language B. The responsible authorities decided during the project that this rule would now apply to electronic signs for the first time. Sadly, as the city is a Language A area, none of the street names that the signs would need to display existed in Language B. The project unexpectedly had to carry the cost of translating over five hundred street names from Language A into Language B. Additionally, new rules were put in place about language display priorities that required software changes and how names could be abbreviated in each language to fit within the maximum number of characters on the sign.

__The pole.__ The display head is not mounted centrally on the pole but at one end. All poles are pre-cut with a mounting slot, meaning that the display head could only fit on the pole projecting in one direction — back over the pavement. When the project manager of one of the remote installations wrote in the required weekly progress report, "The poles have been put up with the wrong inclination," warning flags were raised. Did this mean that the poles had been mounted horizontally, or at an angle of 45 degrees? It turned out that despite having been shown how to install the poles at the edge of the pavement, the installation team had inadvertently rotated each one of the 50 poles in their city by 180 degrees. This meant that if it the display heads had been installed, instead of projecting over the footpath as planned, they would protrude into the roadway. It was suggested that the effect of this was that you could read when the first bus was coming on the sign and then you could have seen when the bus had gone past as the sign would be lying smashed in the road. The result was that extra time had to be allowed to refit all the poles correctly.

__The electricity supplies.__ Two major issues arose with the electrical supplies to the poles. Firstly, from one city there was a request to include an additional hole in each pole so that the electrical connection could be made more than a metre off the ground rather than at the base, as is more usual. When asked to justify the change to the local installation, the local project manager replied that some of the sign locations were tidal. The project team's resident wit suggested that he could imagine the shiny new sign on top of its stainless-steel pole showing a message that the next bus is due in two minutes and a queue of passengers waiting patiently beside the sign whilst waist-deep in sea water.

The other unexpected and much more serious problem with the electricity supply was a change in policy on who could make the connection to the dedicated public lighting network. Previously, the lead city had included connection to this network as a normal part of their installation of illuminated signs. Enter the monopoly electricity supply company. This company now decided that not only were its staff the only people competent to complete the connections but that they would disconnect some of the signs that were already operational because these had not been installed by said staff. The adverse impact on the implementation schedule was magnified when it became apparent that the electricity supplier would only allocate one team to make the connections for the five hundred planned signs and that the team could only complete two connections per day – adding at least 150 days to the project's rollout timetable. Eventually this issue was escalated to CEO-to-CEO negotiations and the supplier agreed to provide two installation teams – one operating in the morning and one in the afternoon!

You begin to understand why the project manager has more grey hair than would be expected for his age.

In summary, as far as the street signs are concerned, the project didn't spot the potential for any of the problems arising with the installation of the pole or with dual-language regulation changes. These are clear examples of a project plan not surviving the first encounters with real life.

Tying a Bow

There is a simple three-part analysis of risks and issues that can help us visualise what we are trying to achieve, usually called a bow tie analysis. Although this type of analysis can give rise to complex diagrams, at its simplest it pictures an event arising from an issue or risk as having multiple causes and multiple consequences.

Figure 8.2 shows both the causes and events relationships and where different types of barriers to risks, essentially mitigating actions, may be implemented.

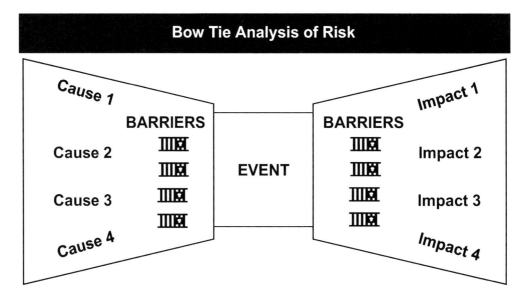

Figure 8.2 Bow Tie Analysis

Using this analysis as a base, our role in risk and issue management can be seen as being twofold:

- Identify and evaluate the series of events which may impact on the delivery of our project.
- Erect barriers (mitigating actions) which will protect the event from its causes or reduce/eliminate its impacts.

It's time to pick up the bio-energy from effluent project again because an email from the company's new head of risk management has just been received.

Re: Bio-Energy from Effluent Project

Dated 9 January

Good morning.

Please allow me to introduce myself. I've been newly appointed as the head of global risk management, and I understand that you've been charged with leading a looking at bio-energy from effluent. I'd like to meet with you to discuss the business and other risks associated with your project. Please can you let me know when over the next two weeks you'll be free for a discussion. Please also suggest an agenda which picks up on the key risk areas that the project is facing.

Kind regards,

Naz Kumar

What sorts of things should be on the agenda for this meeting with the head of global risk management, and how could they be quantified?

Think back to the first meeting with the boss and to the seven questions that we learned to ask and get answered over the subsequent days. There is, in fact, an eighth one, actually a full set of questions, that we need to think about in relation to risk. These questions come steaming right behind the other seven. The new questions involve What If:

What if the process that R&D have developed doesn't scale up to a full production plant?

What if a competitor is more advanced than we are and gets to the market first?

What if we cannot get a funding partner on board?

What if we can't find initial clients to trial the solution?

What if . . . ?

And so on.

Now, consider how you would analyse the risks and issues contained in the following short briefing note from the development team that arrives in your inbox the very next day:

Re: Briefing Note from Technical Development Manager

Dated 10 January

We've noticed that there's potential issue of increased gas pressure in the first effluent fermentation vessel of the preproduction prototype if the reaction temperature rises above a maximum of 35.5 degrees Celsius. This increase in pressure seems to increase the energy yield in the entire process by 0.05%, but in order to maintain safety levels, we would need to increase the ability of the fermentation vessel to withstand a long-term increase in operating pressures, and the extra cost is unlikely to be offset by the additional energy yield. We therefore propose to take action to manage the temperature in our first fermentation vessel to keep it below 35.5 degrees Celsius and to protect the plant by installing an automated pressure release valve.

Jay Stevens,

Technical Director

Two things immediately stand out from this briefing note. Firstly, this is about an actual issue. In our world it has to be either a risk or an issue; there is no such thing as a "potential issue" as referred to by the technical manager.

Secondly, we always need to ask if this is a closed issue or if it raises any red warning flags about where further action might need to be taken. The answer is that this briefing note should have sounded loud alarms bells and set red lights flashing. The issue was not adequately dealt with, and a huge risk remained.

Now let's consider the voicemail from the on-site manager of the first pilot rollout site several months later.

Voicemail from Wadi Plant Installation and Commissioning Manager

13 October

"This is Tom from the Wadi Implementation site. It's Thursday morning here. Everything had been running great for the first week since we got the mobile plant trucks in place, as you know from my report last Thursday, but now we have a major problem. There's been an unprecedented heatwave here – the daytime temperature has been hitting the mid-forties Celsius for the last three days, and it's affected the plant operating temperatures as we were doing our pilot test runs on the plant. The first effluent fermentation vessel started running far too hot, the pressure increased, and it blew the safety valve. This caused a large amount of the barely fermented contents to spray out everywhere, it's going to take some time to clean all this up.

"Sadly, the blowout happened at the worst time – just as a government minister was doing a live interview on local TV about the importance of such technical advances as ours in creating new jobs and combating global warming. It's all there in the live broadcast. The minister, the glamorous interviewer, and the bystanders all covered from head to toe and gagging at the smell. I expect the video is already going viral.

"We're currently decontaminating the visitors, their equipment, and the plant. The minister is widely tipped to take over from the prime minister next year, and her loss of dignity means that to say the least she's not seeing the funny side of things."

The original briefing note shows a technological risk and a proposed mitigation which fails to account for potential operating conditions. The voicemail is an issue which not only indicates that the proposed mitigation was not sufficient to remove the technological risk but also that there was an unanticipated reputational issue which must now be managed.

How should the technology risk have been assessed and mitigated?

Would a bow tie analysis have helped following the briefing note?

What can now be done to deal both with the immediate reputational issue and the prevention of a recurrence of the technological issue?

Would a bow tie analysis help now?

Such risks do mature into issues, and usually at the worst of possible times. First, consider the case in 1830 of William Huskisson, an English member of Parliament. During the opening event for the world's first passenger railway, whilst standing and reportedly talking through the door of his carriage to the Duke of Wellington (then UK prime minister), Huskisson was hit by Stephenson's Rocket locomotive. He thus achieved immortality as the world's first recorded railway fatality.

Consider now how another similar issue two decades later might have changed world history. A young Otto von Bismarck, not yet Germany's famed Iron Chancellor or the driving force behind German unification, visited the soon to be completed alpine Semmering Railway as his government's representative. A temporary gangway over a ravine collapsed under him, and young Otto reportedly only survived death by clinging to a cliff ledge (Wolmar, 2009). Surprisingly, this did not reduce his lifelong enthusiasm for railways.

Timeliness and Resilience

Timeliness catches us from both sides. Figure 8.3 indicates how our ability to manage risk will reduce as the project progresses because our earlier decisions mean that there are now fewer options available to us as countermeasures.

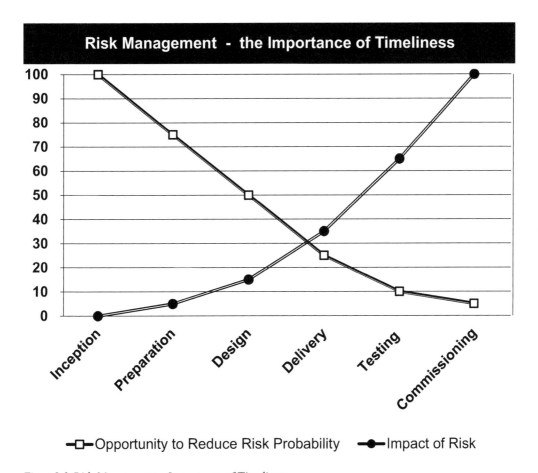

Figure 8.3 Risk Management – Importance of Timeliness

If, for example, we have contracted with only one supplier (giving the project a single point of failure), we are now locked into that supplier's ability to deliver. If, at the start of the project, we had chosen to design a strategy including contracting with multiple parallel suppliers of the same item, we would have both reduced the impact of a single supplier failure and widened the scope of actions available to us – for example, by increasing volumes purchased from another supplier.

This principle of taking timely action means that our first objective in managing risk is always to create a project and a delivered solution that, from the outset, designs out as much risk as possible and creates a resilient project and solution. When we have designed in this resilience, the residual project risk is what will remain for us to manage during the whole life of the project.

Achieving the objective of maximising resilience from the start of the project means we must ensure that each of the decisions taken sets us on a path to being as resilient as possible within the usual confines of time, cost, quality, and regulations that we will have to operate. Consequently, as project managers, our first steps in managing risks and issues should begin as soon as we start thinking about the project and its outcomes and certainly before we have any risk management experts engaged or processes in place. We need to formulate and ask our What If questions from the very outset of the project and continue adding to them at all times, paying additional attention to them during our planning and initial solution design.

Some methodologies also value risk and calculate the cost of mitigation strategies. We can see a simple direct cost of managing risk by taking the example of either single or multiple suppliers of identical products one stage further. If placed with only one supplier, the initial contract should achieve a lower total price based on higher volumes supplied by said supplier. The difference between the lower single supplier total cost and the higher total cost with multiple suppliers represents the cost to the project of mitigating the risk of supplier failure.

How Do We Manage Risks and Issues?

We have identified that risk management is not a one-off event at the start of our project. Our processes need to support the continuing monitoring of risks both at an individual level and at an overall project level. We can also be asked by business sponsors to reassure them by reporting on how the total level of risk in the project is changing over time – and, of course, they are looking for us to ensure them that it is decreasing! Our risk management evaluation processes will also need to support such reporting.

Irrespective of the measures chosen to evaluate risk, at the core our judgements will always be subjective. They will, of course, be based on the best advice available and, we hope, using the best comparative data that we can find. However, they will still be subjective judgements.

Figure 8.1 set out the basic steps we need to take to ensure that we manage risk effectively. These are, in essence, quite straightforward and worth reemphasising:

- **Plan**. Establish and document our approach in a risk management strategy.
- **Identify**. Review the project and identify areas where risks may potentially arise.
- **Document.** Document individual risks in accordance with the standards in the risk management plan.
- **Evaluate**. Consider each risk and decide whether actions are required to deal with the risk. Also consider overall project priorities for risk management actions.
- **Action planning**. Preparing a co-ordinated series of appropriate actions to tackle all risks and issues, including triggering change control processes.

- **Take action**. Act on the prioritised risk management actions.
- **Assess and report**. Evaluate changes in issues and risks during and following completion of the actions.
- **Update project documentation**. Ensure that the changes that we have put in place are properly reflected in the current versions of project documents.

In addition, we need to ensure that we have in place adequate continuing reviews of the effectiveness of our actions in tackling issues and reducing exposure to risk.

Step 1: Prepare Risk and Issue Management Strategies

The risk management strategy and issue management strategy document how we intend to manage risks and issues throughout the life of the project. They are critical documents that support a major area of our ongoing management of the project. Each needs to cover all of our seven standard questions of Who, Why, What, Where, When, How, and How Much, as well as the specialised What If:

- **Who** will be responsible for activities related specifically to risk or issue management? Some of these – for example, project board responsibilities, should already be included in the business case but should be restated here. We should, however, still summarise all responsibilities for identifying, evaluating, reviewing, and agreeing on the risk. This needs to include reference to the roles of the project manager, project board, and business sponsor. For some projects a risk manager role may also be required. We could also include details of external advisers on risk if the nature of the project would justify it.
- **Why** is it important to manage all of our risks or issues? Not all business sponsors and members of the project board will fully understand why risk management is important and why the project needs to invest time and resources in preparing for things that are not going to happen anyway.[2]
- **What** criteria will be used for assessing the level of probability and the impact of individual risks? All criteria to be used for evaluating each include two factors: its probability of occurring and the impact on the project if the risk does occur. In the risk management plan, the explanation of these criteria should include descriptions of the types of impacts that will be considered and how probability levels will be assessed. We also need to specify the criteria to be used when treatment actions are required immediately in response of a critical issue.
- **Where** will risks and issues be recorded and reported? Our records should include detailed descriptions of the risk or issue and the actions taken and set out our summary reporting of risks, issues, actions required, and actions taken (often via higher-level risk and issues registers). A note of warning: where the risk register is used as the sole record of the management of risk, but without supporting detail, it either can become a horrendously large spreadsheet with extremely small type or must be summarised and essential detail can be lost.
- **When** will individual risks be assessed (risk records), and at what points will all risks be reviewed together (risk register)? This timing of risk management activities should include, for example, reviewing our risk register as part of each project board meeting and additional processes for dealing with emergent higher-probability risk between meetings.
- **When** will issues be reported, and what rules will govern their escalation?
- **How** can we design and implement procedures and methodologies for the overall management of risk and issues? These procedures may be based on a corporate or other standard, and we should identify if and why these have been tailored to fit the project. The processes for recording, evaluating, and reporting risk should include the processes for raising a notification

of a potential risks or actual issues by anyone engaged in the project and for recording and monitoring risks and issues.

- **How** will progress on individual risk or issue as well as the overall level of risk on the project be reported?
- **How** will the release of any contingency funds be authorised?
- **How much** funding should be allowed for the management of risk? This funding is for the process of risk management rather than any contingency funds that have been set aside to deal with issues arising.
- **How** are the levels of criticality set? These levels will determine whether a risk or issue requires further action or escalation – i.e., the rationale for making our action decisions.
- **How** many staff resources are required to manage risk? Depending on the type of project, its size, and its complexity, the project manager may be able to act as the risk manager, or the project support team may need to be supplemented by a dedicated risk management function. The numbers of staff and the skills required may also vary throughout the life of the project.
- **What if** we have to act in a crisis situation because either a risk has escalated or an issue has arisen which requires immediate action? Our processes and procedures will need to allow for taking actions without requiring reference to the full project board, including interfacing with the wider organisation.

All the steps in risk management (planning, identification, evaluation, and treatment) set out in the previous section apply equally to identified issues. The only major difference is that with an issue we are treating the effects of an event which has already occurred, rather than seeking to prevent it. Our issue management strategy also contains the answers to our seven questions and feeds the actions required into our change management process, although normally with a high level of priority.

Step 2: Identify Risk

Identifying risks is, in essence, not different from the discovery methods that we have already used in researching the project's objectives and making our initial guesstimates. Once again, our approach can include one-to-one interviews with our stakeholders and any others who have useful information, perhaps based on their involvement with previous comparative projects. Interviews can also be very useful when information is being sought on a specific topic from an expert and can often be conducted remotely. Risk discovery, however, presents another excellent opportunity to engage our whole project team using a workshop-based approach, including involving representatives of our "volunteer army," if the project supports one, and our stakeholders.

Preparation. As preparation for the risk discovery exercise, it is always useful to learn from other similar projects. Preparation can involve researching available documentation, including published studies, and discussion with people involved. Our objectives in exploring the good as well as the bad experiences of other projects are to:

- Identify the sources of their risks and issues and the potential impacts on different project objectives and stages.
- Understand how risks and issues were managed.
- Map the lessons from the past project to similar points in our project.

If our client organisation has previously undertaken comparative projects, there should be opportunities to:

- Obtain copies of their lessons learned reports (if they were prepared).
- Review their risk (and issues) registers and their risk records.
- Have informal discussions with previous project managers.

Once we have our completed preliminary research, a useful preparatory step is for us to analyse our work breakdown structure and consider the risks that may arise in relation to each element of our project. This analysis should include input from the client organisation's risk manager and, if one has been appointed, our project risk manager.

Whilst both using the work breakdown structure as a starting point and learning from previous experience are helpful in ensuring we cover all areas of the project and apply previous lessons, we still need to make sure that our risk discovery processes will:

- Address the risks arising from novel areas in our project which fall outside the experience of previous projects – for example, perhaps our project needs to work across different jurisdictions or take account of different ground conditions or technologies.
- Take account of changes in external circumstances which will affect how applicable lessons on risk management from previous projects still are. We must bear in mind that the governing legislation, potential market opportunities, required standards, or stakeholder expectations may have changed.
- Identify overall project risks.

Facilitated risk discovery workshop. Using the same sort of facilitated workshop approaches set out in Appendix 1, we can set up group brainstorming sessions to help us identify risks and countermeasures. With the bow tie analysis in mind, a useful way of doing this is to split the attendees at each workshop into two teams. One team is tasked with identifying risks and the other with identifying the barriers or countermeasures to treat the cause of the risk to reduce its likelihood and also reduce its impact. After these teams have completed their work, the facilitator matches up their output during a reporting session. We can even bring in a competitive element and find a "winning" team. We might allow a "score" for the risk team for every risk where no countermeasures are set out and a "score" for the countermeasure team where a countermeasure is identified but the risk team has not spotted the risk.[3]

Usually, the risk and countermeasure identification process can be structured by asking teams to focus their answers on each of the different objectives and project stages. For each project objective, both teams are given these questions to consider:

- What might stop us achieving this objective completely?
- What events or actions would reduce our level of achievement of this objective?
- How would these events or actions impact our stakeholders?

For each project stage, the teams are then asked to consider:

- How could events make us less efficient in completing this stage – for example, requiring more resources, reducing quality, or making us take longer?
- What would our stakeholder reactions be?

Ideally the teams will have five or six members, depending on the total number of attendees, and available breakout spaces. With larger groups, multiple risk and countermeasure teams can be established, with each being given a selection of the project's objectives or stages to consider. Floating observers can be appointed to move between teams, having already been briefed to get the teams to consider some of the key areas of concern identified in the preparatory work. The observers should be briefed to ask open questions to guide discussion, such as: "I see that you've looked at the risks of transporting the effluent processing machinery to site, but what risks do you think might arise in relation to choosing a suitable location?"

Such a workshop, or series of workshops, will usually require detailed preparation by the project manager and a facilitator, including identifying the most appropriate participants and ensuring that they are fully briefed. As the first risk discovery workshops are likely to take place in the very early stages of the project, those attending may be largely unfamiliar with it. Therefore, our briefing to them will need to include information on the background to the project, its objectives, and how we intend to deliver them.

As the expert on everything related to the project, it is appropriate for the project manager to open the workshop and deliver the briefing and set the objectives for the session. Despite the temptation to facilitate personally, it is often better for the project manager to take a step back and be an observer or participant, rather than the facilitator. If it is decided not to bring in an independent facilitator, filling this role is an opportunity to engage an important stakeholder or a senior member of the project team in the risk management process. If the business sponsor wishes to be involved, it may be best for them to open the workshop, leave, and return for a final reporting session.

Using the workshop format, we can also consider whether it is useful to spend some time in developing the attendees' understanding of risk before moving into the discovery process. One way to do this is to present predeveloped risk scenarios, perhaps based on our research of other projects. We can set some teams to project forwards how we might react to a particular event or series of events and other teams to look backwards at how the risk arose and consider how we could have taken positive action earlier to reduce it or avoid it completely.

As with our previous discovery exercises, all items identified, in this case risks, should be recorded for subsequent evaluation without any judgement being made, however slight, tangential, or just plain bizarre the risk they identified might seem to be.[4] We will be agreeing on and addressing the most important of these risk areas in the next stage of the process, so the documenting of *all* risks that we do now is vital. Picking up on our earlier theme of satisficing, the requirement at this stage is for the list of risks identified to be adequate as a starting point, not to be perfect. We will be reviewing, updating, and reevaluating our list of risks as the project progresses, and any risks missed now can be included later. Beware discounting any risk, however trivial it may seem at the time.[5]

Taking such a workshop-based approach to identifying and gathering information risks has a number of advantages over other methods in that it:

- Develops team commitment and sensitivity to a difficult area and allows the project to benefit from a wide range of skills, knowledge, and experience.
- Increases the quality of the output over other methods of discovery through the interaction between attendees.
- Fosters team engagement in risk management, making it easier for team members to identify and report new risks as the project progresses. Their deeper understanding of the process and the risks identified will mean that we are likely to achieve a greater degree of support and acceptance if/when treatment actions are required.

- Allows workshop participants to collectively gain a much deeper understanding of the risk analyses, risk evaluations, and consideration of different treatment options.
- Increases acceptance of risk mitigation actions because, if we have selected our attendees carefully, the individuals to whom we may have to assign such actions later will have been involved in the risk discovery process from the start of the project.

Whatever approach we take to risk discovery, we need to ask (and research and consult as widely as possible in reaching our answer) the following question: "Where do risks hide in our project, and what can we do about them?" In truth, risks can arise everywhere and at all times. As we can see in a bow tie analysis, risk may have many sources. Like the overheating of the primary fermentation vessel in our bio-energy from effluent project, a single risk event may impact different areas – in that case, both technology and reputation. The areas in which risks will arise are always project dependent. We should, however, always start our discovery exercise by reviewing our project for potential exposures based on commonly encountered areas and thinking through some of the countermeasures available, probably using a bow tie analysis for each risk. The most common areas where risk is encountered include:

- Organisational/personnel.
- Quality of project output.
- Technical.
- Commercial.
- Financial.
- Reputational.

Organisational/Personnel Risks

There should always be concerns in relation to a project's personnel and its organisation from the outset of the project. These major initial risks include:

- Our ability to recruit members of our project team, including functional specialists of a high enough calibre to deliver to the quality standards and the timescales that we require.
- Staff retention until the project's work schedule can afford for us to release them. There is a particular twist to this risk where staff have been directly engaged on a temporary basis, perhaps for the life of the project. In order to avoid any gaps in their own work programme, it is possible that they will chose to move to their next role earlier than our project's planned work schedule would ideally permit, thereby leaving us with a resource shortfall to cover for brief but often critical periods.

One way we can try and mitigate both recruitment and early departure risks is through our project resourcing strategy. We can gain resilience by using diverse sources of team members, including "borrowing" from elsewhere in the organisation, using contract or agency staff, and engaging external suppliers of specialist skills, as well as directly employing staff to work on the project. We will then have also established a mixed supply chain and a variety of relationships upon which we can draw if necessary.

There is, however, also another major personnel and organisational risk that we should consider – overreliance on key people. Overreliance is risk in any area of business where resources are limited or broken down into small, specialised pockets of expertise, and this will also apply to many projects. When considering our project's resources, we should include as a "key person" risk anyone whose loss of knowledge and skills from the project would impact its delivery – i.e., anyone who represents a "single point of failure" for the project. Depending on the type of our project, the key people group could include designers, developers, engineers, skilled craftsmen, and, of course, ourselves.[6]

Consider applying a simple bow tie analysis, as in Figure 8.4, to illustrate the risk to a project whose work schedule requires a small team of skilled craftsmen.

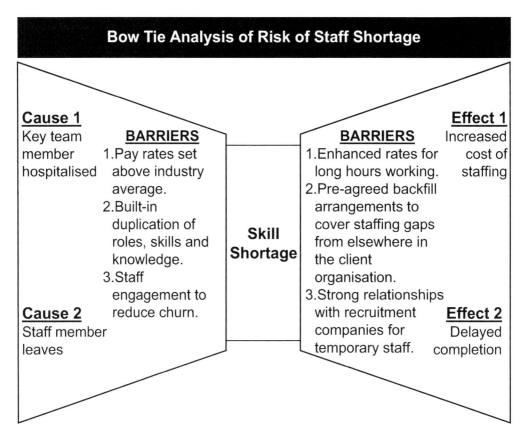

Bow Tie Analysis of Risk of Staff Shortage

Cause 1
Key team member hospitalised

BARRIERS
1. Pay rates set above industry average.
2. Built-in duplication of roles, skills and knowledge.
3. Staff engagement to reduce churn.

Cause 2
Staff member leaves

Skill Shortage

BARRIERS
1. Enhanced rates for long hours working.
2. Pre-agreed backfill arrangements to cover staffing gaps from elsewhere in the client organisation.
3. Strong relationships with recruitment companies for temporary staff.

Effect 1
Increased cost of staffing

Effect 2
Delayed completion

Figure 8.4 Bow Tie Analysis of Risk of Staff Shortage

Perhaps this team was difficult to recruit. What would happen if suddenly one team member is hospitalised and will not be fit for work until after the end of the project (cause 1) or if a skilled craftsman leaves for another more lucrative job (cause 2)? Either cause would throw our schedules into confusion (event). Our lead time to replace the lost team member could be weeks or months, and the shortage of skilled resources for that period would either impact our costs if the project has to pay existing staff overtime or bring in more expensive contractors (effect 1) or if the task's completion and dependent tasks complete later and we have to report a lengthened project delivery period (effect 2).

In practice, issue mitigation resulting from a sudden loss of key staff resources is often difficult. We will usually have to react quickly to replace staff. The effect of the loss of knowledge to the team can be as important as the extra costs or delays. In practice, for example, a skilled mason or electrician may pick up on the work required to complete the scheduled task more easily than a design engineer, construction manager, IT developer, or anyone who requires a detailed and comprehensive knowledge of the overall solution and the current state of its development.

We cannot normally take effective steps to reduce the probability of external events such as a sudden illness, but we do not have to wait until the risk becomes an issue to take action to increase our project resilience. How might we reduce the impact on the project of the loss of the knowledge of a key person? We could seek to build in more resilient solutions for the operation of the project, such as effective documentation standards, fostering buddying and partnering relationships between team members individually, and encouraging the sharing of knowledge and current activities widely across the team. This latter strategy of ensuring knowledge sharing is one in which borrowing from the SCRUM methodology with its daily short meetings can really help. Building such techniques into our project's method of operation gains us the flexibility to both reallocate any non-available key person's workload across other team members and to provide any new starter with effective support both from the whole team and from a buddying relationship.

One further consideration on the loss of key individuals: we should always consider ourselves, as the project manager, to be part of the key people group.[7] As we carry the central knowledge of the project, its stakeholders, and its delivery team, our loss is always a major risk to the delivery of the project. We are, or we should be, the one person who has the comprehensive body of project knowledge and history and can answer any or all the seven questions about every aspect of the project immediately.[8] Consequently, we are always key to the successful delivery of the project and to meeting its plans and schedules. What will happen to the project if we are the ones to fall ill or, worse, accept the offer of a job elsewhere?

The mitigation of our loss to the project should stem from the detailed processes that we should have built into our project. We should have designed our processes to build resilience into project management as well as the project overall. As with other project team members, we should fully document our work at all times, including in our plans, registers, and reports. We should seek to share our knowledge across the project team, the project board, and the business sponsor through engagement and discussions. These activities will naturally reduce the impact on the project of our loss as the project manager. In addition to this resilience, the addition of a long-term business manager to work closely with the project manager will also mitigate this risk, as in the Chapter 6 example.[9] We could also consider whether one or more members of the project management and support team should become a deputy project manager.

Quality of Project Output Risks

This may not be often listed as separate from other areas of risk, but it is inherent in every project. As with the risk of time, technical, and schedule failures, the impact of failing to maintain quality is that the project will be seen to have failed. Maintaining output quality is specific to every different project and runs from ensuring that a computer development will operate in accordance within accepted norms and will be tested and documented correctly before acceptance through to ensuring that the concrete to be used in a construction project is correctly formulated or that machinery to be purchased and commissioned for a new manufacturing production line will operate in accordance with specifications and safety requirements. There is also one area of quality risk that is common to every project, which is the fitness for purpose of our project management and supporting processes.

The mitigation of quality related risks includes:

- Building in resilience to our project and solution design.
- Ensuring our project processes address and reduce quality risk.
- Reviewing and, where necessary, testing the quality of every project process and output to ensure that it meets, or exceeds, the required standards.

As a matter of policy in managing quality associated risks, we should expect key work undertaken by all parts of the project team to be fully documented. We should also ensure that all output is independently checked or peer reviewed as part of its completion process before being formally reviewed and authorised. One engineering design organisation insisted that every document, including every calculation and every subsequent amendment, was signed off by:

- The originator.
- A peer checker.
- An independent reviewer.
- An approver.

The same rigorous approach was applied to every revision. Was this overkill? Some people might think so, and perhaps it is on some projects, but not if we are building a tunnel or a bridge that needs to meet rigorous safety standards. An infamous example of a project failure attributed to a failure to check, review, and approve calculations was NASA's Mars Climate Orbiter mission. This NASA spacecraft disappeared, and probably became a $193 million fireball in the Martian atmosphere, when someone allegedly missed that the spacecraft builder had provided critical data on acceleration from its systems in Imperial units but the NASA mission controllers and systems assumed that they were in metric units. Whoops, a total mission cost of $327.6 million lost through non-conformance with quality checking requirements!

The Mars Climate Orbiter project illustrates that in all areas of the project delivered by our functional specialists, we must build in quality assurance measures and testing of their deliverables to ensure that the project outputs will meet the standards required. Consequently, in designing our resilient project and managing our quality risks effectively, we need to consider:

- Who should we make responsible for designing the technical quality processes which would mandate that designs and calculations will be checked and reviewed?
- Who should be responsible for quality checking any functional expert's calculations, such as when a pesky spacecraft keeps drifting off course (cost of conformance)?
- How and when should we implement higher levels of independent expert checking in critical areas?

We also need to ask how the quality of our work as project manager and the project management and support team can be tested and quality assured. As with the loss of the project manager, the risk of a quality failure by us is mitigated because of the resilience of the processes implemented for project management activities. These activities should all include authorisation, monitoring, and reporting procedures, which build in an independent scrutiny of our actions and thus a measure of quality assurance. Frequently, however, projects also appoint separate quality assurance teams to support the project manager. Such quality assurance teams may be tasked to review and report on deliverables produced by the project manager and by functional specialists. In the UK, the government implemented a set of formal independent reviews (Gateway Reviews) in which the quality of the project's management and output must be assessed by an external panel and a report made to the business sponsor. This concept has now been adopted more widely by different governments and organisations. The UK Gateway Review process itself includes independent reviews and reports to the business sponsor at key six stages in the life of the project:

Gate 0 – Strategic assessment.
Gate 1 – Business justification.
Gate 2 – Delivery strategy.
Gate 3 – Investment decision.
Gate 4 – Ready for service.
Gate 5 – Benefits realisation and operational review.

Technical Risks

Whether it is new product development, research, construction, an IT system, or a manufacturing process rollout, there are inherent dangers that the technology we are delivering will not perform as required or that it will take longer than anticipated to become operational. The now classic example of failing to meet delivery expectations was the novel processes for manufacturing the COVID-19 vaccine by AstraZeneca where the client's (the European Union Commission) expectations for delivery schedules were not met due to reported technical production issues with achieving required quality standards in some of the newly created manufacturing plants.

Mitigation of technical risks can include:

- Ensuring that the project and its outputs follow industry norms and agreed-upon or compulsory standards or regulations, including health and safety.
- Where external contracts are involved, ensuring that the responsibility for managing risk and bearing the consequences of an adverse event are carried by the appropriate parties.
- Implementing design reviews, unit testing, and independent technical assessments in order to identify potential or emerging problems as early in the project life cycle as possible when we have greater freedom of action and, normally, lower mitigation costs.

Commercial Risks

There are sufficient commercial risks in contracting with other parties to help deliver our projects to fill many volumes and to keep vast armies of lawyers in every jurisdiction standing ready with their pencils sharpened. The major commercial risks for a project usually centre around one or more of the following questions:

- Can we find a supplier, or suppliers, willing to supply us within our constraints of time, cost, quality, and contractual terms?[10]
- What will be the impact if a supplier, or suppliers, fails to deliver fully to the time, cost, and quality constraints within our contractual terms?[11]
- What happens if external circumstances and/or our requirements change during the contract period?
- What can we do if a main supplier ceases to trade?

There are some general actions that we can take to mitigate commercial risk – and these underlie the approaches and examples set out in Chapter 10 and Appendix 2. These actions include:

• For any major supply contract, engaging legal and procurement experts at the outset. Such specialist support is *always* worth the additional cost if it enables successful contracts to be let and disputes and litigation to be avoided.
• Making sure that the "golden thread" of our requirements is reflected in explicit standards to which the supplied goods or services must conform and that these are included in both our supplier evaluation criteria and in unambiguous contract terms.
• Doing a due diligence review on potential suppliers, including taking up references regarding financial stability and the supply of similar goods or services.
• Contracting on the client organisation's terms and conditions and not those of the supplier.
• Using industry standard contracts, if possible, but supplemented with any special conditions and data concerning particular needs.
• Maintaining a strong relationship and regular contacts with all suppliers and monitoring their deliverables.
• Taking early action to keep suppliers operating within the terms of the contract.

Financial Risks

Any event that may alter the financial estimates on which our project's budget is based represents a financial risk. Bear in mind the level of our imperfect knowledge when we prepared our initial guesstimates and the assumptions we had to make about the uncertain future. We should have built in some additional contingency funds to cope with the uncertainty – which we can now understand was actually creating a fund with which to mitigate risks and their impacts.

The pattern of financial risk, who carries them under supply contracts, and how they can be mitigated is inevitably unique to each project and can be influenced by the client organisation's financial management policies, but we normally need to consider all purely financial risks. These include at least:

• Inflation.
• Exchange rates.
• Interest rates.
• Costs of change
• Default by supplier.

Inflation. On longer-term projects, how inflation is accounted for and what indicators are used is important. Not only do we have to predict the future rates of inflation when preparing estimates, but we need to recognise that any supply contract negotiations may result in us having to use different indicators. For example, our budgets might have been drawn up using a general rate of inflation indicator, but one supply contract seeks to use an indicator specific to that industry and a second contract at a rate specific to a different industry.[12] We can mitigate the impact of encountering industry-specific inflation rates during contract negotiations by ensuring that we research and consider these whilst building our business case.

Exchange rates. Where supply is on an international basis, there is an issue on who bears the pain/gain from currency fluctuations. The bearing of the exchange rate risk is often implicit in the choice of contract pricing currency. This is an area where the client, especially public sector bodies, will normally want to eliminate risk, and have pricing be expressed using their national currency, whereas suppliers from another country may want the contract priced in their home or another currency – e.g., US dollars or euros. If there are concerns on currency instability, it may be possible to agree on a fixed exchange rate either with the supplier (transfer of risk) or a third party (treatment of risk). Our engagement with the client organisation's financial management team is advisable to ensure that their requirements are met before any contract negotiation with the supplier is concluded.

Interest rates (i). Where the project is being financed through borrowing by the client, the terms of the loan raised will impact on project viability. Therefore, any potential for increases in the cost of the loan represents a risk to the project achieving its financial goals. If the project is to be debt funded, then during the preparation of the financial elements of the business case, we should have undertaken a sensitivity analysis to determine the impact of interest rate changes. Provision to mitigate any adverse interest rate variability is an area we should discuss with the client organisation's financial management prior to finalising our business case. Sometimes additional interest charges can be required to be met from within the project, in which case we should have argued for a provision for this in our contingency funds. Alternatively, additional charges can be met by the organisation adjusting the overall project funding. Both changes, however, will require feeding into our financial justification in the next iteration of the business case.

Interest rates (ii). There may also be a risk to the supplier from interest rate changes where there is a significant time lag between incurring costs and receiving payment. If this gap in cashflow is debt funded, it represents an increase in supplier costs which they may seek to recover from the client organisation, especially if there are any project delays for which the supplier blames the client. Our mitigation is to ensure that under the contract any risk to the supplier's cashflow is carried by the supplier.

Costs of change (i). Not only is change a commercial risk but a financial one as well. There is a well-worn saying that many suppliers "quote low for the bread and butter and make their money from the jam (changes)." A typical example is where day rates are quoted for a mix of staff to deliver the contract outputs, but any changes always seem to be only deliverable by staff at the higher grades and therefore at the higher rates and take much longer than seem appropriate. A real-life example of how change can increase costs for the unwary project manager happened on a software implementation contract. A minor change requested by the client was quoted by the systems integration supplier as requiring detailed customisation and needing 15 days of development and testing. A second opinion was sought from an independent expert from the originating software house. They reviewed the change request and even delivered the requirement by making a single small parameter alteration in under five minutes. Consequently, underspecifying our solution is a particular area of risk leading to higher levels of expensive change. We may need to use comparator solutions to confirm our specifications and experts to review the supplier's proposed resolution. Solution underspecification can occur either because we have genuinely not understood what the solution requires or because the supplier has quoted, accidentally or deliberately, for a solution which does not meet the contract requirements. The supplier then argues that the difference is actually a change (meaning more jam). We can mitigate this by including overall performance standards as a basic requirement in our contract in addition to, or even instead of, a very detailed specification.

Costs of change (ii). Any change to the original scope of the solution can attract costs. We can mitigate against "nice to have" and "wouldn't it be a good idea if" types of scope changes by being as difficult, obstructive, and downright unhelpful as possible when anyone wants these sorts of changes once the specifications have been agreed on and the contracts signed.[13] It is almost inevitable, however, that there will unavoidable changes required under the contract, and once again, as mitigation we needed to have built in a contingency fund to our business case to cope with it. Contingency funds can evaporate fast, and they need to be closely guarded. We should always try to ensure that only changes that are evaluated as "must have" under our MoSCoW priorities are accepted.

Default by the supplier. The supplier's failure to deliver is not only a risk to the time and quality deliverables of the project but also to our expenditure plans. Consequently, we always need to consider the potential impact on the project's finances if a contracted supplier company fails. There is a risk of financial loss when a supplier company defaults, especially if it ceases trading or has to file for bankruptcy. The exact rules will vary in different jurisdictions, but the effect on the project is the same. We have committed to a supplier to help us, and part way through the contract, they cease trading and can no longer fulfil their commitments. We are left having to find a substitute supplier and recover any funds paid out for incomplete delivery of goods and services. Major risks include, for example, when we have been obliged to make some payments in advance of the completion of delivery or when a company's trading is suspended and we have to find a substitute supplier at short notice. We can mitigate the risk of needing to try and recover direct losses from an insolvent company by negotiating a schedule of payments under our contract where we protect the client organisation by ensuring that all payments are made as late as possible in the supply process. Ideally, we would make no payments until at least after receipt of goods (and their transfer to our ownership) but preferably would delay until after any required commissioning and testing are completed and accepted. We may also have to bridge a gap caused by lack of service from the supplier until a substitute supplier can be contracted. In all instances of a defaulting supplier, we can expect our costs to increase, if only because we chose the lower-cost tender originally from the now non-performing or bankrupt supplier. Paying higher prices both for a stopgap and for a longer-term substitute supplier would usually be a case where we could consider a call on contingency funding.

Reputational Risks

The classic question is this: whose reputation will be damaged by an adverse event when a business sponsor (on a public project, often a senior politician) champions a project which fails to deliver expectations but does not necessarily fail to deliver on the terms of the contract? In the EU and AstraZeneca COVID-19 vaccine supply contract spat, it appeared that the highly publicised expectation by the political sponsors was for vaccine dose deliveries against a set schedule. The contract, however, reportedly contained 15 "best reasonable endeavours" caveats to cover the supplier in the event of problems in the development and manufacture of the vaccine; i.e., the onus on the supplier was only to make "best reasonable endeavours" to supply according to the schedule, and when the risk of lower-than-expected production became a reality, a very public argument ensued and the reputation of the leading individual in the EU client was damaged.

There are many examples where an adverse project event has caused a rippling outward of damage to reputations to both companies. Share prices can be hit and organisations can lose credibility, and to individuals, these effects are real even if not always fully justified. The adverse effects of huge cost increases, protracted contractual disputes, late delivery, or a poorly performing project, such as a new public-facing IT system, are keenly felt – often by the chief executive of the client organisation. The protracted inquiries into the UK Post Office's failed Horizon computer system meant that years later, the CEO lost her credibility and, despite having moved on to other roles, had to resign her non-executive directorships and even stood down as an ordained minister of the Church of England.

To mitigate against adverse and damaging publicity, in Chapter 6, we already identified that we need an emergency/crisis communications plan to be put in place. As the project progresses, this plan will need to be reviewed and widened to engage all players whose reputations may be placed at risk by any adverse project event. We also need to realise that even if the adverse impacts of an event may only be internal to the organisation and have no requirement for external communications, our emergency/crisis communications plan will still need to mitigate the loss of credibility of the project with senior decision makers in the client organisation and other stakeholders.

Step 3: Record Risks and Issues

An essential step in our process is the creation of comprehensive risk and issue records. For both risks and issues, there are key records that we need to open at the start of the discovery process and maintain throughout the life of the project. Almost identical records and registers are kept for both, although for issues, the assessment of risk probability is removed. We need:

- An individual risk or issue management record, which allows us to record each risk or issue, potential mitigations, and actions at an appropriate level of detail.
- A risk register and an issue register, which summarise the risk or issue records and form the basis of our reporting, usually to the project board.

The project manager is the owner of both the detailed records and the overall registers, although on many projects, the detailed work of maintaining current records and registers will be undertaken by dedicated members of the project support team, including a risk manager where one has been nominated.

There are software tools that can help maintain risk and issue records and registers; however, many projects will use spreadsheets which can be accessed widely by project team members and may even maintain some element of manual documentation. If this is the case, remember that the records and registers are the project manager's responsibility, and we should keep master copies securely that are only updated under our control. Examples of manually kept risk records and registers are shown Tables 8.1 and 8.2. These records and registers can be recreated relatively easily in spreadsheet and database systems.

An example of a Risk Management Record is shown as Table 8.1, importantly this sheet maintains a history of actions and risk assessments throughout the currency of the risk.

Table 8.1 Risk Management Record

Risk Management Record					
Risk name					
Risk reference no.		**Owner**			
Description					
Mitigation					
Risk creation		**Created by**		**Date**	
Probability %		**Impact** %		**Criticality**	
Action owners					
Action required 1					
Action required 2					
Action required 3					
Risk review 1		**Reviewed by**		**Date**	
Probability %		**Impact** %		**Criticality**	
Action owners					
Action required 1					
Action required 2					
Action required 3					
Risk review 2		**Reviewed by**		**Date**	
Probability %		**Impact** %		**Criticality**	
Action owners					
Action required 1					
Action required 2					
Action required 3					
Action completion		**Reviewed by**		**Date**	
	Name		**Signature**	**Date**	
Actions completed					
Project documents updated					
Risk/issue closed					

The review process may formally close a risk only when:

- The assessment of the **probability** has reduced to zero. This would happen when the point at which the risk was current has passed – for example, a risk that we may not be able to attract a suitable supplier is passed during or at the end of a successful procurement process.
- The assessment of the **impact** has been reduced to zero because mitigation measures have been successfully taken – for example, the impact of a mains power failure has been removed by the installation of a standby generator. This also applies to an issue, which can be closed only when all countermeasures have been taken successfully.

Closed risks and issues are maintained as part of the historical record of the project and for reference in case they recur. Risks and issues, however, can be opened at any time, and those assessed to be critical or urgent would normally be raised with the business sponsor as soon as practicable.

The risk register is essentially a summary report of the state of all current risks within the project. It is possible to use it to record the overall levels of criticality of risks of different types and of the project overall over time. Formal reviews of the risk register are usually undertaken as part of the duties of the project board and are often a standing item on the agenda for their meetings. A copy of a simple risk register is shown as Table 8.2.

Step 4: Evaluate Risks

When the risks facing the project have been identified, we need to assess them in order to understand what actions we need to take and when we need to take them. Our first step is to use the standards agreed on in our risk management plan as a yardstick for our assessment of each risk. This assessment is usually based initially on a qualitative set of assumptions about probabilities and impacts, but further research can sometimes identify metrics which can be used to refine our decision. Such metrics are always specific to the risk and may also be specific to the type of project and industry. Data to support quantitative assessments may be more available in some territories than others – for example, when assessing safety risks, workplace accident rates are heavily monitored and reported in the UK, but in other locations, they may not be as easily available.

There is, however, a further element that we need to consider when making our judgements about individual risks. This element is the current risk control environment. Our project will not be operating in a vacuum, and there may well be existing controls, possibly specific to the client or possibly as part of industry or national regulations, that will moderate both the probability and the impact factors.

Table 8.2 Project Risk Register

Risk Number	Risk Description	Impact Description	Probability 1 (low) – 10 (high)	Impact 1 (low) – 10 (high)	Criticality (Probability x Impact)	Mitigation Actions Needed/Taken	Status Open/ Closed/ Complete
Total Risk Score					0.00		
1.0	**Organisational and personnel risk**						
1.10							
1.20							
1.30							
2.0	**Quality of project output risk**						
2.10							
2.20							
2.30							
3.0	**Technical risk**						
3.10							
3.20							
3.30							
4.0	**Commercial risk**						
4.10							
4.20							
4.30							

Project Risk Register

	Project Risk Register						
Risk Number	Risk Description	Impact Description	Probability 1 (low) – 10 (high)	Impact 1 (low) – 10 (high)	Criticality (Probability x Impact)	Mitigation Actions Needed/Taken	Status Open/ Closed/ Complete
Total Risk Score					0.00		
5.0	**Financial risk**						
5.10							
5.20							
5.30							
6.0	**Reputational risk**						
6.10							
6.20							
6.30							
7.0	**Other risk**						
7.10							
7.20							
7.30							

As an example, consider a theoretical project that involves the construction of a 15-floor building. The construction process will require significant numbers of the project team working at height, and therefore there is a very real risk of a staff member falling and being seriously injured or killed.[14] The company with whom the client has contracted to lead the build company, however, has very stringent on-site safety policies, including special codes of practice, general and site-specific training, risk assessments, monitoring, and provision of fall-arresting safety equipment for staff working at height. There have been no reported incidents of falls on the company's sites in the five years since the safety policies were introduced, and they have also insurance arrangements built into their contract costs which will cover any financial compensation awarded for injury on site. Irrespective of the inherent levels of risk, or risk in the industry generally, our assessment has to be that the risk of injury or death from a fall is heavily mitigated by existing controls. Therefore, we would be right to judge it as being a risk with very low probability and having a financial impact on the project which would also be very low.

A suggested approach for the assessment of the two dimensions of risk, probability, and impact, is set out in Tables 8.3 and 8.4.

Table 8.3 Risk Probability Evaluation

Risk Probability Evaluation		
Probability	**Category**	**Comment**
Over 80%	**High**	The event is almost certain to occur without taking measures to reduce its probability. Measures to reduce adverse impact should be taken.
60%–80%	**Medium/ high**	The event is very likely to occur, measures may or may not exist to reduce probability, and measures should be identified to reduce any adverse impact.
40%–60%	**Medium**	The event occurs frequently, and there are established countermeasures to reduce probability. Research is required to identify if project failures have occurred because these were not taken or if they were not fully effective. Measures should be identified to reduce any adverse impact.
20%–40%	**Low/ medium**	The event may occur, but countermeasures to reduce probability and lessen potential impact are well known to be effective.
10%–20%	**Low**	The event has been known to occur, and there is some evidence of countermeasures having been taken elsewhere. Any increases in probability to be monitored.
Below 10%	**Very low**	Potential risk with little evidence of the event occurring elsewhere. Any increases in probability to be monitored.

Both tables generate a category of risk based on a subjective assessment of the particular risk in the context of our project against a set of broad definitions.

Table 8.4 Risk Impact Evaluation

Risk Impact Evaluation		
Impact	**Category**	**Comment**
Over 80%	**High**	The project will be unable to meet stated goals, and overall vision will not be achieved. Consequently, the project will be viewed as a failure and may be halted immediately.
		Except in limited internal projects, there will be consequent harm to the organisation's reputation and ability to fulfil its objectives and obligations.
60%–80%	**Medium/ high**	Project may not meet many of its objectives, including delivery to time, cost, and quality.
		Many of the anticipated benefits may not be achieved, the targets set out in the approved business case will not be met, and continuing the project will be seen as non-viable without major alteration.
		Except in limited internal projects, there may also be unavoidable harm to the organisation's reputation.
40%–60%	**Medium**	Some of the anticipated benefits may not be achieved, and the targets set out in the approved business case will not be met. Consequently, the project will usually be viewed as being only partially successful, and other measures external to the project will be required to meet the project's objectives.
		Except in limited internal projects, there will be a need to manage communications to ensure that there is no damage to the organisation's reputation.
20%–40%	**Low/ medium**	The project may miss some of its targets and may need replanning. There will also be some loss of benefits, and the business case will need to be reviewed to ensure that it is still viable.
10%–20%	**Low**	Some replanning will be required to reduce the impact and ensure that the project's targets can still be met within acceptable tolerances. There should be little impact on overall benefits.
Below 10%	**Very low**	Impact should be containable within project contingencies.

Our analysis of risks needs to allow us to compare and prioritise our actions in mitigating diverse risks with very different probabilities and different impacts. Essentially, we need a robust way of comparing how diverse risks will affect the project. Table 8.5 suggests a simple approach for scoring the probability and impact of a risk based on the levels set out in Tables 8.3 and 8.4.

Table 8.5 Probability and Impact Scoring

Probability and Impact Scoring					
Probability of Occurrence			Impact on Project		
Likelihood	Category	Score	Evaluation	Category	Score
Over 80%	High	10	Over 80%	High	10
60%–80%	Medium/high	8	60%–80%	Medium/high	8
40%–60%	Medium	6	40%–60%	Medium	6
20%–40%	Low/medium	5	20%–40%	Low/medium	5
10%–20%	Low	3	10%–20%	Low	3
Below 10%	Very low	1	Below 10%	Very low	1

Consider a risk in relation to the level of staff leaving our example IT systems replacement project for other jobs. Our current recruitment and onboarding processes will cope with a certain level of new recruitment requirements – say, based on 12% of staff resigning in a year. Using this figure, we set our probability of exceeding this level to medium and our impact assessment to very low – i.e., below 10%.

There is frequently no quantitative science in these allocated values. They are set solely on a relative basis, and there are many different ranges for scores that can be used. Scoring is not even strictly essential for effective risk management, and some approaches use narrative terms such as high/high or medium/low, etc., for describing the probability and the impact. Whatever means of assessing probability and impact are chosen, it is sensible to document both the category definitions and the risk assessments against them. These should be agreed on with at least the business sponsor and ideally also the project board by including them as part of the risk management plan before moving on to evaluating individual risks.

The final part of our assessment approach is to relate the probability and the impacts of each risk to each other. This is shown in Table 8.6 and uses the scoring suggested in Table 8.5 to develop what is often termed a risk matrix, which shows the level of exposure of the project to each risk. The risk matrix is, in fact, a genuine "heat map" which should highlight increasingly "hot" risk exposures and identify areas where action is required with differing levels of immediacy. Again, it is possible to use a heat map based on terms such as high/high, etc., but assigning scores to both probability and impact and then multiplying them also allows for the calculation of an indicated total risk being carried by the project or in individual risk areas.

Table 8.6 Risk Management Heat Map

Risk Management Heat Map			Impact Scoring					
			High	**Medium/ High**	**Medium**	**Low/ Medium**	**Low**	**Very Low**
		Score	**10**	**8**	**6**	**5**	**3**	**1**
High		10	100	80	60	50	30	10
Medium/High		8	80	64	48	40	24	8
Medium		6	60	48	36	30	18	6
Low/Medium		5	50	40	30	25	15	5
Low		3	30	24	18	15	9	3
Very Low		1	10	8	6	5	3	1
Key to Risk Exposure Scores								
Critical	Very high		High		Medium		Low	
70 and above	50 to 69		35 to 49		15 to 34		14 and below	

(Left side label: **Probability Scoring**)

In our example of staff turnover, using the scoring method suggested, we would score 6 for medium probability and 1 for very low impact, giving a score on the heat map of 6. This equates to low exposure.

It needs to be stressed that this approach is still a subjective view of risk. However, if we are consistent in our evaluations, the risk matrix can become a useful comparative indicator of both how the total risk carried by the project is changing over time and in which areas of the project risk is either increasing or being successfully dealt with.

Risk Evaluation – An Extra Trick for the More Sophisticated

Based on our premise concerning timeliness of action – i.e., that early interventions will give us greater scope for taking mitigating actions – there is an additional step that we can include as part of our risk evaluation process. This step will help us reduce an increase in our exposure to a risk by triggering pre-emptive mitigation actions. To do this, whilst we are undertaking our risk evaluation, we can set one or more "tripwires" as indicators of potential change in a level of risk exposure.

Setting such a tripwire means we can identify when a change in a risk's probability or impact level may *in the future* lead to a risk exposure increase above that which we can tolerate. The tripwire triggers an early warning in time for us to take prompt countermeasures to mitigate the risk before the exposure level escalates. These countermeasures, because they are taken early, may be smaller or have a wider range of options than if we acted later. A tripwire can be formulated as: "If indicator A increases by B%, then mitigation action C will be taken to avoid any increase in the exposure category." We can set multiple tripwires at different levels for the same risk, so a second level might be: "If indicator A increases by 2B%, then in addition to mitigation action C, we will take actions D, E, and F to avoid any increase in the exposure category."

Let's pick up again on the example of our risk of staff turnover exceeding 12%. When we monitor the risks, we discover that during the year, we have had 17% of staff resign. The effects of this increase in resignations are as follows:

- Our recruitment team is struggling to find, recruit, and onboard an additional 5% of new members for our project team above that which we originally thought would be acceptable.
- We are currently running different parts of the project at lower than optimum staffing levels, incurring either additional overtime/contractor costs or facing slippage in completion of some tasks. There may be knock-on effects which could adversely impact our final delivery date for the project and require us to call for additional funding.

Our risk review of the staff turnover risk now concludes that:

- The probability score of the risk has risen to our top category high with a score of 10.
- At 17% turnover we are substantially above the 12% original tolerable risk level. Consequently, we have had to raise our impact category to low/medium, which has a score of 5.
- Our resulting exposure score to the risk has risen to 50, which is in the very high category, and mitigation action is now required.

Consider, however, if instead of relying on a review of our staff turnover risk when the annual figures became available, we had set a tripwire based on the level of resignations trending above 1% per month for three consecutive months. This increase would have been flagged up by our reviews of risk during the year. This would have given us time to investigate the reasons for the resignations rising and to put in place mitigating actions to retain staff or increase recruitment activity long before the exposure to the project of the staff losses increased to a more critical level.

Step 5: Prepare Treatment Action Plans

When treating a risk or an issue, we are essentially spending resources to increase our levels of certainty of achieving a favourable outcome for the project. As we increase our certainty levels, the costs usually increase – and can sometimes increase exponentially. In consequence, before deciding upon any course of action, we need to establish the point at which the benefits achieved by our treatment, or combination of treatments, will represent the best value for money. In many cases best value for money will be self-evident, but we may often have to recommend choosing between nuanced mitigation measures and potential levels of benefit. In our staffing turnover example, we could consider increasing the core staff on the project, and hence increasing costs, to reduce both the probability and the impact of not having enough staff. Another option might be to outsource some key areas of work, transferring our staffing risk to a supplier. A third option might be to put in place arrangements for the short-term supply of staff and wait until the risk actually matures into an issue before taking action. All these strategies might achieve a similar benefit of reducing the risk of an adverse outcome, but with different costs to the project.

We therefore need to fully understand the range of treatments that are possible, identify those that have the best balance of costs and benefits for the project, select those that we are going to use, and prioritise them into a treatment action plan for each risk or issue. Each treatment action plan should answer all of our seven standard questions, including:

- **Why** do we need to take action on this risk or issue?
- **What** treatments have been authorised?

- **Where** is action required?
- **Who** will be responsible for taking and monitoring the actions?
- **How** will progress be reported?
- **When** will actions need to be completed?
- **How much** funding have we allowed to resource the change?

Actions identified in the treatment action plan may also need to be treated as formal project changes. This means that, in addition to being included in the risk and issues registers and on the risk or issues record, the actions need to be fed into the change management process as change requests. This means they also need to be included in updates of the main project plan or stage plans. This documentation and its subsequent authorisation are critical, especially where one agreed-upon treatment action plan may affect more than one identified risk or issue and needs to be prioritised along with other change requests.

Consequently, our next step in the risk and issue management process is to identify if, how, and when we need to take action on each risk or issue, based on our evaluation of the exposure of our project to each one individually. In assessing risks, as opposed to issues, there is an extra process that we need to take in order to do this, which is to set out the criteria we will apply to evaluating each risk exposure. Ideally these criteria should already have been included and agreed on as a part of the risk and issue management plans in step 1. They should reflect the client organisation's appetite for risk and the ability of the project to plan and complete the mitigation actions. If not, these criteria need to be clearly established before we review each risk and decide:

- Can we tolerate this risk exposure?
- Do we need to take a mitigating action to treat this risk?
- What are our priorities for action across the whole range of risk mitigation actions?

A set of levels of exposure to each assessed risk score was suggested in Table 8.6, and Table 8.7 is an example of how we could use the level of exposure to make a graduated response to each risk. Again, this is a subjective approach, and the criteria to be used are a matter for agreement on a project-by-project basis.

Table 8.7 Risk Exposure Action Required Assessment

Risk Exposure Required Action Assessment		
Exposure		Action Required
Score	Category	
70 and above	Critical	**Action required immediately** Risk requires immediate action to reduce exposure. Plans should be drawn up, agreed on by the business sponsor, reported to the project board, and implemented without delay. Each action should be assigned to an identified individual with a specific time frame for action, quality target, and budget. Project manager to monitor and report progress on actions taken to reduce the risk exposure to business sponsor on a daily or weekly basis. Project manager to update project plans to include the new activities.

Risk Exposure Required Action Assessment		
Exposure		**Action Required**
Score	**Category**	
51 to 69	**Very high**	**Action required in the short term** Risk requires short-term action to reduce exposure. Plans should be drawn up, agreed on by the business sponsor, reported to the project board, and implemented without undue delay. Each action should be assigned to an identified individual with a specific time frame for action, quality target, and budget. Project manager to monitor and report progress on actions taken to reduce the risk exposure to business sponsor on a weekly basis. Project manager to update project plans to include the new activities.
35 to 49	**High**	**Action required** Risk requires action to reduce exposure. Plans should be drawn up, agreed on via an expedited process involving the project board, and implemented. Each action should be assigned to an identified individual with a specific time frame for action, quality target, and budget. Project manager to monitor and report progress on actions taken to reduce the risk exposure as part of project board reporting. Project manager to update project plans to include the new activities.
15 to 34	**Medium**	**Limited action and monitor** Mitigation actions to reduce risk exposure should be identified, evaluated, and agreed on with the project board. Exposure should be carefully monitored by the project manager, and any increase which would place it in the high-risk category should be drawn to the attention of the business sponsor and the project board.
14 and below	**Low**	**No action yet – monitor** Risk does not need action to be taken yet, but exposure should be monitored as part of the regular reporting processes.

Such an assessment process is not necessary for issues, which all default to an "action required immediately" status.

We can now complete our example of evaluating the risk of staff resignations. We had a risk exposure score of 6, which fell into the low category. Consequently, under our standard, we need to take no action yet and just have to monitor the risk exposure going forwards.

When identifying risks and issues, we may have also identified some potential treatment methods. In taking any of the initially proposed treatment methods forwards, we will need to validate, refine, and possibly expand our range of options before completing an options analysis to decide on the most appropriate treatments. This process may need us to engage in a variety of different consultations, including with experts in the relevant area, our project team members, the project board, and significant stakeholders.

Our treatment options are usually grouped under a few broad headings, which also form a preferred list for actions to be taken. It runs as follows:

- Avoidance.
- Mitigation (i) – Probability reduction.

- Mitigation (ii) – Impact reduction.
- Transference or sharing (partial transfer).
- Tolerance.

The first two treatment options apply only to risks, whilst the last three apply equally to issues.

Tolerance. Although shown as the last option, risk tolerance is both our first option and our last option to treat a risk. It is our first option because if we consider that the exposure to the risk can both be managed within our project team resources and remain within our acceptability criteria, then we should tolerate the risk. In our original staff turnover risk example, we had decided that our existing processes for recruitment and retention would allow us to tolerate a 12% rate of staff churn but that we still needed to monitor the risk to ensure that it did not rise beyond that 12% tolerance criterion. Tolerance is also our last option because, after taking all actions possible, some residual risks may still remain, and we may have no other choice but to tolerate them. On an IT change project, despite everything that we have put in place to mitigate the risk, there is a chance that a new computer system will not become operational on time. On a factory relocation project, whilst we can take some countermeasures, we cannot fully protect against a site getting hit with a once-in-five-hundred-years flood event.

Similarly, we may have to tolerate an issue for which we have no viable treatment options. To take a particularly dramatic example, consider the suspension of working and the global shortages in supply of many commodities and components following the start of the COVID-19 pandemic, which caused many projects to be delayed all over the world and to be hit by increased commodity and shipping costs.

Risk avoidance. If a risk cannot be tolerated – i.e., it exceeds our tolerance criteria – then we should investigate how we can avoid it entirely. Under the heading of risk avoidance, we have a number of different tactical responses which we can employ:

- **Finding an alternative approach.** Building resilience into our project may mean that we can completely avoid the risk. For example, centralising manufacturing and distribution at riverside site A has a once-in-ten-year risk of disruption due to potential flooding,[15] but choosing site B, which is at a higher level and above the flood risk area, would avoid the risk entirely but may cost more. Other alternative approach strategies include selecting an existing solution, rather than building a new one, and using existing tools, methods, and techniques for delivering the project rather than implementing different ones.
- **De-scoping.** We could decide to reduce our scope to avoid the risk. For example, if granting the public new online access to our data, as envisaged in the project scope, could give rise to the risk of a security breach, we could remove the public access from the project's scope. To take another example, if there is a risk that the costs of the electrification of a railway line will rise too high, we could decide to shorten the length of track being electrified.
- **Improving project capability to deliver**. If the risk is identified as a lack of capability within the project to deliver with the existing level of knowledge and expertise, we can avoid the risk by dealing with that lack of capacity. For example, a risk arose on a very complex software project that the new server security could not be set up and maintained by the existing staff. The risk was avoided through a mix of bringing in an expert contractor for the initial set-up and in parallel by training and mentoring existing staff. In another example, a rail electrification project was originally to be delivered using a single new type of machine to install

all supports for the overhead lines, but in operation it was found to be too slow and supports had to be installed using other methods as well.

- **Improving project knowledge of solution requirements and possibilities**. The better our understanding of the required solution, and the better our ability to communicate to ensure that this understanding is shared and will meet stakeholder expectations, the more we can avoid the risk of a poorly implemented solution which is not well received by the stakeholders. Included in this category are techniques such as the use of prototypes and improved user engagement with design, development, and testing. A simple example of this would be, in an IT system, ensuring that the system's users first get the opportunity to set out layouts of the screens that they will use, identify the navigation between different functions, and are then able to test these ideas through prototypes. Another example is in construction, where the use of computer-generated images, virtual reality models, and "fly throughs" presents enhanced opportunities to validate the designer's ideas against user requirements at an early stage.

Mitigation (i): risk probability reduction. If we cannot either tolerate or avoid a risk, then we should seek to proactively mitigate our potential exposure to it if should it occur. If we cannot avoid a risk entirely – i.e., reduce its probability to zero – we may be able to take actions early which, whilst not avoiding the risk entirely, will reduce the probability of it occurring. All the strategies in the risk avoidance category remain applicable, but our expectation of the result is less. Examples of actions in this category from our previous discussions could include using a multiple supplier strategy or raising the level of the new building on our riverside site A to the level of a once-in-a-250-years flood.

Mitigation (ii): impact reduction. Impact reduction actions are highly task specific and can be taken early – i.e., alongside a risk reduction activity – or triggered if an issue arises. Proactive actions could include:

- Avoiding downstream delays by cross-training staff to enable cover if a key member of staff leaves.
- Completing manufacturing and delivery of required supplies as early as possible and accepting that we may have to draw on contingency funds to cover any additional warehousing, insurance, and warranty costs.

Reactive actions could include:

- Immediately searching for additional staff from whatever source as soon the departing team member gives notice.
- Finding supplementary supplies if one supplier seems to be not meeting delivery schedules.

Risk transference. If we cannot tolerate, avoid, or mitigate a risk, we may seek to transfer the risk to others. Risk transference has several tactical approaches, but our first issue to consider is how much risk is to be transferred and to whom. Sometimes risk can be shared between the client and the supplier or in some cases between several suppliers. A pragmatic rule to follow regarding transference is to consider transferring the risk to the party that can best manage it.

- **Contractual arrangements.** As with any other contract terms and conditions, we can negotiate who will bear the impact of any risk.[16] So, if we contract with a supplier to manufacture and deliver goods on a certain date and they fail to do so, the agreed-upon costs to us of their failure will be repaid to us by the supplier.[17]

- **Insurance.** Risk can be transferred to an insurance-providing third party by either the client or the supplier. Such insurance arrangements are often found in engineering and construction projects. Involvement of an experienced insurance company can also increase the expertise in risk management available to the project.
- **Sharing between multiple suppliers**. Supply arrangements can include multiple parties delivering through joint ventures, partnerships, and other co-operative arrangements. These arrangements may give additional opportunities for risks to be transferred or shared which the project manager should explore.

Step 7: Assess and Report Changes in Risks and Issues

Included in the risk and issue management records are opportunities for us as project managers (or other designated individuals) to undertake a review of progress on the implementation of the agreed-upon actions. The timing and number of such reviews will be action specific, but they are critical to ensuring that we remain in control of changes and are able to report properly on both them and their impact on our overall delivery of the project. There is also provision in the record for the project manager to sign off on the change as being completed and the documentation as having been fully updated, another essential control.

Risk Reviews

In our allocation of responsibilities to the project board, as shown in Figure 8.1, their role included reviewing the risks and issues registers. Logically, there are two different types of review that we can expect the project board to undertake:

- A review of progress on the identification, evaluation, and treatment of risks and issues as part of their oversight of the management and progress of the project at each project board meeting.
- As part of an end of stage review, or a comparative point for agile projects, to conduct a more detailed examination of those risks and issues that have been treated and also those risks and issues that may affect the delivery of the next stage.

For some larger or more complex projects, it is useful periodically to reengage with the attendees from the original risk discovery process. We might perhaps call another facilitated workshop to undertake a formal review and update of the project risk register, prior to submitting it to a project board meeting.

Communicate, Communicate, Communicate

As with all our reporting, we should seek to establish a "no surprises" culture and engage in two-way communications with our stakeholders throughout the project about risks and issues. This engagement improves our ability to take their views into account when formulating our solutions and courses of action, leading to smoother implementation of more acceptable outcomes from our project. This engagement applies equally to our risk and issue management activities. One of our objectives should be to bring our stakeholders along with us throughout the process reducing the scope for subsequent disagreement and misinformation.

Step 8: Update Project Documentation

Whilst shown as a separate step, documentation updating and its consequent version control are essential elements of the process for risk and issue management and can be both time consuming and repetitive. On larger projects the oversight of this documentation can usefully be given to a nominated member of the project support team; otherwise, we need to undertake it ourselves. This does at least have the advantage of ensuring that we are fully connected with every nuance of the management of every identified risk and issue.

Risk Awareness and Risk Management

Now it's time for another of our guest contributors and a chance to look at some major lessons drawn from a real-life project. As a highly experienced project manager working across countries in northern Europe, this contributor advises major corporations, national, and regional governments on improving the ways they manage projects, especially construction projects. Many of his comments on risk management reflect the journey we have taken throughout this chapter, and his experiences underline the lesson that we should leave no risk, however small, unexamined. In this example, like in our fictional bio-energy from effluent case study, the risk may not have seemed that large initially, but life was to prove otherwise . . .

Introduction

Risk management is an established method for delivering projects on time and to budget. Identifying risks at an early stage in order to start mitigating risk, long before the risk can arise, is a central element in project or programme management. Using risk management enables the project to define risk budget and then to apply it in a highly controlled and rapid manner.

However, implementing risk management is not a substitute for fostering the risk awareness that an entire project team should have, and a risk register does not solve the challenges posed by the risks that the project will encounter. It is common practice for a risk manager to:

- *Identify the most important project risks in a series of small sessions with parts of the project team.*
- *Sort and rank the risks in isolation from the team.*
- *Report risks via the risk register and in project board meetings and project reports.*

Very often only the "top ten" risks appear in the risk report, together with details of the thinking of the project and risk managers about them. Mitigation strategies, however, are often only developed when it is certain that the risk will arise.

After decades of cost and time overruns in projects, many more projects now recognised the need for a more robust budget and are delivered within it. This overall project budget may also include the calculation of a separate risk budget, normally focused on meeting the "top ten" risks faced by the project. But there is always a danger in the creation of a robust (that is to say, higher) budget that this actually entices the project to spend it. There is a wider economic issue in the public sector if planned project budgets are now becoming too high and spending is greater than strictly necessary. As project financial demands grow in size, the overall budgets of the individual German states are struggling to meet the increased demands of fewer projects whilst trying to avoid having to completely scrub others.

Real risks also arise mainly in tiny formalities such as applications, obligations, permits, interfaces with third parties, public participation, etc., with low costs associated with managing them but a high risk of delaying entire projects for years. Risks on this small scale cannot be covered by a lone risk manager and a risk register. The entire project team needs to be aware of all these challenges, and we need to take action well before these risks can occur.

The Case of the Common Bat

There is an additional problem if risk management concentrates on the "top ten" or the "10 to 30" main risks in that it may not always cover the actual risks. For construction projects, the big risks are related to the ground, construction failures, and handover, but these risks are strongly covered by the consultants' and the contractors' experience and by established mitigation processes that are usually already listed in the tender or tender book.

This is a real-life example of the effects of a small-scale risk, and although it involves the impact of having to meet detailed legal requirements, at no time is there any criticism of any regulation, obligation, or law. These are all in place to protect nature and our way of life. As you will see, there are many of them, and often they are interdependent.

In Germany one such regulation requires the resettling of bats when any construction crosses a bat's territory. The actual handling of bat resettlement is simple and inexpensive. You have to "scare off" the bat and offer it a new nest site close by at a cost of about €32. This resettlement is often done in co-operation with local conservation organisations to make sure that it is done correctly.

A major energy project in the northern part of Germany was tasked with the construction of a 200-kilometre (120-mile) high-voltage power line, which is partly underground. The project established a risk management strategy and engaged a risk manager who developed a risk register and estimated the risk budget. Together with the project leader, ten main risks were identified and closely monitored. Nine of the risks concerned the power cable, and one concerned the converter station.

At the very end of the planning approval process, after a lot of natural, archaeological, and soil investigations, the project aimed to start some of the preparatory work on the cable corridor. It was mid-February, and the first task was to cut down all the trees on this cable corridor. Another German regulation only permits tree felling during a winter tree-felling season, which runs from 1 September to 1 March. But it was quickly found that common bats were living in some of the trees, and so the work had to be paused. In a series of frantic last-minute actions, the project tried to convince the authorities to grant an exception regarding bat resettlement so that the tree felling could be completed during the permitted period, but the authorities insisted that the project had to provide the alternative nesting places and allow time for the bats to resettle before the trees were felled. Consequently, the project lost the winter felling season, and then the project learned that the next bat resettlement window would not be until the coming December-to-February period.

The project accepted this delay, identified it as a mistake, and no further risk investigations were done. But, in December, they learned that the €32 resettlement nesting bat box had to be built from a special wood and would need an independent certificate (a result of many breaches of this obligation in the past). The main supplier of this wood for the boxes was a Finnish company, but they had sold out. Once again, the project tried to get an exemption from the authorities, but this was rejected, and the project lost the felling season for another year.

This situation was accepted by the project as a funny mistake, and again no further risk investigations were carried out. Now, two years on, with the stylish, special Finnish wood €32 nest boxes ordered, the project learned that they needed to apply for access to private land to place the boxes on the other trees near the cable corridor. Access negotiations began in another hectic operation, but the landowner followed the formal, lengthy procedures for granting access, and consequently, the project lost the felling season for another year.

Finally, three years late, the bats were relocated and the felling work was carried out at the last minute, just before the felling season ended. The impact of not controlling a small-scale risk was not only the delay to the project but tens of millions of Euros in overrun costs. The mitigation costs if the risk had been tackled at an early stage would have been €32 per bat box and some cups of tea for the local nature organisation.[18]

The entire company learned from this experience, and it has changed its top ten risks on such projects. One risk is still about the ground, and one is about the cable, and the other eight are now environmental and licence/permit application risks.

A Method of Detailed Risk Management to Establish a High Level of Risk Awareness

A number of major projects have adopted an approach to risk management which stresses that it is important to identify most risks in detail and mitigate them at an early stage. But it is equally important to ensure that each team member is aware of all risks and is willing to take mitigation measures at any time in advance of the risk event occurring. This approach has a positive side effect, in that every member of the team needs to read the project baseline and understand the project's scope, which often does not happen in real-life construction projects.

*   **Step 1** is to create a detailed baseline that describes all tasks in as much detail as necessary to deliver the project. The baseline should focus not only on the construction stage but should include similar levels of detail, and effort, for the planning and approval periods. This is not always easy today as there can be hundreds of applications for consents, often from different bodies, different obligations, and the associated interfaces between them. Sometimes major projects try to make it easy by categorizing and compressing tasks, and searching for the magic tool that manages everything together. But we still have to go step by step when we build the pyramids, and if we miss one, we will fail.*

*   **Step 2** is to go through each baseline task chronologically with the entire team and identify and describe each risk that has been discovered in each task. Experience shows here that the risks in the construction tasks are the ones that are easiest to recognize and mitigate. Much more challenging are all the agreements, approvals, obligations, and interfaces in the planning period. In every project, of course, we hope that these processes will work well, but these processes are often not as logical as the construction tasks. The approval processes may be somewhat defined, but nevertheless we have to agree on many details and changes with humans. This means that we very often have a misunderstanding about what is needed, leading to more discussion, and perhaps to us learning that in the end, the wrong person has been contacted because they are not the one able to decide the issue. This becomes much more critical when the project needs to adjust approvals between different authorities and stakeholders. The human risk factor of these interfaces is massive, as the progress of the project is completely dependent on the personal statements of key actors even without any applicable regulations and processes.*

*   **Step 3** is to go through all the identified risks again and again with the entire team and discuss and describe risk mitigation in each task, including assigning responsibility for leading the mitigation. This step is the beginning of very proactive thinking on risk. The detailed review allows the team to start simple and cheap mitigation measures in the very early stages of planning, or simply remove risky tasks from the scope, and can also be used to optimize the project and start early value management. At the beginning of these workshops, the task looks endless, but it will accelerate once the team has got the feel and the rhythm of the process.*

*   **Steps 4 to 6** follow the general risk management approach. The risk manager, the project manager, and the programme control staff add the risk-related costs to each task, define the probability per task,*

and finally evaluate the overall risk probability and cost approach using methods such as Monte Carlo calculation.[19]

Step 7 *is the final determination of the risk budget by the steering group or the project board and should be based on the results of a special joint risk workshop for the project board members. In this workshop, the 30 to 40 top cost and time driving risks will be discussed again and then ranked. This determination is ultimately the result of the personal experience of the members of the steering group or project board.*

These seven steps are very intense, and the project team could be blocked for a reasonable time from other tasks, but the benefits to the project are much greater than the effort involved. Team risk awareness will be at its highest possible point, and everyone will understand the project scope. In the mitigation development, every team member will understand responsibilities and related actions and processes based on real examples. This will overcome the problem that many projects, be they simple or complex, still trust that every team member understands them fully by quickly reading some documents on their morning commute.[20]

Research and experience show that most of the risks that could be identified in these early stages will seem simple. If a project identifies them too late, or once the risk arises, a lot of effort and personnel must frequently be invested to mitigate them. The results are often unsatisfactory, a foul compromise about what is left to be done.

Major projects, such as Berlin's new Brandenburg Airport Willy Brandt, have suffered greatly from an approach that ignored looking into the details of each task and identifying the related risks. The progress of the Berlin Airport project was halted hundreds of times because risks "emerged," often in a chain reaction with other associated risks. Hundreds of emergency meetings and workshops were held, often with the result that mitigating one risk was seen as triggering other risks elsewhere. This was one reason why the project was delayed by nine years and overspent by several billion euros, most of which was to cover extended time- and overhead-related costs. The terminal itself is still essentially the same as was planned in 2004. Good people adopted a culture of "see what happens first," and the impact of this "risk-ignoring approach" on the project was enormous. There was low motivation to bring the project to a successful close, and any good ideas in value engineering or simplification of construction methods were stopped because of another risk appearing.

Reflections

Choosing your project, or using the bio-energy from effluent case study:

- Create an agenda and an attendee list for a one-day risk discovery workshop.
- Identify ten risks that may arise from such an event and categorise them into broad risk types (e.g., finance, commercial, technology) and categorise their probability and their impact.
- For each risk, conduct a bow tie analysis to identify its causes and its effects upon the project.
- Complete the risk record for five of the risks, including detailing the treatment required, and prepare an overall risk register.
- Prepare a prioritised risk treatment action plan for each of the risks.
- What actions would you have taken to mitigate against the problems arising from an issue similar to the Case of the Common Bat?

Notes

 1 Moltke was born in 1800, technically the last year of the 18th century, and his is reportedly the only surviving recorded voice of a person born in that century.
 2 Almost a direct quote from a project board member – but obviously things were arranged so that they never got first choice of the biscuits again.
 3 Running a workshop in this way often means that the project manager has to personally put up the prize for the winning team. The choice of prize, is, of course, culturally sensitive, but in many settings, either champagne or chocolate seem to be great motivators.
 4 How many project managers would have responded positively to an off-the-wall suggestion in a risk workshop in 2019 that precisely because we had not had a global pandemic for a century, we were now overdue and the risk ought to be recorded?
 5 Spoiler alert – watch out for the example of *die Fledermaus.*
 6 If, by now, you do not feel that your loss would be a major blow to the project, you must feel that you have mitigated this risk superbly well. Feel free to take a month's holiday and relax. After all, nothing could possibly go wrong that they can't fix without you . . .
 7 This is not the time for false modesty.
 8 Please repeat: Who, and Why and What and Where and When and How, and How Much.
 9 In fact, during the later stages of one project, the business sponsor decided that its management was now running so well, and the plans and processes were so clear, that they could save money and have the business manager take over the project manager's role as well. Obviously, the risk mitigation had been too good. There wasn't even an invitation to a launch party. No, I'm not bitter. No, really, not bitter at all . . .
10 Note the addition of the phrase "contractual terms" here – more about these in Chapter 10.
11 There is always a risk that we fail to translate the requirements of the project clearly and unambiguously into the conditions of the contract. This can make two sets of lawyers very happy when the different parties start to pursue each other in court.
12 For some reason industry-specific rates invariably seem to be higher than the general rate of inflation.
13 In fact, we say, "That seems to be a great idea, please could you do a small cost benefit analysis of the change and I'll try and get it on the next project board agenda" or "I'll put that forwards for inclusion in phase 2."
14 In the UK during 2019, 38 deaths were reported from falls from height on construction sites. Other countries may have very different accident rates.
15 Never forget that in 2015, when Storm Desmond flooded a biscuit factory in Carlisle, it created a UK-wide shortage of ginger and other favourite biscuits, thereby no doubt adversely impacting many project meetings.
16 Experience of negotiating contracts from both sides indicates that a supplier will always seek to add in to their price cover for any risks being transferred.
17 Lane rental is a mechanism to deal with supplier delays which both transfers the risk of delays and avoids claims for damages under the contract. This system was originally designed for highways construction and requires the supplier to pay a rental sum for the construction site, which is both a bonus for early delivery and a penalty for each day that contract completion is delayed.
18 I would obviously have provided biscuits as well . . .
19 A mathematical technique for modelling different outcomes based on assessing a range of different input values for uncertain factors.
20 Or, if they are working from home, then over breakfast immediately before their early morning video call to discuss the documents.

Bibliography

Burns, R., 1785. *To a Mouse on Turning Her Up in Her Nest, with the Plough.* Kilmarnock: John Wilson.
International Standards Organisation, 2022. *ISO 31000:2018 Risk Management – Guidelines* [Online]. Available at: www.iso.org/standard/65694.html.
Moltke, H.v.D.H., 1995. *The Art of War.* Toronto, ON: Presidio.
Wolmar, C., 2009. *Blood, Iron and Gold How Railways Transformed the World.* London: Atlantic Books.

9 Managing Change

Coming to Grips with the Real World

Aim of this chapter: To introduce the importance of change control and configuration management and to understand how changes may be managed.

Learning outcome: Understand and apply the basic principles of controlling change during the life of the project and be able to use simple tools to support doing so.

What Is Change?

In project management terms, change means any alteration to what we are seeking to achieve or how we are seeking to achieve it – or, indeed, often to both simultaneously. We have already looked at risks and issues, which may generate demands for change. Other demands for change that we face can arise from any or all of the following principal causes:

- Comprehension of the project.
- Changing internal demands.
- Changing external demands.

Comprehension of the Project

This category can be thought of as being internally generated change and may be unavoidable. Often, at the outset of the project, it seems that no one really fully understood the problems and opportunities involved in delivering the project in the first place. As the project progresses, we can see both our destination and our route to achieving it more clearly. A clear example of this lack of comprehension of what the project needed to achieve was in a bespoke IT development project bringing together massive amounts of data every day from more than a dozen different sources. When preparing their specification of requirements documents, the client *assumed* that these data providers would agree to submit data in a single common format. The contract for the development and operation of the new system was let on this basis. During the detailed design process, it became clear that that submission of data in a standardised format was not technically possible and the client's assumption was wrong. Major changes to the project were necessary not only to research and develop sets of programmes to convert data from each provider into a common format that could be recognised by the system but also to provide additional computer hardware to ensure that the required multiple data conversions operations could be completed each day.

DOI: 10.4324/9781003405344-10

Changing Internal Demands

These often come in the form of "While you're doing this, you could do that as well." In practice, most projects will experience some kind of pressure for such growth, which we call scope creep. Scope creep is not always a bad thing, as it can show that our project team is flexible in its solutions, responsive to change, trying to improve project outcomes, and seeking to innovate. If unmanaged, however, it can also derail a project. The opposite, scope reduction, also occurs, but normally when costs, time, or risks begin to escalate to an unacceptable level.

Changing External Demands

The demands that the outside world places on the project may change or may not be what was expected, and change must then occur to address them. From project inception to project delivery can be a protracted period. The world does not wait for us, and changes may be forced upon the project. You will recall our site visit to meet the newly sacked project manager of a public transport fares project. This is an example of change generated by changing external demands. In this case the government had changed the rules surrounding individualised ticket account cards, thereby creating an issue which resulted in a project change.

Another example of imposed change could be to do with third-party approvals. A scheme for constructing a development of 150 houses included a surface water drainage scheme designed to protect the development and nearby houses from a one-year-in-ten risk of flooding. However, when the scheme was submitted for local authority approval, said approval was only given subject to increasing the flood protection level to cope with a one-year-in-one-hundred flood event. This change in protection level, in turn, required redesigning the surface water drainage and carrying out more extensive groundworks than originally planned.

Why We Need to Control Change

Our responsibility for the effective control of changes is irrespective of who originated the demand to change something that the project is doing. This means that we need to remember to expressly include the client's management structure as one of the sources of change. Experience shows that even small changes can have consequences across the project, and an aggregation of small changes may materially affect the entire project and its continuing validity. Consequently, our exercise of control over changes is critical to maintaining successful project delivery.

Whilst change on projects is inevitable, making any particular change is not something to be undertaken lightly. We need to consider every request for change and subject it to rigorous investigation, evaluation, prioritisation authorisation, control, and monitoring. These processes will largely mirror those we have already discussed in relation to other areas of our role and are perhaps most similar to those for risk management.

Figure 9.1 shows that our basic procedure for controlling change is to:

- Define and agree on our process for managing change in our particular context – i.e., for this project within this organisation and within the parameters that the project will have to operate. The output from this step is a change management strategy highly tailored to the needs of our project.

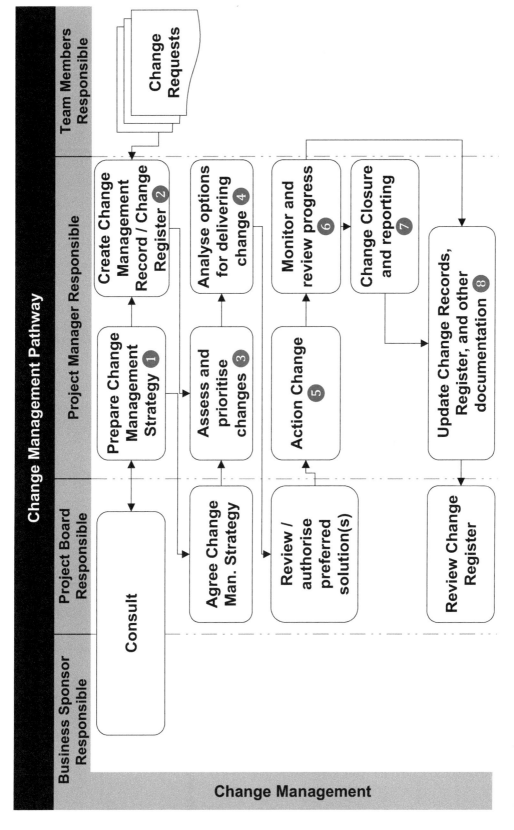

Figure 9.1 Change Management Pathway

- Capture all requests for change from across the project and stakeholder community. As with risks, these change requests may be generated by anyone engaged in or with the project. There is always a clear and present danger of change being defined and agreed on without it being properly captured, and making the request process as easy as possible will help avoid this.[1]
- Define why a change is needed, assess both why it fits within this project and its impact on it, and prioritise the change for delivery.
- Identify the options for both delivery and potential solutions and any differences that the various options may have in their impacts on the project.
- Implement the preferred option for delivering the change.
- Monitor progress of the implementation, including reviewing the impacts of the change across the wider project.
- Maintain updated records of the change and its impacts.
- Prepare a change closure report, including an assessment of the effectiveness of the change and any recommendations for further action.

Time to slot back into your role as project manager of our world-leading bio-energy from effluent project.

It's been a long week. Not only have you had to finesse that incident with the local politician at the Wadi pilot site, but two key staff resigned, and then you had to prepare the monthly project board progress report for next Wednesday's meeting. To cap it all, this afternoon you had a long and difficult emergency contract meeting with the company manufacturing the trailers for mounting the mobile units at the next three pilot sites. They seem to have interpreted the specifications wrongly, and you may not be able to mount all the different processing units on their trailers. It looks likely that lawyers are going to have to be involved on Monday. With all that to deal with, time was getting on this afternoon before you even sat down to do the weekly progress report. Now it's finished, and at least most of the Friday night traffic had cleared before you left the office.

It's already late in the evening as you arrive home. You're looking forwards to a quick dinner, a long shower, and a good night's sleep. Sadly, it's not to be. As you walk through the door, your phone rings. It's the boss.

As ever there's neither introduction nor opening pleasantries.

"I've been speaking with the CEO and the board chair. They've been talking about your approach and think that it would be a good idea if you could do a ninety degree turn on your planned rollout. The CEO and the chair think it would be better to showcase the company and the capabilities of our new technology by building a single full-scale plant on each continent, with the aim of delivering immediate benefits and convincing governments to contract with us to build full-scale sites servicing whole cities. We'll put in state-of-the-art executive visitor centres, offer decent hospitality, and fly people in and out as needed. Of course, we'll have to drop your plan of placing small independent trailer-based units in multiple countries."

In one short conversation, over a year of project planning, approvals, and development has been upended. Your first instinct is to say "No chance," but you're also sensitive enough to have noticed that the boss referred to "your planned rollout," "your approach," "you doing a ninety-degree turn," and "your solution." All this despite everything being agreed on and signed off by the boss as business sponsor, the project board, and even the main board of directors. So, instead, you breathe deeply, count to five, and reply:

"That's interesting. I'll set up a team leaders' meeting first thing on Monday, and we'll examine in detail what this radical change would mean for the project. In the meantime, and as it's now formally the weekend, I'd say we need to get the team leaders' full attention and stress the importance of this development to them. I think that it would be best if an email came directly from you, telling them that you've asked me to arrange a meeting to discuss it. It might also be good if you could include a formal note of your conversation with the CEO and the chair."

Nicely done! You're learning how to manage the boss; not only have you regained your weekend, you've just promised to arrange a meeting when you get in – not necessarily to have one immediately. You've pushed some work back on the boss over the weekend and got him to confirm in writing the request to change the project's direction. Congratulations!

Solutions will naturally immediately start forming in your mind. One possible way to approach this would be to examine how the new proposal could be integrated with the current plans, essentially bringing forwards the first city-scale plants. But how would they be funded? Getting the first large-scale sites agreed on, built, and running will take much longer than the current plan, so perhaps another idea would be to look at the proposal as a parallel project, which will be informed by the current one.

And now, stop. STOP! These thoughts are for next week. First, you need to recharge. You need that dinner, that shower, and that good night's sleep. You need to enjoy your weekend so you can return refreshed and revitalised.

So, come Monday morning, start by answering this question:

What do you feel will need to be done first to prepare to deal with this major request for change?

Project Baseline – What Are We Changing From?

Our project baseline is not necessarily another separate document. Instead, it is the collection of documents we have already produced, including the descriptions of our project's costs, its delivery schedules, its specifications of quality and scope, and its outputs. Our control of these types of critical documents is referred to as configuration management. Configuration management is often also extended to other things – for example, designs, prototypes, computer programmes, and hardware. It is usual to refer to anything subject to configuration management as an item.

Configuration management is an integral part of our wider change management process, and any change to the project that requires a change to any one or more of these configuration items needs to be subject to configuration management. At its most basic, all this really means is that we have to go through an approval process for the amended document or item in the same way as we did for the original, and we have to check its ramifications for other documents and items across the project. For example, changing the order of tasks in the project plan may have an impact on our plans for the drawdown of finances for the project and supplier contract schedules, or changing the part of a design – say, for a piece of machinery, might require us to alter the specification of the power supplies for it or the strength of the floor where it will be located.

Consequently, it is useful to include in each configuration item the details of all those other configuration items on which it depends and all those items dependent upon it. These details then provide a chain which indicates where other configuration items may need updating and reauthorisation.

Consider the implications of a simple request from a supplier with manufacturing issues regarding a conveyor system for your new production line. They request a change in the delivery schedule of their contract, slipping the on-site delivery date by a month from the end of September to the end of October. Whilst this is a blow, we accept their reasons for the request and agree to it.

Following our chain of configuration items, this change will certainly mean us having to change the stage plan, and possibly the overall project plan, to reflect the new date.

We would then have look further at the subsequent tasks in the project plan and consider any need to amend the timing of our testing, acceptance, training, and implementation tasks, making sure that we still avoid difficult times such peak holiday, workload, and illness periods.

We will probably have also to amend our financial plan, especially the cash requirements for the finance department, because if delivery is to be a month later, presumably our payment for the goods will be as well.

We will need to review and possibly also change the dates in installation and commissioning contracts, insurance contracts, warranties, storage contracts, and so on.

Sources of Change

Whilst all change may be generated from across the project, and the wider environment in which it is being delivered should result in the initiation of change requests as in our process in Figure 9.1, there are two areas of change generation which need to be expressly included in our processes and consideration.

Risks and Issues

As we identified when considering the management of risks and issues,[2] there is an overlap with our change management because the treatment actions we identify in our risk management processes will normally result in project changes. These actions could include treating the issues we encounter, as well as risks we have anticipated and are seeking to avoid. If any of our actions require changes to our project activities, then they will give rise to change requests and be subject to our change control processes.

To recap, our risk and issue treatment actions could have arisen from externally imposed issues or risks. For example, consider the necessary changes to a project resulting from the following scenario:

A design team leader emails the project manager responsible for developing and launching a new product which slots within planned changes to a manufacturing production line:

Re: New Safety Regulations

Hi John,

As I said when we last met, the government-mandated safety regulations have changed. We've completed an impact assessment across our whole portfolio of products, and from 1 January next year, instead of the outer covers on your new product being left as plain metal, everything will need to be painted bright red with standard-sized hazard warning markings. We estimate that complying with this will cost an extra £2,000 in design and development effort and take an additional four days to complete, given that we have to create a suitable place for the markings to be attached. There will need to be some differences in the production process, because in addition to spray painting and attaching the markings, you'll need to allow time for the paint to harden before the stickers are attached. We'll advise on the best paint to use to minimise this time.

Yours,

Jackie

Project changes can also arise from a totally unexpected issue. Let's now consider a second scenario:

The site manager of a building contractor calls the project manager responsible for refurbishing a 16th-century pub/restaurant:

"Hi, Chris. I'm on site at the moment, and we have a problem. When we drilled into the wall between the bar and the restaurant to start opening up for the new doorway, we found it was not a solid wall but one filled with rubble. I've stopped the team for the moment to check how you want us to proceed. When we cut through the doorway fully, we anticipate that the entire rubble filling between the two outer walls is going to come away, and judging by the size of the wall, this will be several tonnes. It's also not going to be such a clean job, and when the rubble does come away, there'll be choking 16th-century dust everywhere. Either we're going to have to erect additional screening to the rest of the pub or you're going to have to close for a very major deep clean afterwards. It's also going to take extra time get the rubble out of the building and do an initial clean up before we can continue. I estimate we're going to need an additional five days of effort over and above our quote to deal with this. Also, unless you're going to ask us to build a massive rock garden somewhere, you'll now have also to arrange removal and disposal of the rubble, which is again extra cost and time.

"I need your decision about what to do, otherwise I'm going to have to move the team to jobs for other clients, and once we start those, I'm not sure when I can get everyone that we need back on site again. You'll also have to think about the impact on the electricians, plasterers, and decorators that you've got booked to follow on from us."

Both the changed safety regulations and the rubble-filled wall mean that we have to accept changes to the projects. In the second case, which is an issue with critical priority, we also have to decide very quickly on what actions we need to take and how our project plans have to be changed.

Evolution of Requirements – Creeping About

The second area to consider is changes in scope. We set out the scope of our project in all of our launch documents, such as our business case, our project plan, and our four document set.[3] We also understand that circumstances beyond our control, as in the changed safety regulations and the rubble-filled wall examples prior, may force us to make changes to the scope of our project and mean we need to update those baseline documents. Some changes to our scope, however, are internally generated and are completely voluntary and unforced. Consider, for example, the following two informal conversations between project team members:

> We'd get much faster take-up if we modified the design of part A to make it appeal to a wider customer base. We could also replace the steel washers, which will eventually rust and look unattractive, with white plastic ones. There's only a trivial extra cost and no additional time would be needed. I don't think we need to bother anyone with the formal processes for such a tiny adjustment.

and

> I was chatting with the client's wife the other day, and it emerged that she wanted to add some small additions to the work programme. Whilst you're on site putting in the footings for the house, please could you put down some footings for a treble garage which they will build in future. I have a rough plan of what's needed. It should only take an extra hour or so to excavate and a little extra concrete being poured. When the concrete is set, please could you cover it over again with the excavated soil. There should be enough scope in the estimates to cover this – and think of the customer goodwill that it will generate.

Both of these are examples of voluntary change, which would change or extend the original scope of the projects as we defined them in our work breakdown structure, and possibly in other documents such as drawings, requirements specifications, contracts, and work instructions. As project manager, we should always recognise that such suggestions are requests for change, and if they are to be considered, they must be raised by the initiators as part of our formal change processes. We should also always work with all our project team members from the time that they join the project so that they adopt the necessary rigorous attitude to identifying and recording change.

Many of the continuous, and often very small, incremental changes that we face in our real projects may not be quite as obvious as these two examples. Often small changes to our agreed-upon tasks, like the modifications or the replacement washers or the garage foundations prior, are often seen merely as "minor tweaks," but taken together they can add up to an unacknowledged and unauthorised creep in the scope of the project. Such "tweaks" may be perfectly good suggestions for improving the final outcomes of our project, but the dangers from scope creep arise from its uncontrolled nature and the unanticipated effects it can have elsewhere in the project and on our overall performance.

One other common cause of uncontrolled scope creep that we need to consider is the timeliness of our response to a change request. This timeliness of response is also often prescribed in supplier contracts but may be completely at variance with the need for speed and pace in our project.

Consider our earlier example of opening a doorway in a wall that turned out to contain a rubble-filled space rather than solid stonework. The builder has identified the issue and requested a change. Until we respond there may be workers and equipment sitting idle, often at the supplier's cost. The temptation for the builder is to keep his team working and extract the rubble and insert a changed design of doorway, but this will be an unauthorised change – i.e., scope creep. On major contracts, notification periods can be many days, and the impact of these periods is to cause major delays or result in unapproved changes to our project plan, with tasks being reallocated or rescheduled by a team leader or a supplier in order to keep their team working usefully.

One very frequent first impact of unreported scope creep is that suddenly we, as project managers, find that we are facing unexpected issues with maintaining our planned completion schedules and controlling our costs. This leads to us needing to spend time analysing why our project is costing more or running late and subsequently having to take measures to justify or rectify the unauthorised changes.

Scope creep can equally impact our suppliers and stakeholders. Suppliers can also find themselves delivering to us late because of such unauthorised changes and having to try and either cover the increase in costs associated internally or seek to recover them from us after the event. In our doorway example, we may have a perfectly valid contractual argument that the builder implemented changes without our prior consent and hence should bear the additional costs. We may also recognise that the builder responded to the situation in such a way that project costs and delays were minimised, and we may not wish to enforce our contractual rights. This can lead to difficult discussions with the business sponsor and the legal team.

When scope creep starts to generate poor project performance, we may find that through their association with it, our business sponsor, stakeholders, and others who have supported our project suffer a reputational cost. As we have seen, it is not unusual for business sponsors, and even chief executives, to lose their jobs when a major project starts to underperform. There is, however, good news about managing scope creep, as many of the management approaches and measures we have already taken to build a controlled project also help control scope creep. These include:

- **Fostering engagement.** Engaging widely in framing our project and its outcomes should allow us to capture a wide view of the requirements for the project very early on. This engagement then reduces the potential for new requirements emerging from stakeholders, users, and business sponsors during our project delivery period.
- **Ensuring fine granularity of our instructions**. Using our work breakdown structure to define the scope of individual tasks, which we can develop into individual work packages, means that we have defined what the project will deliver on a very detailed level. Where scope is poorly defined and remains at an overview level, there is potential for ambiguity and changes in the delivery at a task level. As a rule, the finer the level of granularity of our instructions and the more comprehensive that we can make them, the less opportunity there will be for scope creep. We will also have far better and earlier opportunities to spot any points where the scope of what is being produced exceeds our brief for it.
- **Opening up the change request process**. Empowering anyone to submit a formal change request seems simple, but opening the process for obtaining an authorised change to everyone on the project team reduces the temptation for any of them to create an unauthorised change.[4] It is also a very useful point on which to focus when onboarding team members.

Step 1: Develop a Change Management Strategy

Setting out how we will manage change is an essential step to ensuring that change requests and their delivery will be properly controlled. The change management strategy will have to answer the same questions that we posed for all our other strategies, although our answers will be somewhat different:

- **Who** will be responsible for activities related specifically to change management? Some of these should already be included in the business case but should be restated here – for example, project board responsibilities. We should, however, still summarise all responsibilities for identifying, evaluating, reviewing, and agreeing to the change. This summary needs to include reference to the roles of the project manager, project board, and business sponsor. It should also make clear that anyone engaged in the project is able to initiate a change request. For some major projects, a change manager role may also be required, or in a project anticipating multiple changes, it may be necessary to set up a separate change authority, which operates in support of the project board. We could also include details of external advisers on change if the nature of the project would justify it.
- **Why** is it important to manage all of our changes? Not all business sponsors and project board members will fully understand why change management is important and why the project needs to invest time and resources in ensuring that change is properly controlled. We may need to underline the message that failing to control change will usually lead to problems in delivering our project in accordance with our carefully documented and agreed-upon budgets, timetables, quality standards, and scope.
- **What** criteria will be used for assessing the level of priority of individual changes? The criteria we use should include many of the business-related factors from our options analysis in the business case. In the change management strategy, our explanation of these criteria should include descriptions of the types of impact effects on the project that will be considered relevant and how we will assess them, including the weighting to be assigned to each of the identified criteria. These criteria and the weightings that we assign to them are project specific. In selecting the criteria against which we should evaluate any potential change, it is useful to consider our "golden thread" of success factors and requirements which we built into the project at the outset. In addition, we will need a further catch-all category into which both our examples of issues of regulatory change and the rubble-filled wall will fall. This category is critical issues – essentially change in which we have no choice about acceptance, although there may be choices in the way the change is actioned. We also need to establish the conditions for enabling an emergency escalation to authorise changes arising from an urgent and immediate issue. Some possible general areas for inclusion are:

- Increase in project benefits:

  - Improved return on investment from the project's solution.
  - Decreased operating costs after project goes live.
  - Reduced forwards maintenance load.
  - Improved operational performance.
  - Increased client revenue after project goes live.
  - Higher levels of long-term profitability.
  - Earlier delivery of financial benefits.
  - Improved customer service/satisfaction.

- Improvement in project delivery:

  - Reduced time to complete delivery.
  - Decreased project costs.
  - Reduced resources needed to complete project.

- **Where** will changes be recorded and reported? We should set out details our change management records and project change register, including where and how they will be stored and who will have the right to create, update, read, and delete them. As ever, we may choose to record the changes in dedicated software, in a safeguarded electronic filing system, or manually. The important thing is that our documentation recording changes should include detailed descriptions of each change and the actions taken in one secure place.
- **When** will individual changes be assessed? The timing of change management activities should include, for example, reviewing our project change register as part of each project board meeting and additional processes for dealing with change when action between meetings is required.
- **When** will both requested changes and the actions taken be reported?
- **How** can we design and implement procedures and methodologies that enforce the overall management of changes? These procedures may be based on corporate or other standards, and we should identify if, where, and why these have been tailored to fit the project. Each process should clearly specify the processes for recording, evaluating, and reporting change, including:

  - The process for raising requests for change by anyone engaged in the project.
  - Recording and monitoring change requests.
  - Reporting progress on individual change as well as the overall level of change on the project.
  - Change authorisation mechanisms, including the release of any contingency funds and emergency changes.

- **How** will the levels of priority be set which will determine whether a change requires further action or escalation – i.e., the rationale for making our action decisions?
- **How** many staff resources are required to manage change throughout the life of the project? Depending on the type of project, and its size and complexity, the project manager may be able to act as the change manager, but in others, the project support Team may need to be supplemented by a dedicated change management function.
- **How much** funding should be allowed for the management of change? This funding is for the process of change management rather than any contingency funds that have been set aside to deal with issues arising.
- **What if** we have to act in a crisis situation because a change has escalated or requires immediate action?

Step 2: Record Change Requests

As with recording risks and issues, we need to both document the details of every change request raised and track its progress through the approval, delivery, and monitoring processes. This documentation should enable us at any time to be able to report at the level of both an individual change and a summary project. Using the same methods as for risks and issues, we can achieve this recording through the use of a detailed change request document, a change management record, and a summary project change register. The latter two should be updated with progress details during the active life of the change request.

Whilst project software can help by reducing the administrative load of doing this, be aware of the risks of constructing a single spreadsheet to act as both a set of change management records and the project change register. If we do, it is best consider separating your data input areas into one sheet in the workbook and picking up the information needed from there into two report areas, one a series of detailed change requests and the other the summary change log. It is necessary to always keep backup copies in case of file corruptions.

The chosen format for both the change management record and the project change register should include space to record the prioritisation of the change request. Table 9.1 shows a basic change management record.

Table 9.1 Change Management Record

Change Management Record				
Change name				
Change reference no.		**Change raised by**		
Description of required change				
Cost and time implications				
Change creation		**Created by**		**Date**
Priority assessment				
Project documents affected	**Changes needed**		**Updated version no.**	**Date updated**
Business case				
Stage plan				
Project plan				
XXX doc				
XXX doc				
XXX doc				
Action required 1				
Action required 2				
Action owners				
Action required 3				

Change Management Record						
Change action authorised	**YES**	**NO**	**Authorised by**		**Date**	
Reason for acceptance or rejection						
Signature						
Assessment of progress						
Action required 1						
Action required 2						
Action required 3						
Action owners						
Change progress review			**Reviewed by**	**Reviewed by**	**Date**	
Assessment of progress						
Action required 1						
Action required 2						
Action required 3						
Action owners						
Change completion			**Reviewed by**		**Date**	
	Name			**Signed**	**Date**	
Change completed satisfactorily						
Project documents updated						
Change completion report issued						

Table 9.2 shows a blank project change register, divided up into different areas in the same way we partitioned the risk register. These section headings are completely subjective and project specific.

Table 9.2 Project Change Register

	Project Change Register				
Change Number	Change Name	Change Description	Priority	Actions Needed/Taken	Status Open/Closed/Complete
1.0	**Personnel change**				
1.10					
1.20					
1.30					
1.40					
2.0	**Technology change**				
2.10					
2.20					
2.30					
2.40					
3.0	**Commercial change**				
3.10					
3.20					
3.30					
3.40					
4.0	**Financial change**				
4.10					
4.20					
4.30					
4.40					
5.0	**Other change**				
5.10					
5.20					
5.30					
5.40					

Step 3: Assess and Prioritise Change Requests

Change requests represent additional competing demands for our project's resources – i.e., finance, staff, and time – and may also demand alterations to scope and quality. Remember, our change management strategy needs to have set out the criteria to be used in evaluating and then prioritising each change request.

Assess Acceptability

Before we start considering the priority of a change request, our first task is to review it and decide whether the change is acceptable within the broad objectives of our project and our capacity to deliver it. It is always useful to keep the business sponsor informed of any change requests coming in, not least because they may have a wider view of developments in the client organisation which could affect our view of the change requested. It can also be useful to include change requests generally as a standing item for business sponsor briefings.

Consider the following scenario:

> *The project manager of an internal business change programme receives the following email from the director of customer services:*
>
> #### Re: Customer Ordering Enhancement
> Dear Jane,
> Thank you for your update on the progress of the customer ordering workflow element of the wider financial system project. I'm glad to see that you're making such good progress with the supplier and that you intend to run a six-week pilot of this module at the beginning of the next quarter.
> My team and I have looked at the images of the data input screens for our call centre agents. We think that if you could make these screens directly available to our priority customers through a dedicated portal, we could start offering them a 24x7 ordering capability. This simple change alone would mean that we could refocus our staff to offer more in the way of problem resolution rather than just order taking over the phone. Such a change will make a massive improvement in the service we offer to our most valued and profitable customers. I've spoken with Chris in his capacity as finance director, and he's broadly supportive of the concept.
> Please can we have a meeting to discuss how you can bring this important development about.
> Best regards,
> Kate

> *Whilst on the surface Kate's suggestion for developing a new customer channel, the customer ordering portal, is reasonable in business terms, and as the finance director, who is also the project's business sponsor, is "broadly supportive," it looks certain that will merit further investigation. Project manager Jane, as the guardian of the project's objectives, has to ask if developing a client-facing portal could be fitted within the scope of the existing internal workflow project. Unfortunately, the answer is that it probably would not fit, as it would:*

- *Require a new area of systems development.*
- *Demand at least some engagement with the customer base to ensure that the new portal would meet their needs.*

- *Lead to additional cost and time and require different IT design and development skills to build the web-based portal.*
- *Extend the completion date of the project by a period of potentially several months.*

A tactful reply by Jane is required:

Re: Customer Ordering Enhancement

Dear Kate,

Thank you very much for your email.

I agree that your idea of a customer ordering portal for priority customers could be a major benefit to the business and is worthy of further investigation. I can confirm we do have existing facilities within the software package which could allow for a custom-built web portal to be interfaced to it, although we cannot for technical and contract reasons just reuse the internal screens. Also, there may be many other customer-facing services that could be delivered using such a portal, for example, submissions of shipping documentation and invoicing, and these need to be fully explored as well. I've talked to the developers, and on the face of it, a web portal would be possible. It could be developed and implemented as a follow-on once the internal workflow has been delivered and shown to be working successfully.

I think that we should meet shortly, probably when the pilot has been running for a while, by when we will have had some results coming in which we can discuss as well.

Best regards,

Jane

A masterly response, and it looks like Jane is already lining up her next project management role.

As in the example, our acceptability test for a change request has to be not only about the business validity and desirability of the request but also about its degree of fit with the project and the capacity of the project team to deliver.

Assessment of Impact of the Change and Prioritisation

Being accepted as a viable change is a necessary but not sufficient test to enable the change request to be actioned. Once the viability of the change request is accepted, we also need to undertake and assess each requested change against criteria established in our change management strategy. This should include at least:

- Fully assessing the contribution of the change to achieving the project's and the business's objectives.
- Judging the results of this evaluation against our MoSCoW-based assessments.
- Prioritising the change request in relation to other claims on the project for additional resources.

These three processes are illustrated in Tables 9.3, 9.4, and 9.5.

Table 9.3 restates our basic MoSCoW evaluation process which we used when determining the scope of the project and which we now need to apply to each change request. There is one suggested additional category that can be useful at this stage, "not required." This additional category, usually in consultation with the business sponsor, gives us the chance reject a change request outright at this point.

Table 9.3 Change Priority Criteria

Change Priority Evaluation Criteria	
Category	**Comment**
Critical issue	Must be addressed before the project can continue. Usually this will have been raised under the issue or risk management strategy and is being passed into the change management process for prioritisation and action.
Must have	The planned solution will not be viable without this change being made.
Should have	The project can deliver a solution without this change but may have to introduce some workarounds and compromises if it is not included in the project scope.
Could have	This change is desirable and can be delivered but not at the expense of other financial and time commitments. If resources are not available, this change can be held over for inclusion in any subsequent development of the project solution.
Won't have this time	This change is still desirable, but there are more important things to achieve first. It is likely that this change will have to be held over until after the completion of the main project and then can be considered if there is to be any future development.
Not required	There is insufficient business justification for the change.

Changes are prioritised for action as set out in Table 9.4. It is useful to tailor the information in this table and in Table 9.3 to meet the specific needs of the project and obtain pre-approval for them by including them in our change management strategy.

Table 9.4 Change Actions Prioritisation

Change Actions Prioritisation		
Priority Scoring		**Action Required**
Category	**Score**	
Critical/Must have	75 and above	**Commence planning/action immediately** Change requires immediate action. Plans should be drawn up and agreed to by the business sponsor, reported to the project board, and implemented as required. • Each action should be assigned to an identified individual with a specific time frame for action, quality target, and budget. • Project manager to monitor report progress on actions taken to deliver the change part of project board reporting. • Project manager to update project plans to include the new activities.
Should have	61–74	**Action/planning desirable** Change requires short-term action. Plans should be drawn up, agreed to by the business sponsor, reported to the project board, and implemented without undue delay. • Each action should be assigned to an identified individual with a specific time frame for action, quality target, and budget. • Project manager to monitor report progress on actions taken to deliver the change part of project board reporting. • Project manager to update project plans to include the new activities.

Change Actions Prioritisation		
Priority Scoring		**Action Required**
Category	**Score**	
Could have	**41–59**	**Planning desirable** Change requires action. Plans should be drawn up, agreed by the project board, and implemented if time and resources permit. If implementation is to proceed: • Each action should be assigned to an identified individual with a specific time frame for action, quality target, and budget. • Project manager to monitor report progress on actions taken to deliver the change part of project board reporting. • Project manager to update project plans to include the new activities.
Won't have this time	**21–40**	**Planning may proceed** Actions required to deliver the change should be identified, evaluated, and agreed on with the project board. If time and resources permit, full planning and implementation may proceed and be submitted for project board approval in due course. If approval to implement is given: • Each action should be assigned to an identified individual with a specific time frame for action, quality target, and budget. • Project manager to monitor report progress on actions taken to deliver the change part of project board reporting. • Project manager to update project plans to include the new activities.
Not required	**20 and below**	**No action** Unless the situation alters, the change does not need to proceed. Change request to be closed as refused.

Table 9.5 shows an example weighted scoring matrix recording the assessment of a series of requested changes to a project, labelled CR no. 1 to CR no. 5. The matrix draws on agreed-upon criteria for judging the success of the project, to each of which a weighting has been assigned (see the evaluation techniques in Appendix 1).

Table 9.5 Change Priority Matrix

Change Priority Matrix													
	Criteria												
Change Requests	**Critical Issue**		**Reduced Cost**		**Increased Solution Performance**		**Better Customer Service**		**Increased Revenue**		**Faster Delivery**		**Total Weighted Score (max = 100)**	**Rank**
Criterion weights		10		8		6		5		3		1		
CR no. 1	10	100	0	0	0	0	0	0	0	0	0	0	100	1
CR no. 2	0	0	6	48	3	18	0	0	0	0	0	0	66	2
CR no. 3	0	0	1	8	3	18	2	10	0	0	1	1	37	4
CR no. 4	0	0	0	0	7	42	1	5	1	3	1	1	51	3
CR no. 5	0	0	0	0	2	12	1	5	1	3	0	0	20	5

As ever, this scoring is subjective, but it allows the overall desirability of the changes to be established and the different changes to be ranked in order of priority. In this particular example, it has been decided that the total maximum weighted score will be 100. Each change request has been scored against relevant criteria and weighted scores calculated and totalled. Critical issue change requests are given the maximum weighting and a maximum score – 10 in the example – and so have received the maximum score of 100.

Resulting judgements of if, and when, to progress with any change will normally be ratified by the project board. Smaller or urgent changes may be delegated to a designated change authority, often the business sponsor and/or the project manager for decision usually with subsequent reporting to the project board. In this case we would be recommending CR no. 1 for immediate action, CR no. 2 for planning to commence for short-term action, CR no. 4 for planning to commence if time and resources permit, CR no. 3 for identification of actions required, and CR no. 5 for rejection.

Step 4 Determine the Options for Delivering Change

Once a change is accepted as being suitable and has given a priority rating, we are faced with choosing how it should be delivered. Depending on the type of project and the type of change, there may be many different ways of delivering the change. Our task is to apply our satisficing process once again to select the option that has the most achievable balance of costs, risks, and fit with the remainder of the project delivery plan.

Consider the following scenario:

A request for a change to the purchased equipment required by a project has been received, in this case for handheld scanners for use by mobile staff. Overall, the project requires one hundred scanners, and the detailed specification of this equipment was included in the contract with the supplier. The scanners have now failed the user acceptance testing process because they were judged to be:

- *Heavier than desirable for extended periods of use.*
- *Inadequately powered, with too short a battery life such that it cannot cover a full eight-hour working shift without recharging at least once. This would require staff to return to base and exchange the scanner for a fully charged one.*

The equipment is essential for the solution rollout, and the project will have to pause if an acceptable solution is not available before user training commences. Consequently, the change request was assessed as being critical. To complicate matters, the supplier claims that in their view, the units meet the contracted specifications.

The project manager is now faced with a series of options, each of which has different risks and impacts:

- *Purchase additional units to allow users to change equipment for fully recharged units during a shift and have a custom designed shoulder strap fitted to overcome the weight problem. The users of the equipment will not be fully satisfied with this solution, additional provision will need to be made for recharging the units, and costs will increase. The advantage is that the timing of the project will not be affected.*

- *Work with the supplier to accelerate the launch of a new type of scanner which they have in develop-ment and which they claim will resolve the problem. The supplier has offered to provide the units at the same cost as the original ones, but as the scanner is not yet operable, even as a prototype, there is a risk to the timing of the delivery of the project.*
- *Run a new procurement to purchase the equipment from other suppliers, and work with the existing supplier to interface the new scanners to the overall solution. This will lead to a contract dispute with the current supplier, may result in additional costs as the cheapest supplier was chosen originally, and could delay the rollout.*

Such choices are not easy to make, and even the simplest of changes can present implementation options with different balances of risks and impacts – including to our four old friends of time, cost, quality, and scope.

Step 5: Actioning Change

Once the changes required have been approved by the project board, the project manager is responsible for updating *all* impacted project documentation – including ensuring that con-figuration rules are observed. Communication is vital to ensure that different members of the project team, contractors, and stakeholders are fully informed that a change request has been approved and the implications that it will have for the project. It is also advisable to feedback to the person raising the original change request that it has been approved, or in some cases, rejected.

Updates will be required to project documents, including project and stage plans, work breakdown structures, and detailed instructions to team leaders and staff carrying out the changed activities. Where necessary, contract variations may also have to be agreed on with different suppliers.

Step 6: Monitor the Change Process

Included in the change management record are opportunities for us as the project manager, or another designated individual, to undertake a review of progress on the implementation of the change. The timing and number of such reviews will be change specific but are critical to ensur-ing that we remain in control of all project changes and are able to report properly on them and their impact on our overall delivery of the project. There is also provision in the record for the project manager to sign off on the change as being completed and the documentation as having been fully updated, another essential control.

In our allocation of responsibilities to the project board as shown in Figure 9.1, their role included a periodic review of the project change registers. Logically, there are two different types of review that we can expect the project board to undertake:

- A review of progress on the identification, evaluation, and treatment of changes as part of their oversight of the management and progress of the project at each project board meeting.
- As part of an end of stage review, or a comparative point for agile projects, to conduct a more detailed examination of changes that have been implemented as part of this stage and to include those changes that may affect the delivery of the next stage.

On a regular basis, and necessarily in advance of any project board review, the project manager should assess the progress of the change with the assigned action owners and record the results of the review in the change management record.

Step 7: Change Closure and Reporting

A change request is closed either when it has been successfully delivered to the satisfaction of the project manager or at any point in the approval process where it has been decided that the change will not be progressed further. Once the closure is initiated, all further work on the change request ceases and the change management record and project change register are completed with the closure information.

We should also complete a summary closure report showing how the change was achieved, including additional costs and the impact on project timings, quality of deliverables, and project scope. This should be reviewed by the project board, and ideally a copy of the approved closure report should be sent to the instigator of the change. If the change is closed because of a decision not to pursue it, then this decision should also be notified to the instigator with an explanation of why it is not being accepted.

Step 8: Documentation Is Everything

Throughout the process of identifying, assessing, and delivering the change, we have looked at the record keeping needed to document every step in dedicated change management records and to ensure that any necessary amendments are made to other configuration items – for example, updates to the project plan, the business case, and especially to contracts. As part of our closure of the change request process, we should review the affected project documentation to ensure that it still complies with the project's configuration policies, including:

- All documents have all been updated to reflect the completion of the change.
- Version numbers have been correctly updated.
- Appropriate documentation approvals and sign-offs have been recorded.

Reflections

Choosing your previous project, or using the fictional bio-energy from effluent case study:

- Consider ten major changes that could be requested for the project and categorise them by broad headings that are appropriate to the project.
- List four criteria that you could use to decide on the allocation of resources between different change requests.
- Select five changes of different types, complete a skeleton change management record for each, and populate a Project Change Register.
- Map the five changes against your selected criteria and prepare a priority list for approval by your project board, including details of the MoSCoW prioritisation of the changes.
- Identify what existing configuration items will need to be updated for each of the five changes.
- What information will you need to communicate regarding each of the changes, and to whom?

Notes

1 It certainly makes it much easier to point out the error of someone's ways (gently, of course) if we have a clear, open, and fully explained notification process.
2 See Chapter 8.
3 See Chapters 3 to 5.
4 It also gives us a very good backing when (sensitively and politely) pointing out the error of a team member's ways if they have made an unauthorised change.

10 Supplier Management

Your Supplier Is Your Supplier – Not Your Friend

Aim of this chapter: To introduce the supplier-client relationship management life cycle.

Learning outcome: Understand and apply the basic principles of supplier management during the life of the project, using simple tools to do so.

We have an overriding objective which we must bear in mind during the often onerous process of sourcing, engaging, and managing a supplier to help us deliver our project. That objective is to achieve our stated requirements on time, within scope, and while meeting the quality standards we have specified, at a price which is demonstrably fair and reasonable. Supplier management is another area in which we often need to draw on various kinds of specialist expertise as we navigate its perils and pitfalls.

Not all projects will require engaging with external suppliers to provide goods or services under a formal contract. Some projects may be completely internal to a team. Some may involve using an external supplier but only require simple purchases of materials. The remainder of projects will involve engaging a supplier under a contract to help deliver some part of the project; either goods, or services, or both. The contract may not always be a formal commercial contract but could be an internal arrangement where support is provided by a different part of the client organisation, or, if in the public sector, the arrangement may be with another public body. Even commercial contracts range from a straightforward agreement to purchase a number of items for a fixed price to a production or construction contract running across multiple years, many sites, and different countries.

It's now time for the bio-energy from effluent project to start thinking about engaging suppliers. As its project manager, you're now at the point where you've just had a very successful project board meeting. You worked hard to prepare all the members of the board first, and although you got asked some tough questions, in the end they all agreed to support both the business case and high-level project plan.

In six months, you'll have to locate the first of the planned processing plants on the Wadi site. This site will use a design scaled up from the latest iteration of the plant developed by the research and development team. Essentially, it will be a preproduction prototype which the project will use to demonstrate the commercial effectiveness of the proposition to local and national governments. The plant will run initially for three months. Thereafter the project plan requires siting two installations per month in different countries until total of ten units are in operation. As these units may be moved between different demonstration sites, it has been agreed that each plant will be mounted on a series of trailers connected both to each other and to the local electricity network.

DOI: 10.4324/9781003405344-11

As yet there is no manufacturing capacity in-house, and the project will need one or more suppliers to build the first mobile plants. The supplier, or suppliers, will also need to work closely with the research and development team, who will be modifying and improving the design based on on-site experience. Any design changes will need to be included in the later models as they are produced and may require retrofitting to the existing plant in-country.

You've obviously already done some background research around how this plan may be achieved, but what do you think are the next steps you need to take in order to be able to meet your deadlines?

To be clear, this is after calling the client organisation's head of procurement to arrange a meeting. if you have not already done so!

Using Suppliers

Supplier management does not only apply to those situations where we have to engage with a contracted external supplier. Even if our project draws solely on other departments or teams in the client organisation for support, we still need to manage them as suppliers in the same way by:

- Drawing up our brief to describe the project to them, and set out an unambiguous statement of the work we need them to carry out.
- Agreeing on the terms with them on which their support is provided.
- Managing the project's relationship with all suppliers to ensure delivery to time, cost, quality, and scope.

The controls that we need to put in place when working in a non-contractual relationship with internally provided services have to broadly mirror those we require when working with an external supplier. There is, however, one additional initial action that we need to take if we are using external suppliers. This is establishing whether the client organisation commissioning the project already has a long-term relationship, and even applicable contract arrangements, with a suitable supplier.

Supplier management can be seen as a process involving three main areas of activity:

- **Procurement**. Defining and delivering a procurement process to engage a supplier able to deliver for the project.
- **Contracting**. Shaping and agreeing on a contract around the project's critical success factors and the client organisation's corporate requirements.
- **Contract management**. Applying a range of techniques and strategies to manage and support the supplier in successfully delivering to contract and taking action if they don't.

Whilst each of these areas is separable, and often involves different parts of both the buyer's organisation and the supplying company, they build a single continuous relationship between the client and the supplier organisations. That relationship starts to take form during the procurement process, is written into the contract's terms and conditions, and thereafter is always managed through the mechanisms set up in the formally signed contract.[1] As I have found with many aspects of my project management practice, these three elements are not completely sequential, and there are major overlaps and co-dependencies. For example, large elements of the briefing information that we prepare for suppliers will relate to our requirements both for:

- What we require from the supplier – the number and types of physical deliverables. This information should subsequently appear in the contract, often as a schedule, and against which will we review actual deliveries as part of our contract management.
- How the contract will operate and the performance levels the supplier should achieve. These details will also appear in the contract with key performance indicators attached to them. Again, we will also assess the supplier against these as part of our contract management.

It is key to consider appropriateness when determining how we will seek to create and manage our supplier relationships. As in most other areas of the project, we need to build in processes and controls that are appropriate to the objectives that we are trying to achieve. We will need to balance the need for our control of a relationship against over-bureaucratisation. The relationship required with each supplier will be different, and our need for controls will also vary. If our project seeks only a one-off purchase in a single transaction with no requirement for any other relationship with the supplier, we will need to establish certain specific controls regarding timeliness, cost, and quality. If continuing support and maintenance are required, which may extend well beyond the life of the project, we will need a range of additional process and payment monitoring and co-ordination mechanisms.

Before enmeshing ourselves in the details of each of these three elements of supplier management, it's time for another virtual field trip. We're going to see what can happen if we don't get our procurement, contracting, and contract management exactly right from the very beginning.

There are many sensibilities around this project, and to protect all the parties (both the innocent and the guilty), I can't tell you where we are. I'm afraid I'm going to have to blindfold you both on the way in and the way out on this trip. I can tell you that we're going straight into a meeting room in an office building, so you won't need a coat, but, as you're about to discover, a hard hat and maybe even a flak jacket might be appropriate.

We're going to observe part of a three-day contract management meeting. We slip in unnoticed as the person serving the coffee and biscuits comes out. Removing the blindfold, you can see that this meeting room is modern, long but not wide, and very light. One end of the room is dominated by large video screens and the other by an equally large window. Once again, the room is too small for the number of people in it, and, as it will turn out, the table down the centre is much too narrow for comfort.

Already in the room is a legal team representing the client and another representing the supplier. The project manager, members of the project support team, and the supplier's CEO, together with his project team leaders, are present. There is also a team from the organisation funding the many millions of euros required for the project, including their finance, technical, project, and legal experts. It is a truth almost universally acknowledged that when you get this sort of representation in a meeting room, something very, very bad is already happening to the project.

"Give me my money!"

The supplier's CEO is on his feet, pale with anger, shaking, shouting, and jabbing his finger at the head of the funding team, who is sat directly opposite, utterly impassive, and less than half a metre away across the far-too-narrow table. He continues, "Give me my money! It's not even anything to do with you, it's between us and our client. Give me my money!"

The head of the funding team is unimpressed. He's obviously handled difficult situations before, and he replies in calm and measured tones. "You know very well, because we've discussed this before, that under your contract with our client, as the project funders, we have the right to vet and approve all requests for change. As we showed in the presentation earlier, our project and commercial teams have analysed each of the twenty or so items in the change request in very great detail . . ."

"Give me my money!"

"As I was saying, we have analysed each of the twenty or so items in the change request in very great detail, and our conclusion is this: some of the items are already covered in the contract, some items are for services that you have to deliver at your cost in order to meet the contract requirements, and some of the remaining items are not calculated correctly. To take a very simple example, why does every change item need to be delivered by staff at the top daily rate and not by a mix of senior and junior staff as in the main contract? Even to me, and I'm not an expert in this area, this looks most unusual, and you've not justified it at all. Overall, the judgement of our experts is that the great majority of your change request is not valid. We can see justification for fifteen percent of it, and we are willing to sign off on that amount immediately."

The CEO's face now flushes red, and, furious, he continues at the top of his voice, "I reject absolutely what you and your so-called experts are saying. It's all untrue! Every last cent that we've listed is justified and outside the existing contract requirements, so you must pay for it and I want my money now! I'm going to take this further. We're stopping all work right now, and I want to consult with my legal team privately for a few minutes."

The supplier's staff leave the room to have a meeting in the next (not very soundproof) office.

The head of the funding team nods to his own contract expert to take over. This expert is a curious mix of incredible, painstaking attention to detail and extreme excitability. He addresses the client's project management team and lawyers with rapidly rising volume, accelerating speed, and increasingly wild gesticulations. He states, "Simple. You know that they've significantly failed to deliver under the contract. I told you last year when we looked at their first change request that they would be difficult. They're way behind schedule. And now they have the barefaced cheek to ask for more money to deliver what they had contracted to do. You have only three options left now.

"Firstly, you can go straight to a breach, trigger the termination of the contract, obtain damages from them, and find someone to finish the project for you. In the long run, this would be my chosen option for you, and we can help with that.

"Secondly, you can support our view of their requested changes and pay them a little extra, which we will support. Be warned, though, based on our experience with them, they will probably still keep overpromising and underdelivering. They will keep seeking unjustified change requests like this one. If you chose this route, I can promise you that we will be having many, many more discussions like this about dubious change requests designed to increase their profits.

"Lastly, you could pay them for all the changes that they are asking for, but we cannot support that. So, you would have to fund it separately — and that, in turn, will impact our contract with you. We would need to take the whole situation through to our regional director. I see any conversation at that level as being extremely difficult for all of us. Potentially, we could look to being told withdraw all of our support because you would have breached your loan agreement with us.

"We need to call in your CEO, right now, to discuss which option you're going to choose."

That was hostile, wasn't it? Take a deep breath, we're leaving them to it to try to work through this problem amongst themselves. You'll need to put the blindfold on for the journey back.

To complete the story of this contract: the project managed to work its way through this particular contract crisis, and the next, but the lender's commercial contract expert was right. Within a year the local papers were reporting that the situation between the parties was so bad that the client had terminated the contract for failure to deliver. The project then became a protracted legal dispute.

What is really of interest to us here is understanding how we can avoid situations where relations break down to this extent and we get to this type of shouting match and the almost inevitable outcome.

Part 1: Select the Supplier – Procurement

Figure 10.1 shows the broad steps in the first part of supplier management, the procurement process. This ends at the point of having an agreed-upon tender. The following section concentrates on the tasks in which the project manager is engaged either individually or jointly with the procurement team, as shown in Figure 10.1

Procurement is usually a highly controlled process, designed to ensure fair competition and obtain the best result for the client organisation. This best result is defined as a balance between costs and the quality of the deliverable. As the project manager, procurement is an area where we must understand the processes and regulations to a level where we can guide the project. It is, however, also an area which is complex and legally regulated. Specialist expertise is often required, and as project manager, we should make sure we draw upon it to:

- Guide us through the legal minefield of running a procurement. In every jurisdiction and even in different organisations, not only are there are different laws and rules governing procurement which must be followed but also often alternative pathways for implementing those rules. For major procurements, international rules, such as those of the World Trade Organization, the EU, or bilateral trade agreements, may also apply.
- Aid us in achieving our objective of reaching the best result at a fair price.

Many of the steps required in the procurement process produce documents which will build on ones produced earlier in the project, as well as documents prepared during earlier stages of the procurement process itself. For example, the tender documents to be issued to suppliers include information based on updates of:

- Critical success factors.
- Supply market analysis.

The documents also include information taken from the following:

- Statement of work.
- Procurement approach.
- Delivery relationship design.
- Draft contract terms and conditions.

Procurement Step 1:

One of our early tasks in our initial fact-finding stages of the project was to ask our stakeholders, "What must the outcome from this project do well?" And possibly even, "What must the outcome from this project not do at all?" We described the results of these questions as critical success factors in our options analysis, and they now come into play again. They will help us determine both the requirements to which potential suppliers are going to be asked to respond and the criteria against which their responses will be evaluated. It is therefore extremely useful to go back and

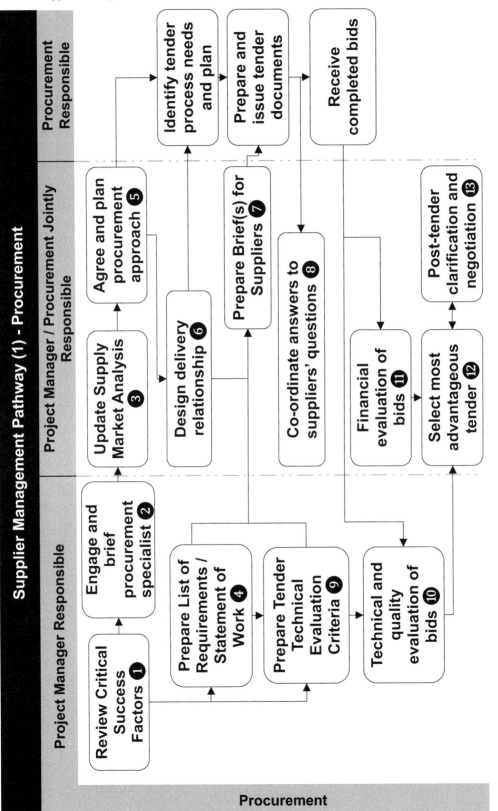

Figure 10.1 Supplier Management Pathway (1) – Procurement

review these early documents in the light of our current state of knowledge about what we are trying to achieve from the project. We should ask:

- Are all the original critical success factors still appropriate?
- Can we quantify them more exactly?
- Do we need to add any additional factors?
- How might we reprioritise them?

Having reviewed the critical success factors for the project, we should also ask:

- Which of them, if any, are relevant to each of the contracts that we wish to procure.
- Is there an opportunity to share our risks for not meeting them with the suppliers? IT service provision contracts, for example, may include provisions to transfer the risk of meeting operational service targets to the supplier under a shared risk/shared reward mechanism.

We should also now be starting to consider how supplier performance against our critical success factors should be assessed and what data we will require to be able to make such assessments.

Procurement Step 2: Engage and Brief Specialists

Project managers are not normally procurement experts, and engaging suitable advice is often a key protection both for the client organisation and for ourselves. Many large organisations have their own in-house teams (as ever often operating under a range of different names such as procurement, purchasing, buying), and other clients may have nominated external experts. In either case we need to engage with these procurement experts as early in the project as possible, not least because we will have to build a major part of our project plan around the procurement process that they recommend. If the project is a "one-off" for the organisation – i.e., in an area where the internal procurement team has no direct experience – it is worth opening a frank discussion with them and the business sponsor about whether the team will require further specialist support from external advisers with recent experience of relevant procurements.

Our first task as project managers in engaging with the procurement specialists is to ensure that they are fully aware of what the project is trying to achieve and the constraints that are being placed upon us in trying to deliver it. Whilst we will need to use their expertise to help us focus what the project needs into a form the market can provide, we also have to be clear with them on our objectives for the procurement and check that they fully understand them.

Our very early consideration of how we could include critical success factors in any supplier contracts is one area in which it is advisable to engage both our procurement experts and our stakeholders. When the shape of these requirements becomes clearer, we should also consider opening discussions with the client organisation's legal team about how such demands could be built into enforceable contracts.

Procurement Step 3: Supply Market Analysis – Update

Early on in designing our solution, we had the option to undertake a supply market analysis. This was part of the research to inform our thinking and help us understand risks, issues, and the potential costs of developing our solution (see Chapter 3). Since that time, our understanding of the required solution, planned timescales, and allocated budgets will have become much firmer. It is therefore useful to consider now whether the supply market analysis requires updating, and if it was not undertaken earlier, whether it is now appropriate to do so.

The update at this point is another opportunity for an information exchange and can be used to inform the market that a procurement is now pending and to give more details on the proposed solutions that will be sought. We can also seek feedback on the shape of potential tenders and contracts that would be attractive to the market, such as the use of different forms of partnership working, pain/gain payment mechanisms, and minimum periods of operational support. It is also another extremely useful way of engaging with the procurement and possibly the legal teams to help further build their understanding of the project and its objectives for the procurement.

Procurement Step 4: List of Requirements

We need to think carefully about how we shape the definition of what we intend to procure. We must ensure that we are going to buy what the client really wants! The list of requirements is not always a simple and straightforward list of items describing the solution that the suppliers are required to provide. At its very simplest, our list of requirements should consist of both:

- **Functional requirements**. The detailed description of the solution that the supplier is required to provide, including timescales, quality requirements, and regulatory standards.
- **Non-functional requirements**. Covering the way the relationship with the supplier will operate.

Lists of requirements can be written in many ways. Consider three basic types that we can develop:

- Detailed specification of every item and or service which instructs the supplier exactly how our requirements are to be delivered.
- Time and materials requirement, essentially a statement of the inputs to be provided, detailing hours of work required plus any materials needed to deliver the work. This approach is often used in service contracts.
- Business-based performance requirement, where the specification lists the business outcomes required and leaves it to the supplier to decide how those outcomes are to be provided.

Sometimes lists of requirements contain elements of all three, but many projects still seem to favour the detailed specification approach. The project acts as if the client organisation wants to buy a kit of parts from the supplier, and, very importantly, the client will therefore carry the risk if the supplier(s) cannot make the different parts work together effectively in the client organisation's environment. Whilst this approach may be appropriate in instances where the client is genuinely responsible for building the solution, it can often place the business risk associated with a procurement with the client, which will be unable to manage it, rather than with the suppliers, which can. Consequently, we may also want to consider an alternative arrangement for supplying the client organisation's requirements, especially where either a mix of goods, services, and longer-term operational support is required or where we may be engaging with many different suppliers (see Appendix 2).

In our list of requirements, we need to consider the basis upon which we want to transact with the supplier and how we want them to behave. Figure 10.2 sets out how an analysis, using two

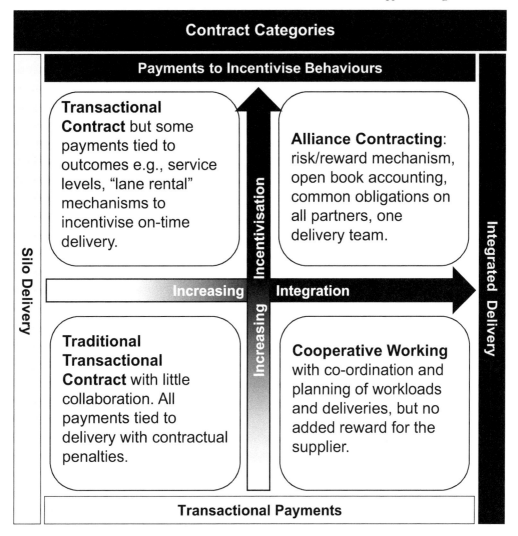

Figure 10.2 Contract Categories

dimensions, incentivisation and integration of working, gives rise to different contractual and contract management requirements.

Consider the example of two actual procurements, both of which were for public bodies in the EU but in different countries. Although dissimilar in scale and complexity, both projects involved procurements for the supply and continuing operation of IT systems. Each procurement required a single supplier to provide on-site centralised hardware, an integrated software system, and peripheral hardware. One client was a hospital in the UK (still an EU member at that time), and the other client was a public body in central Europe. Both organisations had decided that they did not have the expertise in-house to provide and run the required systems and would need a single supplier responsible for:

- Supplying the IT software and hardware.
- Providing the staffing to install, test, and fully commission the software and to train staff in its operation.
- Operate the entire solution over a fixed number of years, with an option to extend the contract period.

The hospital. The hospital designed a procurement strategy based on a delivered operational service that would have to meet certain performance standards, rather than one based on delivering hardware and installing software. The hospitals had asked their potential system users the question, "Do you really care if you have a new digital solution or an army of clerks, as long as they deliver the service you want?" The answer received was, "We want the business solution that meets our needs and is available when we need it."

The procurement was, therefore, written in terms of providing the business solution required. This included specified requirements of functionality for the user, as well as system performance and availability. The procurement, and eventually the contract, expressly did not detail how these should be achieved. This left the various potential suppliers to design their integrated solutions and to price in any risks of failing to deliver to the required performance standards. The funding was granted to the hospital on this basis. A procurement design was agreed upon under which the supplier had to meet a schedule of essential business functional requirements and would be incentivised to deliver and maintain a high-performing service. The final contract between the hospital and the supplier would include a pain/gain mechanism which would provide for:

- Additional payments for overperformance, measured on a quarterly basis.
- Reduced payments for not meeting quarterly availability targets.
- A floor of 90% system availability, below which no payment would be made for the quarter. This availability level had been seen as a critical success factor for the operational system which the project must deliver, as it was judged by the users that below this level, the hospital would have to run parallel manual procedures, negating any benefits from the system.

In summary, the chosen solution for the UK hospital procurement meant that the selected supplier would be motivated to provide, and would be paid for, delivering the operational business solution required by the hospital, not just for installing its component parts.

The public body. The central European public body, which was procuring a much larger solution totalling tens of millions of euros, chose a more traditional procurement approach with an envisaged split between payments for the:

- Supply of goods.
- Supply of standard software.
- Development of custom software.
- Implementation services.
- Running of the operational system.

The requirements for the supply of goods and services were specified by the client organisation in great detail. The huge volume of pricing information provided by the selected supplier was extremely detailed, right down to the required numbers and unit prices of every connector, every plug, and every metre of cable, as well as the numbers of days by each member of their staff to deliver each task required, with differing rates for each grade of staff deployed. The chosen supplier was therefore motivated to supply hardware, implementation services, and custom software as quickly as possible.

Both suppliers experienced difficulties in fulfilling their obligations. The hospital's supplier had problems after the system went live. It took two quarters before it resolved these system problems and finally achieve the required 90% availability. In accordance with the contract, no payments were made for these first two quarters. By contrast, the public body's supplier was motivated to deliver and be paid for:

- Core hardware, which it did very early, and incidentally years before it would be required by the project even if the main system was delivered according to plan.
- Peripheral hardware, which was delivered well in advance of the integrated system being implemented and available for the client to use.
- Staff time as it was being incurred.

In addition, the public body had to pay hardware operating costs and software licences from early in the project and for warranties which were run down during a prolonged software development delay. On our last site visit, you saw that this second contract went very badly and it ended in termination by the client.

The lesson from these two contracts, which were at the extremes of what was possible, is that buying what we want to procure often means thinking laterally and looking beyond the "kit of parts" approach. It might mean, for example, including specifying our requirements to include business solution outputs rather than only supplier's inputs. As in the hospital example, this approach can enable the transference of at least some of the risk of not meeting the client organisation's business requirements to the supplier who should be best able to manage it.

Procurement Step 5: Agree on and Plan the Approach

The updated supply market analysis and our list of requirements are precursors to deciding how we should design the procurement. Even within a strictly regulated process, client organisations have choices and preferences in the way they wish to proceed with procurements. We therefore need to discuss the procurement design as early and as widely as possible. Contrast the approach of two public bodies in different countries to our previous example, both running large procurements under EU rules:

- One client organisation, a European central government agency, took the approach that in order to be seen to be scrupulous and open, everything had to be conducted in writing. All briefings were written, answers to questions from suppliers had to be shared in writing with all suppliers immediately, and no meetings, telephone calls, presentations, or other contacts between the suppliers and the client or project staff and managers were permitted until after the procurement decision had been made and communicated to all parties.
- The other client organisation, a (pre-Brexit) UK local authority operating under the same EU rules, selected a "competitive dialogue" approach. Under this route, once the shortlist of

suppliers was selected, in order to further explore the client's requirements and to shape the suppliers' offers, each supplier had equal access to the organisation and its management through a long and exhaustive series of bilateral meetings. After this process was completed, which took many months, each shortlisted supplier redefined its solution and submitted a "best and final offer."

Both approaches complied with local, national, and the same international procurement rules, but each had very different procurement processes. From our standpoint as project managers planning and controlling our project, each approach required:

- Different completion timescales for the procurement.
- Dissimilar work breakdown structures during the procurement process.
- Very different demands on project and organisational resources.

As project managers, our first step in planning a procurement must be to familiarise ourselves with the way or ways the client organisation wishes, or is able, to procure all the goods and services we need for delivering our project. When designing our procurement, we also need to make an assessment of how much interest we are likely to generate when requesting tenders. Our objective is to stimulate enough response to thoroughly test the market but not so many that we are overwhelmed by the procurement management and assessment processes. In some procurements a two-stage approach is adopted:

- Stage 1 is a qualification stage designed to ensure that suppliers have both the capability and the capacity to provide a solution. Sometimes this may take the form of preparing a shortlist for a single contract, and sometimes it means creating a panel of suppliers under a framework arrangement who are then able to bid for a range of subsequent opportunities.
- Stage 2 is limited to the qualified suppliers, who are invited to compete for the contract and submit their detailed proposals.

This two-stage approach can reduce the amount of work for both us and suppliers by reducing the overall number of detailed tenders that have to be prepared and evaluated.

Determining the exact steps in the tender process and then managing them is usually best carried out on consultation with experts in the procurement team. It is possible for it to be undertaken solely by the project manager, but ensuring that we have expertise guiding and supporting what we do is likely to:

- Improve the quality of the tender responses we receive.
- Protect the client from claims of bias or unfairness if inadvertently deviating from the required process. This protection includes the protection against legal challenges being upheld. These are not uncommon, especially when letting public sector contracts.
- Allow us to move much of the administrative burden of preparing tender invitation documentation and dealing with multiple suppliers to a team experienced in working in a manner which safeguards the integrity of the procurement process.

In Procurement, as in Comedy, Timing Is Everything

When setting the timings for the submission period, we need to reflect the complexity of the information that we are asking for. We may want to set a period that is longer than any applicable statutory minimum response period to enable this. We should always include enough time in the planned submission periods in the tender process to ensure that suppliers have a sufficient period to:

- Review our tender documents.
- Raise any pertinent questions with us about our requirements or about the tender process (including sufficient time for us to prepare a respond to them).
- Prepare a high-quality submission, including taking into consideration our answers to their questions.

However, in scheduling the procurement, we also need to be able to respect our own necessary timeline for the:

- Preparation and authorisation of the tender documents for release.
- Review of documentation received, including taking up references and arranging reference site visits if these are required.
- Getting agreement on the final selection decision.

Unfortunately, it seems that we often get to the "authorisation of tender documents for release" stage immediately before a major national public holiday, and if we don't work through the night, some or all of our decision makers won't be contactable for a week or two. The kind project managers amongst us will take into account our intention to send out the tender documents at 6 p.m. the day before the holiday season when setting the date for receipt of tenders. The less generous amongst us will expect potential suppliers to work through the holiday period. If we are expecting responses from suppliers in a variety of countries, we cannot take into account all the various national holidays anyway, but we should consider this as a factor in setting the length of the response period. Be warned, there is a sting in the tail if we choose to send out tenders immediately before a national holiday period. Someone still has to be available to answer the suppliers' questions, and that person may well be the project manager!

Procurement Step 6: Design Delivery Relationship

Procurement is the point at which we define not only solution requirements to be provided by suppliers but also how we will relate to and operate with them during the delivery of the project and in some cases for continuing support during the life of our project's solution. Increasingly, large projects can involve multiple suppliers, frequently operating from different countries, and these suppliers may exist in hierarchical supply chains of prime contractors and sub-contractors. Appendix 2 looks at how this has evolved in the UK construction industry and the approach of partnership working between clients and suppliers that has developed. Although its origins are complex, the partnership approach has a number of lessons that can be applied to help us better shape our desired delivery relationships in most projects.

When designing our supplier relationship, we must focus on motivating the supplier to continue to want to supply us. This is especially true if the supply relationship for the project and live operations are to continue over an extended period. This relationship will include how we intend to organise both payments and other aspects of the supply arrangements in areas such as change management, management of risks and issues, reporting, and quality control.

Even when we are in a position of strength in relation to the market, in determining our relationship, we need to avoid creating situations where a chosen supplier ceases to be able to deliver, perhaps even ceasing to trade, or deprioritises our contract in favour of other work. Such situations can arise from issues such as:

- Having accepted artificially low prices, which can result in firms either ceasing to trade or deprioritising our contract relative to other more profitable contracts.
- Imposing onerous conditions on the supplier – for example, unrealistic performance targets which result in payments being withheld. One procurement for an outsourced service provider in a European country required accepting the transfer of up to nine staff under EU regulations, each of whom currently performed part of a service being tendered, but the proposed reimbursement equated to five full-time equivalents, thus making the level of return unacceptable to the supplier.
- Creating processes with unwarranted delays in accepting reasonable requests for change. For example, one client insisted on a 30-day response period for all changes, and this could be reset if the client asked a question about the request, meaning that a project could be indefinitely stalled.
- Making unreasonable changes to terms and conditions which alter the balance in industry standard contracts. In one case a dominant buyer changed the entire balance of their responsibilities by inserting one word in a sub-clause of an industry standard contract. This sub-clause had a closed list of five permitted exceptions which the supplier had to deliver at no additional cost, but just adding the word *including* immediately before the list meant the list was now merely five examples and the purchaser could claim anything to be an exception to be delivered at the supplier's cost.

Setting the Payment Approach

Following an approach to contract delivery which engages with the supplier as a partner is a desirable goal,[2] but as we saw in the field trip, what our suppliers really want is payment. At a minimum, their requirement is usually for a prompt payment for the goods and services on which they have already incurred their costs – i.e., they wish to minimise the amount of time for which they are funding our project costs. Ensuring our payment schedules include providing suppliers with realistic and fundable cashflows is an integral part of designing a sustainable supplier relationship and an area where we need to be very sensitive to the market in order to achieve the results we want. Ideally, we also want to try to avoid disputes and the sort of confrontational scenes we witnessed on our last field trip.

As project managers we often have to reflect the project's client organisation's desire to pay for only demonstrable results – i.e., that they should not be required to pay a supplier before whatever goods or services we have purchased are completed and (if appropriate/necessary) installed, tested, and shown to operate correctly. We may immediately find ourselves in a position of potential tension between the objectives of the supplier and the client. On top of this, we may find that the client organisation has internal rules requiring delays of often another 30 or even 90 days after receiving an invoice before payments are made.

In defining the relationship we are going to have with our supplier or suppliers, we must understand the effects of the client organisation's wish to pay later in the project on the cash-flow of the supplier's business and how these effects may rebound back onto the project. This "cashflow conundrum" is illustrated by reconsidering the two previous examples of a client purchasing a project solution requiring a single supplier to manufacture hardware, develop custom software, and provide implementation services. In order to meet our project plan, our supplier will necessarily incur costs according to the profile set out in Table 10.1, which are fixed. The supplier will also receive a final single payment to cover a fixed profit element, which we have also agreed to in the contract, equalling 10% of costs – i.e., €104,000 – at the end of the contract.

Table 10.1 Supplier Expenditure Projection

Supplier Expenditure Projection				
Month	**Manufacturing Costs**	**Software Development Costs**	**Implementation Costs**	**Allowed Profit @ 10%**
Month 1	–€10,000.00	–€10,000.00		
Month 2	–€30,000.00	–€30,000.00		
Month 3	–€50,000.00	–€30,000.00		
Month 4	–€100,000.00	–€30,000.00		
Month 5	–€100,000.00	–€30,000.00	–€50,000.00	
Month 6	–€100,000.00	–€30,000.00	–€50,000.00	
Month 7	–€50,000.00	–€30,000.00	–€50,000.00	
Month 8	–€10,000.00	–€30,000.00	–€50,000.00	
Month 9		–€30,000.00	–€50,000.00	
Month 10		–€30,000.00	–€20,000.00	
Month 11		–€10,000.00	–€10,000.00	
Month 12		–€10,000.00	–€10,000.00	
Month 13				–€104,000.00

The supplier asks for a payment schedule that reflects the profile of their costs, and suggests a payment profile where all costs are reimbursed in the month after they occur. Figure 10.3 shows that this would mean that the supplier's cumulative cashflow requirements peak €180,000, and are as shown by the shaded area:

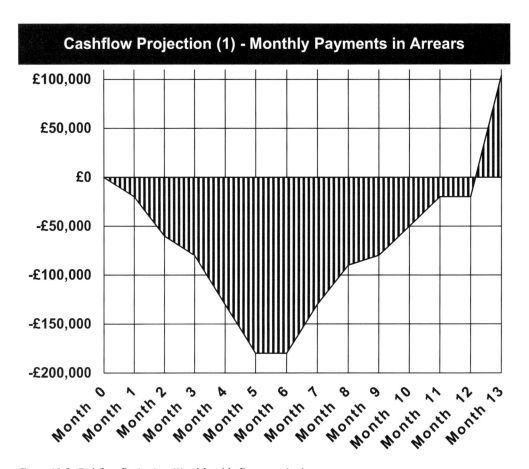

Figure 10.3 Cashflow Projection (1) – Monthly Payments in Arrears

Our problem is that we have learned from our business sponsor that our client organisation's CEO claims to have been caught out by suppliers before on similar projects that failed to deliver and wants to pay for this "thing" only when "it's shown to be all up and running; not a single cent should be paid before that." Using exactly the same cost profile, the adverse impact that this would have on the supplier's cashflow is shown in Figure 10.4.

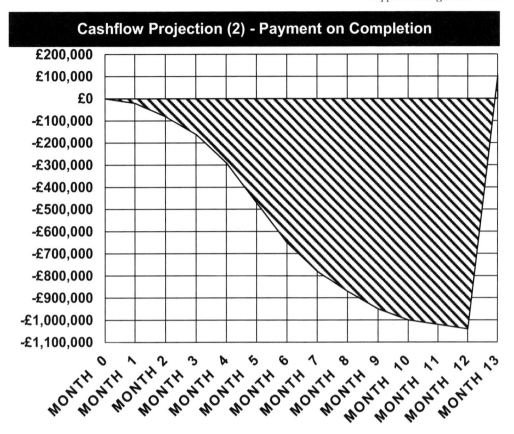

Figure 10.4 Cashflow Projection (2) – Payment on Completion

In the first case, the monthly average cash shortfall that the supplier would have to fund is approximately £87,000, but in the second case, this rises to nearly an average of £611,000 and peaks at £1,040,000 in the last month.

For the supplier there are two critical issues:

- **Cost.** The cost of funding the extra cashflow, which, if not met from borrowing, will have other opportunity costs to their business.
- **Risk**. In the first example, the supplier's maximum exposure to a risk of failure to receive payment on time (or at all) is the unreimbursed costs from the previous month, which is at a maximum in months 6 and 7 at £180,000, plus the loss of any profit earned to date. In the second example, it is all the costs incurred to date and the lost profit, which by the end of the contract totals £1,040,000.

To add a flavour of real life, consider what happens to the supplier's cashflow if the custom software element of this example project is not completed on time.

- With monthly payments this may be painful for the supplier, and the project timeline will be extended. The major impact on the supplier's cashflow is that of lost profit caused by funding the extra non-recoverable costs of the software development team.
- With contract completion payments, the supplier still has the same extra software development team costs to fund, but in addition they will have to meet the extra cost of funding the total cashflow deficit for the period of the contract extension.

The impact of the client CEO's wish to pay only when delivery is completed is therefore:

- To make our contract less attractive to all suppliers as it will force them to carry a much higher level of financial risk.
- To potentially increase the contract price, as the supplier's cost of funding the additional cashflow deficit and the additional risk will have to be built into their pricing model.
- To make the contract untenable for smaller suppliers who cannot fund the additional cashflow requirements.

As with much of what we do in project management, there is no one-size-fits-all answer to the cashflow conundrum. In working with our procurement advisers to set out how we wish to design the contract payment mechanisms, we need think through our suppliers' concerns and the impacts that our choices will have on them. The results of our design of the payment part of our supplier relationship may affect both the number of suppliers responding to our procurement and the prices they wish to charge.

Procurement Step 7: Sharing Information

There is one common factor that is critical to our success whenever we engage support from a team or organisation outside of our core project team, whether it is on a commercial basis or not. This common factor is the appropriateness of the supplier briefs we have to prepare for our supplier or suppliers. Our supplier briefs will change depending on exactly what we require and who is going to deliver that part of our project for us. A supplier brief could range from an email covering a single short task to be carried out by another team in the client's organisation to a full, detailed statement of work running to many hundreds of pages of specifications for the prime contract for a major infrastructure development.

The technical element of each supplier brief is another essential part of our "golden thread" of objectives and requirements for the project. Even if it is only a short email, the information it provides must come from, and be in full alignment with, our:

- Critical success factors for the project.
- Approved business case.
- Work breakdown structure.
- Agreed-on stage plans and overall project plan.
- Proposed contract or other agreement with the supplier.

The supplier brief's contents should be traceable backwards and forwards through all these documents, as well as any agreed-on changes made to them. It is always worth our time to pause and review every supplier brief (and, later in the process, every contract and any work instructions that we issue) to ensure that it truly aligns with our "golden thread" and that the following items are always specified unambiguously:

- **What**. Deliverables we require, maybe including our complete statement of work and change control processes.
- **Why**. Brief statement of the project and its critical success factors.
- **How**. Quality and other standards required.
- **When**. Dates for delivery that align with our project plans.
- **How much**. Payment principles and terms.

We also have to specify any other conditions that we require for this particular task, including:

- **Who**. Supplier personnel to be responsible for the task and our personnel responsible for managing the contract and accepting the work performed.
- **Where**. Locations where the work is to be completed or outputs delivered.
- **What if**. Management of risks and issues, incentives, and penalties for late or non-delivery.

In drawing up the supplier briefs for a commercial procurement, we will usually include input from the procurement experts on the process being followed. We will also include input from the legal team on their required contractual terms and conditions, possibly even including a full a draft contract, and sometimes a requirement that the supplier accepts these terms and conditions in full as part of their bid submission.[3]

As part of the brief, we will also set down the rules for the tender submission. There are a number of these rules that we can consider putting in place which will help the selection process, including:

- Limiting the amount of text the supplier is allowed to provide in the tender. The word count limit should be enough to allow adequate answers to our questions but not so much that the tender has to be put on a pallet and delivered using a forklift.
- Requiring a questionnaire to be completed which addresses every item in the statement of work. This allows technical and quality evaluation to be progressed more easily than attempting to find and extract answers from unstructured text spread across a lengthy document. Using a questionnaire makes it easy for the successful supplier's response to be included as a contract schedule, thereby unambiguously binding their delivery team to provide what their sales team have promised and allowing us to monitor and enforce their performance.
- Asking for bidders to separate the financial and the quality parts of their tender response, as it is easier for the suppliers to submit separate documents than for us to have to divide up a document later. This separation is to ensure that our technical quality assessment will not be influenced by any financial factors, as our process would be undermined if quality evaluators modified their scoring because the price seemed low or high. We will bring together the technical and financial assessments only at step 12 and under strict control.
- Deciding how and whether non-compliant or variant bids can be considered. We may decide, for example, that provided that the supplier submits a conformant bid, they may also submit suggestions for contracting on a different basis, including perhaps different payment profiles, methods of working, and terms and conditions. Or we may not.

Procurement Step 8: Answering Suppliers' Questions

The period between completing the supplier brief and the closing date for tenders is often a quieter period for our project team. This is because the major activity passes first to the procurement team to despatch the tender invitations and then to the suppliers to prepare their responses. However, despite having carefully prepared our supplier brief to be as fully rounded a document as possible, and having had it reviewed and approved by experts in its technical elements, its procurement aspects, and the client organisation's legal requirements, someone is going to ask a question.

Usually, the procurement team should manage suppliers' questions, distribute them to the people best equipped to answer them, and track the responses. It is typical, and in some jurisdictions mandatory, to distribute all questions and responses to all tenderers. The quicker that we, or the procurement or legal teams, can give suppliers an appropriate answer, the better the tender responses that we are likely to get – even to those foolish questions that prove they have not read the brief carefully. Having the right people available to answer questions promptly is the main reason to avoid setting the tender response period to coincide with peak periods or public holidays. If we fail to plan for a sufficient time to respond to suppliers' questions, it will put time pressure on the whole tendering process, including on ourselves. It will also put pressure on the individual suppliers to reflect our response to both their own questions and to those from other suppliers in their submissions.

Procurement Step 9: Technical and Quality Evaluation Criteria

Although we only require them to be finalised immediately prior to the evaluation process, technical and quality evaluation criteria are often completed much earlier and can be included as part of the briefing pack for suppliers. These criteria should be based on our critical success factors and our list of requirements, possibly as later reformulated into the statement of work. They are not the sole basis of our tender selection, but they are a major part of it. By now the details of our technical requirements will probably have been discussed and reviewed many times, so compiling them into a format to be used by our technical evaluators should be straightforward. However, we should always bear in mind that:

- Not all of our evaluators will have been through the requirements development process, so we need to make sure that we are very explicit in the language we use and that we work to remove as much ambiguity as possible.
- In any review of the project by quality assessors or auditors, the technical and quality evaluation criteria will be seen as an important document in establishing that the procurement process was fair and reasonable.

When we identified our requirements, we used the MoSCoW approach for prioritising them, and we need to ensure that these priorities are reflected in the relative weights we assign in our evaluation criteria. As far back as Chapter 3, we looked at a method for weighting scores (see, for example, Table 3.5), and again in Chapter 4 (see Table 4.3). So, here is a quick refresh on how it works in selecting a tender. Essentially this approach splits a single subjective judgement about how well the tender meets our needs into a series of smaller evaluations, some of which will still be subjective and some which will be demonstrable matters of fact. For example, if we have specified that it is a requirement for the supplier to demonstrate that it has an approved health and safety policy or is accredited against certain quality or other standards, then our assessment is objective and will be based on simple provable facts. If, however, a judgement is required about how well a criterion is met, then it remains a subjective assessment.

The weighting method works by assigning a value to the importance of each of the technical evaluation criteria – its weight – and, during the evaluation process, scoring each supplier on its perceived performance. The resultant two numbers are then multiplied to give a weighted score. Weighted scores can be added to give a total value degree of conformance to our requirements.

Procurement Step 10: Technical and Quality Evaluation of Bids

Even when we have taken our procurement professionals' advice on the best evaluation models to use and how to document and communicate everything so as to avoid potential legal challenges, there is still the difficult problem of evaluating the technical quality of the suppliers' final proposals.

The challenge here is bringing together sufficient expertise to review proposals and then arrive at a method whereby different views can be synthesised to produce an agreed-upon score, in response to each criterion that we have set. There is also a problem of familiarity of the evaluators with the bidders leading to subconscious bias. This is one of the reasons for keeping the technical evaluators away from suppliers except at controlled events such as demonstrations and reference site visits. As an example, two local authorities, A and B, were co-operating to procure a similar professional engineering support service. Only two suppliers were shortlisted: Company Y was the existing supplier to Authority A, and Company Z was the existing supplier to Authority B. The two authorities' quality evaluation panels met separately. In response to nearly every requirement, each authority had marked their known supplier higher and the other's supplier lower. The two score sets were virtually mirror images. Bringing these together into one synthesised view proved difficult. The impasse was settled by weighting the scoring according to the value of the contract that each authority was letting, meaning that the larger authority's views took precedence.

There are, however, some things that we can do to help increase the objectivity of the technical and quality evaluation process. Note that not all of these will apply in all cases:

- **Mandatory means mandatory.** When we set out our requirements in the tender documents, we should have identified those that we categorised as "must have" under our MoSCoW prioritisation, and these "must haves" should have included all of our critical success factors. At the harshest level of evaluation, failure to meet any one of these criteria will eliminate the tender. Some tender evaluations allow some flexibility with a requirement to meet at least a high fixed percentage of the "must haves." A separate scoring weight is then applied to mandatory items as opposed to those that come further down our list of wishes. This flexibility can lead to problems if, for example, failure to meet a "must have" would mean that the tenderer's offered solution does not fully meet every critical success factor.
- **Engage a wider team**. When considering stakeholder engagement, a good example to follow might be that of an information systems procurement, which included demonstrations by suppliers to several hundred users. Each user was required to submit an evaluation of each solution, and their scores were included in the technical and quality evaluation.
- **Test the supplier and their solution**. This may require collecting written references from existing customers or undertaking reference site visits. In both cases, it is necessary to create a template of questions on which the evaluation team needs to be satisfied and apply it equally to each supplier. Reference site visits have to be handled carefully, and, within the limits of the

site being visited, it is useful to include several members of the project team and stakeholders, with each targeting a different area of interest. The project manager and the procurement team leader should always be present and can usefully engage with the supplier's representatives, freeing the rest of the evaluation team to link with their opposite numbers in the reference site's organisation. It is advisable to retain some healthy scepticism when reading references or when on a site visit. Remember that following up any concerns you have is always useful.

- **Break up the evaluation into manageable sections**. In our tender requests, we ask for a great deal of information, and the result is that many tender proposals can be voluminous documents. Assigning sections to different panels of evaluators can reduce the time that each individual requires to undertake their review and scoring, and it also meets our aim of engaging a wider team.
- **Spend time resolving issues**. Evaluators may feel that a tender response does not explain the proposal fully enough, and we have to make a decision as part of the procurement process on whether to allow for post-tender clarifications.
- **Spend time resolving scoring differentials**. In their view of any response to a requirement, different evaluators will often arrive at different scores. Our evaluation process, therefore, needs to allow sufficient time for a moderation process to arrive at an agreed-on score. Our options in achieving a moderated score range from the least desirable of merely adding scores together, or averaging them between evaluators, to the usually preferred solution of facilitating a discussion to arrive at a consensus score. Given that part of our objective is to develop an agreed-on solution with support from across the client organisation and from our stakeholders, this latter approach is useful but can be time consuming.

Procurement Step 11: Financial Evaluation of Bids

Sadly, especially in complex bids with different options involved, this is not just a case of selecting the supplier offering the lowest headline price. We should instead consider the financial impact of the proposed solution, including tender prices, payment profiles, and other ancillary costs to be carried by the client, and also take into account the impact of any timing proposed by the supplier on achieving financial benefits. This basically means that we should rework the net present value calculations, but this time for every supplier and every option in their proposals.

When we have either the supplier's quoted price – for simple, easily compared tenders – or the calculated net present values, for more complex ones, we need to use a fair and reasonable scoring mechanism. This is another area in which procurement expertise is required. In different jurisdictions, different approaches will be preferred. Particularly in relation to public sector contracts, some methods may not be legal. For example, courts in France ruled against a particular contract award decision based on an evaluation method in which the lowest price got the maximum points and the highest price received zero points, which is the basis of a method used elsewhere.

Procurement Step 12: Select the Most Economically Advantageous Tender

Traditionally, when looking for a supplier, one question assumed paramount importance: "How much will they charge?" Using the answer to this question as the sole rationale for selecting our successful tender seemed simple and required no justification. However, this approach did not always deliver the best solution. The Victorian artist and writer John Ruskin is reputed to have summed up such attitudes like this: "There is hardly anything in the world that some man cannot make a little worse and sell a little cheaper, and the people who consider price only are this man's lawful prey."

The thinking on the best approach to supplier selection has moved on from the cheapest supplier to the concept of the most economically advantageous tender, or MEAT, which is also known less memorably as the best quality/price ratio method. This approach includes a weighted evaluation of how the tender would meet our requirements, as well as its cost.

This concept is included in the UK and European regulations for public procurement. These regulations include both a useful definition and a guide to how to select such a tender. The regulations tell us that the most economically advantageous tender is assessed from the view of the contracting authority and is to be identified on the ratio of the best price (or cost) to quality. Quality is to be assessed on various criteria, which could be qualitative, environmental, or social aspects linked to the subject matter of the contract. These regulations also require that the weightings to be used for each of the criteria are set in advance and included in our tender documents (UK Government, 2015). Designing our weighting and evaluation process is another point where it is essential for us to engage with the project's procurement advisers, and possibly also our legal advisers, in order to make sure that the supplier selection will demonstrate transparency and fairness and minimise the risk of legal challenges to the final decision.

Whether legally required or not, sharing our evaluation model with the suppliers helps them structure their response to meet our priorities.

In designing the final selection process under MEAT, the key elements of the tender evaluation which we need to consider are:

- Qualitative evaluation of supplier responses to the tender requirements. This can include our assessment of the technical ability of the proposal to meet our specified requirements (as discussed previously), plus the more general aspects of the supplier's proposal such as aesthetics, accessibility, environmental, and innovative characteristics. Perceived staff quality and organisation is also a major area of evaluation in contracts which include buying in of services from the supplier. We are concerned with the ability of the supplier to organise the staffing of the service and with the experience and qualifications of the staff being proposed.
- Meeting non-functional requirements. In most procurements we will also include a statement of our needs other than those for the solution itself. These needs may include items such as delivery dates and processes, continuing support from the supplier during and after the implementation period, and meeting any other supply conditions that we require.
- Mathematical evaluation of the prices and other financial impacts. We must take care, when formulating a method of evaluation, to ensure that it is seen to be fair (and legal) in all cases.
- Use of a formula to bring together the financial and qualitative evaluations. This is another area where care has to be taken to achieve the correct balance between the two elements. For example, a 70/30 price/quality split will confer advantages on any supplier who offers a low initial price and plans to use overpriced change requests or fees for additional services to increase their income, as we saw in our last site visit.

In bringing together each of these areas into one selection decision, there are also choices on how comparative weights, calculations, and evaluations are to be made. The final selection process is again one of establishing and agreeing on the rules to be followed, weighting the criteria to be used beforehand (in some jurisdictions, procurement regulations require that these are sent to suppliers as part of the tender documents), and evaluating the suppliers' responses.

Procurement Step 13: Post-Tender Clarification and Negotiation

When completing the evaluation, it may be necessary to ask for clarification of some aspect of the bid which is unclear. This is not an opportunity for the supplier to change their bid but merely for them to explain their proposals more clearly. Some procurements will also allow for post-tender negotiations, in which many of the details of the bid may be negotiated. Matters covered in post-tender negotiations may be areas included in the statement of work in which it may be considered that suppliers could improve their offer, including, for example:

- Proposed completion dates.
- Warranties and guarantees.
- Payment terms.
- Ongoing support, including maintenance, repairs, and spares.
- Escalation procedures and crisis response and management.
- Quality standards and processes.

In the case of either post-tender clarifications or negotiations, it is usually best if this process is led and recorded by the procurement team. They will be used to being able to demonstrate that this is being undertaken in a fair manner which does not distort the tender process. This is particularly true of post-tender negotiations which may give rise to legal challenges, especially if the result of the negotiations could alter the final choice of supplier.

Part 2: Engage the Supplier – Contracting

It has been said that on the best projects, once agreed on, the contract can be placed in the bottom drawer and never referred to again. If this was ever true, it was on an extraordinary and very simple project, because the contract is the lens through which the relationship between the client and the supplier is going to be focused for the length of the project and for any subsequent period of warranty, support, or maintenance. The contract contains the full details of what the supplier is going to deliver, when, how much they are going to be paid, and what terms and conditions both sides require to observe during the relationship.

As with the procurement process, when completing a contract, the project manager must understand the principles but must also draw on specialist expertise to ensure that the contract which is agreed on both meets our needs and is fair and reasonable to both parties. It is usually preferable for us as the purchaser to determine and stipulate terms and conditions in advance of a tendering process and not to trade on the supplier's terms. Consequently, there is a need to allow time for a process to develop a suitable contract, including:

- Discussions with the procurement team on the requirements to be included in the contract.
- Discussions with the legal team on how our requirements may be translated into a workable contract.
- Discussions with the supplier and obtaining their agreement to the contract terms.
- Obtaining the contract signature by the client organisation, often referred to as "contract execution." This process may include internal reviews and analysis of the final draft contract prior to execution. The selected suppliers will usually have to be allowed time for a similar process between agreeing on a final draft of the contract and contract execution.

Although we may not have control over the timing of many of the necessary steps, much of the work required will be carried out in parallel with the procurement process. A model set of steps for the contracting process is set out in Figure 10.5.

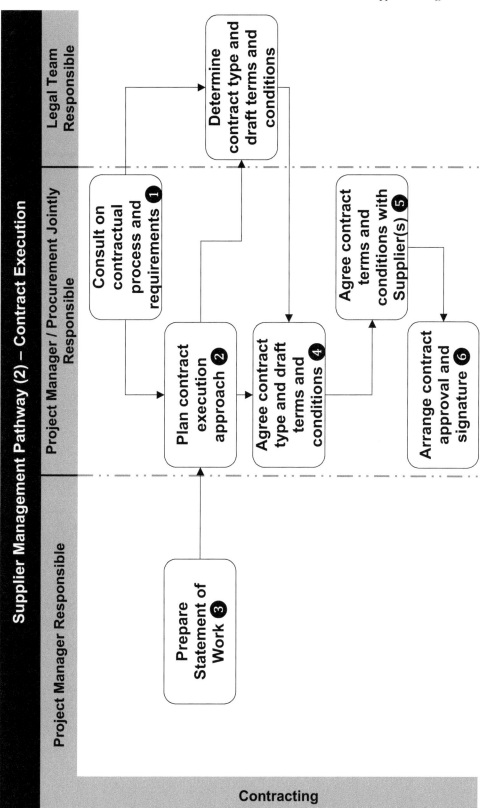

Figure 10.5 Supplier Management Pathway (2) – Contract Execution

Contracting Step 1: Consult on the Process and Legal Requirements

As project managers, we will be responsible for managing contract deliveries. Therefore, we may have views on the contract and its terms which we wish to discuss with the client organisation's legal team, and we will want to test whether our views can be built into the contractual framework. Even if this is not the case, it is nearly always useful to engage with the client's in-house legal team or external legal advisers as early as possible, not least because they may have views which will influence the preferred procurement approach. As we have already discussed, they may have detailed experience of the client's attitudes in key areas such as risk sharing or payment terms which could require us to modify our plans and projections. They may also have preferences for different types of contracts, templates that they frequently use, and standard terms and conditions upon which they insist. They should also be able to advise on the contracting timetable and the steps required to obtain the client's authorisation for contract execution (signature).

Contracting Step 2: Plan the Approach

We need to work with the legal team to understand their views on the sorts of steps and time periods required to complete an agreed-upon contract of the type we need. This will enable us to prepare what is essentially a stage plan. There are usually crosscutting dependencies with our procurement plan that we need to take into account, particularly if we intend to send out draft terms and conditions with the briefing pack for suppliers. As ever, we need to pay particular attention to the dates being projected for agreeing on any key stages with decision makers, and especially for arranging for contract execution. It can be a frustrating experience to have completed a successful procurement and contract negotiation only to be held at the execution step because a key signatory is not available for some weeks and we have not made alternative arrangements.

Contracting Step 3: Prepare a Statement of Work

The list of requirements prepared in the procurement workstream can be seen as a client organisation–facing document, which has been built based our understanding of the client organisation's needs and our updated analysis of the project's critical success factors. In the procurement workstream, we also needed to take account of our supply market analysis and convert the list of requirements into a supplier-facing document, our statement of work, sometimes also called a specification of requirements. There is a crosscutting requirement in the contracting workstream for this document as it should become an integral part of our contract with the successful tenderer.

Although both our statement of work and our list of requirements will have the same basic content, there can be differences which stem from their use and their readership. However, if only a single document is prepared, it must be capable of being added as a schedule to the contract and must be a clear, unambiguous set of instructions about what a supplier is to deliver, when, and to what standards. In particular it should provide:

- **Clarity.** Terms used throughout should be clear, consistent, unambiguous, and not capable of being misunderstood. The consequences of not doing so can lead to contract disputes of the type seen on our last site visit.
- **Detail**. The detail provided should be adequate for the supplier to understand exactly what is required and should include all relevant reference material.

For many procurements, and especially those looking for complex solutions, it can be worthwhile to engage a technical expert to review a draft statement of work to ensure that the level of detail is appropriate and reflects the project's needs. It is also worth asking for a legal review to ensure that we have not created any potential loopholes, which we may regret later.

Contracting Step 5: Agree on Contract Type and Draft Terms and Conditions

Although we may specify our requirements for the contract, the responsibility for drafting and approving the contract structure, and its detailed terms and conditions, rests with the client organisation's legal team or advisers. Contract terms, conditions, case law, and client preferences will vary between different countries and jurisdictions. In some cases, it may be advisable to use industry standard contracts (such as the NEC family of contracts, originally known as the New Engineering Contract but now including contracts covering a wide variety of services and high value goods) with terms and conditions which have been agreed on at an industry level (or wider), supplemented by additional clauses. In other cases, it may be desirable to develop a new contract; the legal team, for example, may wish to prepare a contract based on the client organisation's normally preferred terms or possibly draft a completely new contract from scratch.

By whatever means the contract has been developed (in-house or externally, bespoke or a customised version of a standard contract), the main principles of our engagement with the contract development process as project managers, and that of our procurement advisers, is to ensure that the contract as finally drafted at least:

- Accurately reflects the supply of goods and services that we require.
- Explicitly includes all performance criteria that we wish the supplier to meet.
- Sets out our requirements for delivery schedules.
- Contains payment schedules and payment conditions which align with those in our business case.
- Treats risk in the ways we have agreed are appropriate for this project and for the supply of goods and services we are procuring.
- Clearly sets out the details of how the relationship between the client and the supplier will operate, including performance monitoring, review processes, change control, risk management, and dispute resolution procedures.

Contracting Step 5: Agree on Terms and Conditions with Suppliers

After the tender acceptance, there may be legal discussions and exchanges of drafts between the client's and the supplier's legal advisers. This is frequently called the "battle of the forms" and arises because the client organisation wishes to purchase on the basis that we agreed on in step 4, prior, but the supplier often wishes to supply under their own terms. It is not unusual for there to be rounds of negotiation between lawyers on both sides to ensure that all the details that we require and any other terms and conditions are mutually understood and acceptable to all parties. Our contracting stage plan should have allowed for this to happen. The winner in the "battle of the forms" is usually the side that made the last submission which was not explicitly rejected.

In some cases, this battle can be avoided when clients insist on the proposed contracts being used without alteration and the supplier being required to accept the contract in full as a mandatory condition part of the tendering process. This tactic, as discussed earlier, may reduce the attractiveness of our procurement to potential suppliers. We can always use this approach to try to get the term "best endeavours" inserted as a description of the supplier's responsibility. This is normally interpreted as meaning that the supplier has to stop at nothing, including bankrupting their company, in order to meet their obligations. Any other wording – such a "best reasonable endeavours," as it seems that European politicians publicly failed to understand with their first COVID-19 vaccine contracts – is ultimately a matter of opinion.

Our role is to ensure that the results of such negotiations do not conflict with our goals for the procurement or affect the principles agreed on at the drafting stage in ways we find unacceptable.

Contracting Step 6: Arrange for Contract Approval and Signature

We should already have planned for the approval process, made sure we know who the contract signatory or signatories have to be, kept them informed during the procurement and contracting execution processes, and ensured they will be available when required. We also need to know what documentation they require in order to be able to sign the contract and must make sure we have that prepared for them as well. The signatory's requirements will vary between projects and between organisations. In general, the more remote the signatory is from the project and the larger the contract, the more information and supporting documentation they will require to ascertain that correct processes have been followed and that the contract terms have been approved by the required managers. In some client organisations, this may require sign-offs by managers of various teams involved in or overseeing the procurement, including legal, finance, risk, and procurement itself.

Part 3: Ensure Delivery – Contract Delivery Management

The importance of controlled management of client and supplier relationships cannot be overstated. We are human, and we often build close personal relationships with our suppliers' managers, especially where we are working in a partnership arrangement or are co-located. These strong relationships can enormously facilitate the routine work of co-ordinating and controlling the delivery of our project. They can, however, also lead us into an area of danger when we fail to remember at all times that, whatever our personal relationships, the relationships between the client organisation and its suppliers are governed only by the contracts between them. This can lead to falling into the trap of failing to use our contractual mechanisms properly and then regretting it later.

In the process of drafting the terms of the procurement and then the contract, we should have set out a number of conditions that we require the successful supplier (or suppliers) to meet. This is our baseline against which the supplier and their contract performance will be monitored. The measures that we use and the frequency with which we review supplier performance will vary greatly between projects, types of supply, and the risks to the project associated with them.

Figure 10.6 sets out the types of process steps involved in the effective management of a contract and monitoring supplier performance within the context of a project:

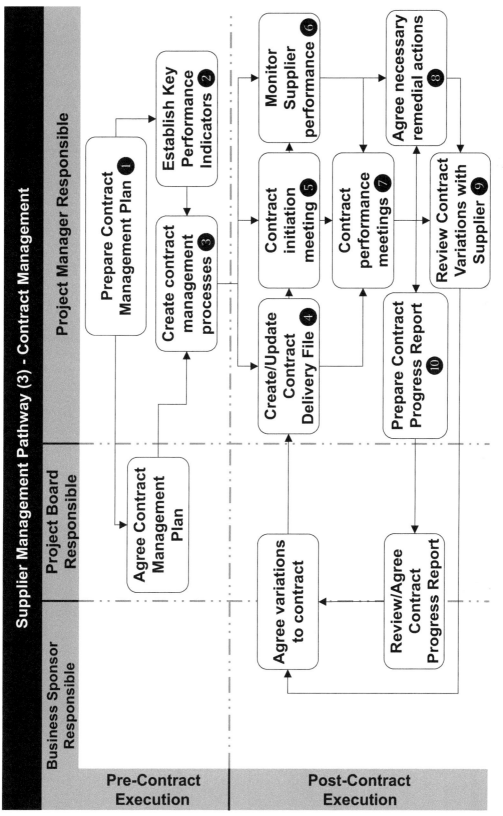

Figure 10.6 Supplier Management Pathway (3) – Contract Management

A Lawyer's Story

There are a significant number of project managers whose careers straddle both project management and working as commercial lawyers. This can sometimes be a blend of knowledge and experience which is immensely useful when difficulties arise in managing supplier relationships. Before looking in detail at contract management, the following contribution from one lawyer and project manager looks at the importance of applying those contract mechanisms that we spent time thinking through, took advice on, and built into the contracts with our suppliers.

Before we start, it is worth mentioning that a contract notice is a provision which obliges our supplier to send a notice to us about circumstances that will necessitate changes to the cost, timeline, or delivery schedule in the contract. We have already seen an instance of changes in circumstances in Chapter 9, when we discovered that a wall was rubble filled rather than solid.

But let her tell you her story in her own words:

After many years in commercial litigation as a building and construction lawyer, and in project management in oil and gas, I have come to the conclusion that the most critical action for a good project manager to perform is ensuring the timely and correct issuing of contractual notices for time and money.

A project manager may feel significant pressure from sponsors and project directions at times (even most of the time on some projects), but a good project manager will ride out the storm and press for the notices to be issued within the time limit stipulated in the contract. The principal [client]/contractor relationship doesn't need to be compromised at all – when delivering a notice, a contractor is simply doing what the principal requires of the contractor in the contract. It is mere compliance with the regime the principal itself put in place, so the principal can't complain when the notices are issued in accordance with that same contract. Likewise, a principal is simply following the regime, and it can be presented as that when being issued.

Being able to deliver (or cause to be delivered) a contractual notice for time or money whilst still maintaining a positive working relationship with the recipient is a prime skill of a project manager. If a notice is not issued, "for the sake of the relationship," then the relationship once it deteriorates (and more often than not that happens when schedule or costs blow out) is going to be fraught with difficulties. In litigation the party that has issued notices correctly and on time will win hands down, regardless of the "moral merits" of the situation. An arbitrator or judge will look solely at the documents, and only in the absence of documents on both sides will oral evidence of "he said she said" be considered. And aside from the inherent and often insurmountable difficulties caused by conflicting oral evidence, it is far, far, far more costly and lengthy a process than one where a notice issued in time answers the question swiftly and conclusively.

Now, that may seem like a blindingly obvious statement, but over and over and over again I have seen a failure to issue notices either on time, or at all, as the root cause of multibillion-dollar disputes that have racked up hundreds of millions of dollars in legal costs and mired companies in litigation/arbitration for at least four or five years (and sometimes longer).

There are many examples where the timely issuing of a notice saved a dispute and the concomitant costs and losses from occurring.

Example 1: Train Derailment Delaying Steel Delivery by Two Months

Construction of a ten-storey mixed commercial residential building with an initial contract price of AUD 30 million was progressing nicely, up to month 5. Vital bespoke steel required for construction was on a train more than eight hours from its final destination near the construction site when a freak derailment severely damaged the rail track and the train itself. Given the nature of the derailment and its location (hundreds of miles from anywhere in the middle of nowhere), it took two months to get the steel offloaded from the crippled train and onto trucks for trucking out to the construction site.

The exact extent of the impact of the derailment on the time of delivery was uncertain at the time it happened, but it was very clear there would be at least a two-to-three-week delay. There was a collegiate relationship between principal and contractor at that time, and discussions were had to the effect of, "Don't worry about paperwork – we'll sort it out at the end, you guys [contractor] are going to accelerate and we're [principal] happy with that a derailment isn't your [contractor's] fault at all."

Now, at my insistence, the notification of a potential claim for an extension of time (EOT) was issued, with an open date. This was issued in accordance with the clear contractual obligation on the contractor to give notification of any event that may cause an EOT as soon as the contractor became reasonably aware of facts that would lead to a claim for an EOT once the extent of the delay became clear.

Further down the line, it became clear the delay was going to be eight weeks and not, in fact, two or three. After some more time had passed, the relationship between contractor and principal became strained when it was clear there was going to be a blowout in time of three months all up and a cost increase of nearly 8% (given the expenditure on trucking and other associated costs of addressing the derailment and delay). The "don't worry" attitude had evaporated, and it soon became "lawyers at ten paces." At that stage, the fact that we [contractor] had issued the correct notice within the time limit required meant we were on a solid footing for the EOT claim together with the associated delay costs and avoided a costly liquidated damages claim. All in all, the mere fact the initial notice was issued, just a one-page piece of paper with three short points on it, saved the contractor more than AUD 2 million in legal fees, liquidated damages, and what would have been unrecoverable delay costs, let alone the soft costs of contractor staff (at all levels) in giving instructions to lawyers and preparing witness statements and undertaking discovery.

Example 2: Freak Flooding of a Basement Requiring Extensive Dewatering Delaying All Work for Three Months

A second example of a timely issuing of a notice of anticipated delay and EOT claim was in the course of constructing a seven-storey mixed commercial residential building with an initial contract price of AUD 22 million. The two basement carpark levels were, in fact, under the water table, and special construction methodologies had to be employed to address that. All was going well, and the slab over the upper carpark level and the lower slab of the ground floor were in the process of being constructed, when a freak weather event occurred and more than four months' worth of rain was dumped in just a few hours. This completely flooded the lower basement level, as well as part of the upper basement level. Damage to the concrete was sustained, and dewatering and rectification of the water damage delayed construction for three months.

At the date of the flood, it was unclear how long the rectification and dewatering works would take. It was initially anticipated to be just three or four weeks, which could have been made up by crashing the trades in the last third of the project schedule. Again, a "she'll be right, don't worry about a notice" *attitude was prevalent; however, I insisted on a notice of an anticipated claim for time and money being issued over the objections of the contractor's directors and notwithstanding the principal's (then) relaxed attitude.*

As the delay dragged on and on, the principal's patience wore evermore thinner until, at month 9 (after work had recommenced) and the delay costs claim went in, the relationship snapped. Again, the contractor saved at least AUD 3 million in legals, liquidated damages, what would have been substantial yet unrecoverable delay costs, and significant opportunity costs of staff being tied up with lawyers providing statement and discovery.

Example 3: A Disaster by Any Other Name Is Still a Disaster

By way of contrast, a USD 35 billion project I was involved in suffered a USD 6 billion blowout in project costs (liquidated damages and lost delay costs caused by delays in delivery of critical modules to site and extensive consequent rectification works to the modules from poor initial construction in the module fabrication yards). The costs of the massive litigation and arbitrations regarding the additional costs wiped out one of the contractors entirely, and it was bought out at a fire sale price by a competitor. In this project I was one of the legally qualified project managers responsible for providing advice on the notices to be issued under the contract to the project directors and project sponsors within the three-part joint venture company, which been created to deliver this contract. As soon as the late delivery of some critical modules was on the cards, I prepared the required notice. It was rejected by the project sponsors as they did not want to "rock the boat" with the client. As time passed, and more trigger events occurred requiring notification of delay and increased costs, I prepared more notices, only to have those held back also, all in the interests of "preserving the relationship," which was day by day becoming more precarious.

However, on the other side, the client was not backwards in coming forwards with its required notices requiring new programmes and estimates of costs increases. Very few of the notices were responded to within time. As yet more time passed, and the delays became more and more apparent, some attempts were made to issue notices for extensions of time, but they were issued only after prolonged internal debates and so were not issued in the tight 14 calendar day time frame but instead were issued late (sometimes as late as 21 days after the deadline).

Suffice to say, when the litigation and arbitration rolled around, the contractor joint venture company was stuck between a rock and a hard place. The documents were completely stacked against it, and the commercial staff who had been around to "negotiate" the "agreement" regarding EOTs and delays costs were no longer working on the project (as it had reached commissioning and handover stage). The costs in running a defence were massive (around USD 2–3 million a month on average for more than five years) because preparation of witness statements was incredibly difficult as the relevant staff were scattered to the four corners of the earth and some refused point blank to assist, and a (very legally difficult and complex) claim had to be mounted in quantum meruit (i.e., a claim for payment for work done, outside of the contract). It is still in arbitration more than five years after the first delay became apparent and more than two years after handover. The prospects are stacked very high against the contractor joint venture company because of the lack of notices on its part and the plethora of notices on the client's part. The contractor joint venture company has no option but to keep fighting, as to lose will wipe it out entirely, together with its constituent partners. And it could all have been avoided had the notices been issued by the contractor on time in accordance with the contract – the delays would have occurred, but the notices would have provided for extensions of time and delay costs and avoided liquidated damages.

Moral of These Stories

Issue your contractual notices, no matter what. Period. Find a way to convince project directors and sponsors to agree to the issuing of the notices as and when required by the contract, no matter what the apparent cost to the client/contractor relationship. A good project manager needs to develop the negotiating and persuasion skills, together with the resilience, to push through the issuing of contractual notices for time and money within the time frame stipulated in the contract. A good project manager will also maintain a good working relationship with their counterpart even whilst issuing notices − it can be done, with the right mix of emotional intelligence and leadership skills. Failure to issue notices in a timely way can cause the contractor to be wiped out in the end in a dispute with the principal.

Contract Management Step 1: Plan

The contract management plan will document how we intend to manage our suppliers throughout the life of their contracts with us, including, where necessary, arrangements for handing over the operational management of the solution at the end of the project. Once again, it is a critical document which we can prepare well in advance and which will support our ongoing management of the project. The contract management plan will have to cover the management of all contracts required by the project and will have to dovetail with other similar plans and processes, especially those dealing with risks, issues, and changes. Like those other plans, it will need to cover answers to all of our seven standard questions of Who, Why, What, Where, When, How, and How Much, as well as the specialised What If:

- **Who** will be responsible for activities related specifically to contract management, and how will this change if the project has to pass on contracts to the business to manage when it closes?

  Included in this area should be not only the people responsible for day-to-day oversight of the supplier's performance but also escalation paths for issue resolution.

- **Why** is it important to manage all our contracts in a unified manner? Once again, not all business sponsors and members of the project board will fully understand why detailed contract management is critical to managing the delivery of the project and why we need to invest time and resources in monitoring supplier performance at a detailed level. Consequently, we need to explain that our contract management processes are to:

  - Protect the interests of the client organisation.
  - Control delivery by the supplier.
  - Ensure that actions taken by ourselves and the supplier are within the objectives and the terms of the contract that we have agreed.

- **What** standards will we be using to evaluate supplier performance?

  There are two basic categories of performance we need to monitor, which will become our key performance indicators (see step 3):

  - Adherence by the supplier to meeting the conditions of the contract.
  - Quality of the solution delivered by the supplier − these standards need to be drawn from our critical success factors.

- **Where** will our main contract record be established, and how much access will the supplier be granted to it? We could consider different software packages that are available to act as a contract record. Their basic functionality should usually include:

  - A communications function enabling both parties to submit and communicate documents in a controlled way.
  - An auditable workflow function to enable control of items such as change notices.
  - Ability of both parties to access key documents to ensure that there is only a single, agreed-upon version being used.

- **When** will regular reporting and contract performance review meetings take place, and who from both sides will be required to attend them? We also need to consider whether additional meetings, often just between the supplier's delivery manager and the project manager, are needed and how frequently. This frequency will depend on the nature of the contract, the type of project, and the current delivery status. It could range upwards from being on a daily basis, as in a truly agile project, the inclusion of supplier managers and staff in daily meetings should be the norm. On a critical project being managed traditionally, where the supplier is behind schedule or having difficulties, daily is probably advisable.
- **When** and on what basis will supplier performance monitoring take place? This needs to be a period that is long enough to provide a realistic view of performance but short enough to allow for remedial actions to be agreed on and implemented within the contract delivery period.
- **How** can we design and implement procedures and methodologies for the overall management of changes to the contract and for risks? These procedures and methodologies will be sub-sets of our plans for risk and change management in the project overall.
- **How** will we review the quality of deliverables under the contract, and who will be able to authorise their acceptance? In relation to acceptance of deliverables, we may also establish different levels of faults and place requirements on the supplier for their remedy.
- **How** will supplier invoices be reviewed and quality checked, and who will be able to authorise them for payment?
- **What if** we have to act in a crisis situation? This may arise because the supplier is deviating from the agreed-upon standards and protocols or is in other difficulties. We will need to set out pathways to be able to escalate issues for resolution rapidly both internally in the client and in the supplier.

The agreed-upon contract management plan forms an important input to the contracting process, and consultation with the legal and procurement teams during its preparation can be extremely valuable as many of our requirements will have to be reflected in the procurement and contracting processes.

Contract Management Step 2: Establish Key Performance Indicators

In our contract management plan, we established the standards that we would be using to assess supplier performance, and our key performance indicators detail the measures we will be using and the standards the supplier will be expected to reach. In general, these need to follow the SMART principle laid out in Chapter 2 – i.e., they should be **S**pecific, **M**easurable, **A**chievable, **R**ealistic, and **T**ime-based. Whilst we need to design the set of indicators we will use to reflect our project and its particular needs, there are a number of common areas in which we could consider assessing supplier performance.

Contract Delivery

Indicators around contract delivery are based on the supplier meeting their obligations regarding their interactions with the project and, if necessary, with other suppliers and the wider client organisation. We should consider measuring areas such as:

- Accuracy, completeness, and timeliness of invoicing.
- Adherence to change control procedures.
- Issue and risk reporting notifications and timeliness, including the timely issuing of formal contract notices and responses to them.
- Attendance at contract performance meetings.
- Attendance at co-ordination meetings with the project team and other suppliers as required.

Solution Delivery

As with any other aspect of the project's deliverables or overall solution, our concern is ensuring that the supplier meets our requirements for delivery to time, cost, and quality. Our indicators could therefore include areas such as:

- Deliveries made according to the required plan, including an assessment of the impact of late deliveries. For example, late deliveries within the available float time, late deliveries impacting other tasks, and late deliveries resulting in overall projection completion delays.
- Meeting the required standards for quality of deliverables.
- Providing staffing with the required level of qualifications and experience to deliver the contract.
- Performance aligns with cost estimates.
- Meeting the criteria for test results at different stages, including unit testing, whole solution testing, and user acceptance testing.
- Performance against health and safety standards – for example, percentage of completed risk assessments or lost time through accidents.

Contract Management Step 3: Processes

Our objective in establishing contract management processes is to bring the functional operation of all our contracts with suppliers under control to ensure that they will meet our business objectives and requirements as defined by the contracts. We need to ensure that our processes will protect the interests of the client organisation and will not be undermined by informal actions taken outside the contract on the basis of personal relationships – for example, unmanaged "small" changes. The processes that we need to design will be unique because they must reflect the operational environment of the client organisation, the way we have designed and are managing our project, and the contractual obligations we have placed on the supplier.

We need to consider how we will design processes to:

- **Manage the contract itself.** The details of how we intend to run processes and who will be the key actors in areas such as invoice processing and authorisation, co-ordinating meetings, ensuring appropriate supplier attendance, and ensuring that appropriate meeting records are kept.

- **Manage the contract record.** The processes required to ensure that our contract record is maintained are critical to ensuring that we have a complete and reliable record of all our interactions with a supplier. In construction-related projects and potentially other sorts of project, we also need to develop processes to monitor the supplier's adherence to both contractual and legally imposed health and safety obligations.
- **Manage the delivery of the contracted service**. This area is sometimes also called obligation management. We are seeking to ensure that our supplier(s) deliver on their supply obligations within the contract – i.e., to the specified quality, prices, and times. Where multiple suppliers are concerned, we may also need to consider how we intend to manage crosscutting issues between them – for example, when a timeliness or quality issue with one supplier has a consequent impact on another. We also need to establish how the information on our desired performance indicators will be collected, processed, monitored, and shared.
- **Manage change**. The contract change management processes that we design need to integrate with our overall change processes (see Chapter 9) and obligations set out in the contract. Our processes need to pay particular attention to the co-ordination of change management where multiple suppliers are involved in our project or where we may need to take accelerated action to protect the momentum of the project.
- **Manage the relationship.** The type of relationship that we have with our suppliers will vary according to the objective of the contract. Where suppliers are providing a commodity item, the relationship is likely to be simpler and more transactional. Where we are working with them over a period to evolve a unique solution, the relationship will be far more complex. For example, supplying the materials for a building or hardware for an IT system is likely to be a simpler relationship than that with architects designing our new headquarters or a software development house building a bespoke system for us. The processes that we set up for reviewing performance, agreeing on remedial action, and negotiating contract variations will need to reflect both this complexity and our objectives for the contract.
- **Review and approve requests for payment**. It seems easy to say that all payments must be in line with the contract, and of course ours always will be. Sometimes this is not the case. Consider the report of state auditors in California in 2019. Of the state transport agency, Caltrans, they reported that because of the value and number of contracts the agency had entered into, it was difficult to analyse all the invoices and make sure that charges were consistent with the contractual agreements. Across the state, the auditors found many instances where contractors overcharged for labour and overhead costs and consulting work, and in others there was no documentation to support the spendings on some construction tasks.

Contract Management Step 4: Create Delivery File

It is often useful to differentiate a contract delivery file from the pre-award contract file covering the procurement and legal processes, which may be maintained on our behalf by the procurement and legal teams. The contract delivery file for each contract is created and maintained by ourselves as the project manager. In practice the contract delivery file may be many different files covering different areas but subject to the same management rules so that collectively they provide the sole record of the contract and its operation, covering at least the following areas:

- A copy of the executed contract and associated schedules (the original should usually be held in a corporate contract repository).
- All post-contract award correspondence with the supplier, including emails, and copies of any relevant pre-contract award correspondence them, which should also be held in the procurement team's file.

- Records of meetings.
- Contract monitoring results.
- Contract performance and progress reporting.
- Formal notifications of variations to the contract.
- Acceptances of the delivery of goods and services.
- Financial records, including payment schedules, invoices received, and payments made.
- Safety requirements, assessments, inspections, and issues.

Although we hope not to arrive at the situation of a contractual dispute with our supplier, as was the case in our last site visit, if we do, our contract delivery file is usually key to pursuing our case. This means that on every project, we need to treat our records of every contract as if it may give rise to a contractual dispute. Consequently, ensuring from the outset that our maintenance of the contract delivery file will deliver an accurate and comprehensive record of the contract's delivery is essential. It is not, however, just contractual disputes where good-quality documentation is required; this includes external investigations such as those by auditors and public prosecutors. A clear public example of the importance of adequate record keeping was in the trial of Balfour Beatty following the collapse of a rail tunnel being built to London's Heathrow Airport in 1994 using a new tunnelling method. The records kept were sufficiently detailed to allow the judge to ask why it was that there was only one engineer qualified in the tunnelling method and that he was at home when the tunnel collapsed.[4]

Creation of the contract delivery file is not always easy. Depending on the type of project and contract, adopting the solution of our main contract record being a traditional paper file split into different sections may not be sufficient or appropriate. A paper contract delivery file will always suffer from all the difficulties in maintaining a comprehensive physical record when most of our communications and documentation relating to the contract are in an electronic form and when different individuals need access to the same contract file at the same time. Certainly, they all need to be accessing the same, authorised version of any document or information.

These problems have been increasingly addressed by contract management software, which is available as a cloud service or as separately installed software. In some client organisations, this is now a corporate system we will have to use. In other organisations we may have a choice of selecting a solution appropriate to the needs of the project. Many contract management software packages will help to manage and track contracts from the initial drafting stages through the entire contracting life cycle, supporting lawyers and procurement professionals as well our responsibility for ensuring contractual compliance. If we use a contract management software solution, our contractual rules should be drafted to require that all communications between the client and the supplier be facilitated by the software. This will mean we can be certain all electronic communications are captured and that there are no "missing" documents or documents that are deliberately withheld, which could impact our contractual management and influence the eventual outcome of any legal dispute.

Contract Management Step 5: Initiation Meeting

Recall our first site visit to that hot, dusty building site in Chapter 1. We were actually witnessing a contract initiation meeting, also often called a kick-off meeting. In practice, as for much else of the project, the contract initiation meeting and its agenda will vary depending upon the type of contract and its criticality to the project. At one extreme is a contract for a one-off supply of a commodity item on a specified date. In this case we probably will often continue to work with the sales team from the procurement process and will not need a separate formal meeting. At the other extreme is a prime contractor and series of sub-contractors who will be deeply engaged with the project for a protracted period and need to be continuously engaged with the project team at multiple levels throughout. In this case, the initiation meeting may only be the first step in a comprehensive onboarding process for the managers and staff of the prime contractor and sub-contractors.

As our newly appointed supplier(s) will now fall within our definition of stakeholders (see Chapter 7), we can determine the level of engagement required with them when developing their stakeholder profile (see Table 7.4 for an example of a stakeholder profile for a supplier), and we should start by with considering whether the type of contract makes a contract initiation meeting desirable and, if it does, what our desired outcomes should be. Our objectives for such a meeting should include at least:

- Ensuring that all parties are aware of our contract and wider project requirements, remembering that there may be multiple suppliers and their staff present.
- Starting to build relationships between the suppliers' delivery teams and the project team.
- Reinforcing the message that the relationship between the project and the supplier or suppliers will be operated strictly in accordance with the requirements and obligations on all parties as laid out in the contracts' terms and conditions.
- Introducing the contract management processes that we require for managing the delivery by the supplier(s), including areas such as performance monitoring, change management, invoicing, authorisations, and escalations.

Contract Management Step 6: Monitor Performance

Our concerns in managing the supplier will naturally change and evolve as the delivery under the contract moves forwards. For example, if we are seeking a one-off delivery from a supplier, then prior to the receipt of the goods, our main concerns in supplier management will focus on scheduling the delivery in accordance with our stage plans. Post-delivery, our focus would probably switch to determining acceptable quality, possibly installation, and agreeing on payment. Even where we have created a fully integrated team under a partnership working approach, we will need to institute a separate process to review the performance of the supplier against the terms of the contract at each stage of delivery. The more unambiguous we have made our contractual requirements, including delivery plans for the supplier, quality standards, and performance indicators, the easier such a process is to operate. Where the contract terms are open to differences in interpretation, there will always be scope for discussions, disagreements, and disputes, and we need to ensure that our monitoring processes highlight these as early as possible.

An example of what can happen if we allow differences of view to drift until they become a critical issue came in a bespoke software development contract. The client was part of a government department, and the contract was to take an existing statement of requirements and use it to prepare and then operate a unique software solution. Although regular project meetings took place between the client and the contractor, it was not until major issues arose with testing elements of the solution that the client started to check in-depth on the supplier's understanding of the statement of requirements. This revealed discrepancies between what the client meant and what the developers understood.

Contract Management Step 7: Performance Meeting

Unlike the contract initiation meeting, at which several suppliers may be present, a contract performance meeting is usually a meeting between a supplier's manager and the project manager – perhaps with support from other project team members and the procurement team. In more difficult meetings, as with our site visit at the start of the chapter, other parties, including lawyers, may also be present. In essence, the objective of the meeting is to jointly review how a supplier has performed their obligations under the contract. Whilst it is tempting to try and ambush the supplier's managers attending the meeting with issues and information regarding their performance, this both cuts across a desirable culture of openness and is counterproductive. This is because one of our objectives should be for the supplier to have carefully thought through responses to the questions we may have. Consequently, we should prepare and share our agenda for the meeting in advance, including circulating details of the performance that we have monitored against our key indicators.

It is also worth using the meeting not only to look backwards at contract delivery to date but to look forwards to the next review period and seek information on how the supplier intends to improve performance – this is especially relevant if we have instituted a mechanism to reward them for exceeding the target performance levels set in the contract or are applying penalties for missing them. We can also look for information on any risks or issues that the supplier has that may impact our project – for example, increasing staff sickness levels, materials shortages, or price changes. During the performance review meeting, we should also pay close attention to progress on delivering agreed-upon contract changes, perhaps stemming from agreed-upon remedial actions.

An agenda for a contract performance meeting could include at least the following items:

- **Contract delivery performance**. This item includes the overall contract status, the status of key contract deliverables, and progress on any issues that have been brought forwards from previous contract performance meetings. Under this section it is useful to include an additional two specific items:

  - Progress on agreed-upon remedial actions.
  - Progress on agreed-upon contract variations.

- **Key performance indicators (KPIs)**. Ideally, details of the performance against contractual KPIs are circulated in advance and should include both current period performance and a trend analysis over the whole life of the contract.
- **Contract administration issues**. This item allows for the discussion of any areas arising out of the operation of the contract, such as staffing, payments, insurances or maintenance, and licences (usually an issue in some form on IT projects).
- **Next period objectives and targets.** So far, the meeting has been focused on looking at performance retrospectively, but we also need to review the supplier's required level of performance over the next period and assure ourselves that plans are in place to make sure that they are met. A simple example might be to establish that staffing or manufacturing plans are in place to deliver on time and that factors such as holiday period closures or potential delays during bad weather have been taken into account.
- **Risks.** As well as sharing information about known risks and discussing areas for consideration as risks to the contract, it is useful to include a focus on current issues being faced elsewhere by the project or the supplier which may impact upon the contract delivery.

- **Contractual and process improvements.** As our joint experience of the operation of a contract grows, especially where the delivery period stretches over a considerable time, either side may identify areas where it may be improved. These improvements can take many forms and could range from changing the frequency of, or the persons attending, the contract performance meeting to modifying KPIs or altering planned deliverables to take account of developing technologies or methods of working.

Contract Management Step 8: Agree on Remedial Actions

Where the performance by the supplier is impacting, or will potentially impact, one of our objectives for the project, we need to work with the supplier to identify how this may be rectified. The initial stage is to discuss the emerging issue with supplier and work to find a resolution that is acceptable to both parties, and this may or not result in a notice under the contract change control process.

Working together to understand the issue is essentially how the differences arising in the last example of a software development contract had to be resolved. An agreement was reached to the effect that:

- Software development work was halted for a period whilst a joint understanding of each requirement was reached.
- Where necessary, previously completed development work would be amended to reflect the new understanding.
- In future, there would be close daily liaison between the client's experts and the software developers.

As another example, on one particular project, a supplier informed the client that they would be unable to complete production of a certain item on time because there was a problem with the supply of a component to them, in this case a toughened glass touchscreen. The project manager explored what strategies the supplier had put in place to both minimise the immediate impact on the project's delivery timeline and to rectify the problem longer term. The supplier explained that they were currently sourcing alternate suppliers of the component and had already drawn in components from their normal stock of spare parts to keep manufacturing going. They also stated that they would build a stock of partially completed machines and run extra manufacturing shifts to complete them as soon as their new suppliers were in place and delivering, which should be within four weeks. Their view was that although deliveries would be impacted in the short term, they would catch up within four weeks of component deliveries recommencing. The project manager, as in most similar situations, then had a choice:

- If the project manager considered the offered solution to be reasonable, then they could agree on a contract variation to change the delivery schedule and submit this for approval.
- If the project manager disagreed with the supplier's solution and required other actions to be taken, this could lead to starting the process of escalation, and this, in turn, could lead to invoking formal dispute procedures.

In reaching such a decision as project managers, we have to consider a number of issues:

- Is the supplier's suggestion credible? We may have to test the solution in more depth. In the example given, was there really an alternative component supplier, would their product also meet our quality standard, and could they really supply our needs within the four weeks?

- What alternatives do we really have to the supplier's solution? Sometimes we can make changes to the project, sometimes we can find alternative suppliers, and sometimes we have no real choice but to either escalate in the hope of getting a better solution from the supplier or accept the current offer.

The "no real choice" situation arises because whatever remedies we have placed in the contract may not be usable in this situation. This is because invoking remedies can rebound on the project. If, for example, we want to replace a poorly performing supplier, we may have to rerun the procurement process and redo work completed to date, thereby extending our delivery timeline. As project managers we will also have to divert staff time and funding to pursuing financial remedies, often through litigation. Our virtual site visit at the start of this chapter was to a project where the supplier was performing badly, but not yet badly enough for the client organisation to want to trigger contractual penalties, and when they eventually did, the project was shut down. If we agree on contract variations to enable remedial actions to be taken, we should consider whether any of our KPIs require adjustment to reflect the agreed-upon changes.

Contract Management Step 9: Review Variations

By this stage we know that a large part of delivering a project is about reducing and managing uncertainty and responding appropriately to events as they occur. Like many other documents in the project, our contracts with suppliers are built around both our requirements for the project and our understanding of the facts at the time the contracts are executed. As we know, change can be expected as the project progresses. Consequently, each contract that we agree on should provide for mechanisms to enable both sides to manage the relationship, including responding to events which make changes in the contract desirable. Although the contract performance meeting provides an opportunity for discussing such changes as they emerge, the contractual mechanism for controlling them is via the issue of a formal change notice. As in our lawyer's examples, these change notices need to be raised whenever either side encounters any event that may impact upon the delivery of goods and services. Where our KPIs as set out in the contract may be affected by the agreed-upon changes, the change notice should also include details of how these are to be updated.

Going back to the example of the pub restaurant doorway, we may have a phone call with our supplier manager along the lines of, "It's OK, you can get on and put that doorway in, and spend the extra time clearing out the rubble. Oh, and can you also get a contractor to haul it away and dispose of it?" However, we still need the relevant change notices issued under the contract to ensure that supplier's company can get paid for it. Working with one construction company that specialised in refitting old buildings, it was discovered that they regularly lost all their profit on jobs by fixing unexpected problems discovered when stripping buildings back without getting relevant change notices approved first. In fact, the supplier's project manager should *always* respond to the supplier, "That's fine. Get the relevant change notice drafted and signed off and then get on with it."

Failing to abide by this principle might be convenient, and might save time in the moment, but, as we have seen, the potential future consequences of not following the agreed-upon contractual processes are increased time taken to resolve problems, increased expense, and potentially even years of litigation.

Contract Management Step 10: Progress Reporting

How we report progress on a contract to the project board again is dependent upon its complexity, the issues being experienced, and its criticality to the project. It is useful to, as a minimum, report to the project board on the overall delivery performance and performance against contractual KPIs, including showing how the supplier's performance is changing over time. This reporting can be added to the standard agenda for project board meetings once the contract has been put in place. Additionally, we should report progress on change notices received and agreed-upon contract variations.

Not All Projects Go to Plan

The examples in the lawyer's case studies reveal that we cannot always expect our projects and our contracts to run smoothly, and despite our best risk management practices, events can overtake our plans or our suppliers can fail to perform as expected.

> *This time our site visit is coming to us, and we have been passed a briefing paper by a business sponsor from another department. Having heard how well the bio-energy from effluent project is being managed, and knowing of your background, he is asking for an independent and confidential view of his project's progress and the difficulties that his project manager is having in managing the contract.*
>
> *The background is that the organisation is replacing an antiquated insurance claims workflow system, which worked only for the main office-based administration staff, with one which, for the first time, will provide online access to all documentation to all of its staff, both those working remotely and those working in the headquarters. In addition, all of the client's delivery partners across the industry will be able to submit and view documents electronically, and customers will gain access on a read-only basis to selected claims documents via a new portal.*
>
> *The contract was agreed on just over a year prior to the preparation of the briefing paper. The prime contractor is supplying and installing all the required hardware and software for the new system, which will replace the existing aged partial workflow system.*
>
> *The prime contractor is using a mix of their own staff and a sub-contracted team from the software supplier to install and configure the software to meet the client's complex, multiple workflows. At the project board's insistence, the supplier's development team is all based on site to facilitate easy communication and foster partnership working.*

Commercial in Confidence

Claims Handling System Replacement
Extraordinary Project Board Briefing Regarding Prime Contractor Progress, week ending 25 September
By Project Manager J. Smith

Management

In order to overcome the previous difficulties in agreeing on the way forwards on multiple issues, a UK main board director for the prime contractor has now been located on site. This individual has been here most days this week, and there has been a noticeable improvement in relationships. It is understood that they are in direct contact with the overseas head office, which should speed up decision making.

Contract Variation

The contract variation order (CVO) that we have been preparing for some time has not yet been signed. There has been considerable negotiation with the prime contractor during the week, and agreement was reached on Thursday with their director, but reference back to their lawyer led to a large number of changes to the sense of the agreement.

These changes were resolved by late Friday and, subject to agreeing to the delivery schedule next Tuesday morning, the CVO will be ready for signature. There is, however, an outstanding problem with this in that they will not commit to a date for delivering the specified alternate workflows. I made it clear to the contractor's project manager and their director on Friday that we required this in order to sign up to the CVO.

Delivery

We believe all scheduled elements were delivered on Thursday, 24 September, to the very limited capacity training/test environment, but the prime contractor failed to deliver to the production environment. This had been previously agreed on as a key milestone and was required to enable the planned testing by a larger number of users. Consequently, we reset the date for delivery to production from 24 September to 25 September. I deliberately set this delivery date and time for a Friday afternoon to give a little leeway to the prime contractor to complete delivery over the weekend without us having to raise yet another non-delivery issue. This will move the end date for testing by a day, but I am not proposing to change the target acceptance date.

The documentation of the delivery was still poor, has many omissions, and is therefore not of a standard that would allow our staff to take over systems operation. This has been formally notified to the prime contractor as a category A fault.

Testing

We will be checking on Monday morning at 8 a.m. that the delivery to the production environment was made successfully and that we can safely assemble the first phase user test team to start at 10. The aim of testing on Monday is to check that the results in the production environment match those from the test environment, with load testing with 30 users beginning on Tuesday morning at 9.

Our test manager has been working this week to ensure that we have sufficient test scripts prepared to commence testing on Monday, and the project team liaison officer has been negotiating with line managers for the release of their staff.

The director of professional services has managed to rearrange forwards workloads amongst that team to free leading professionals for testing of the professional services template in the week commencing 5 October.

Performance Testing

The software supplier has presented the independent performance testing report which we commissioned directly from them, and it contains details of a very concerning bottleneck in performance caused by the web servers. I have passed details of this issue on to the prime contractor for comment, but it seems that the optimum number of users connected to a web server for adequate performance is 20 under the automated test but in actual usage could be up to around 30. This compares to 300 users in our *existing architecture and as described in our statement of requirements. The software supplier's recommended solution is to replace the web servers which we have purchased and installed with multiple latest-generation servers to fit in the same space in the computer room. This would mean that all of the production hardware already purchased will need to be released by the project, but the IT department feels that it could be suitable for redeployment in the medium term to support other applications as current hardware reaches the end of its serviceable life.*

The worst-case scenario is that the new hardware cost is around £X,XXX per server plus the cabinet and the controllers, which, I am given to understand, are necessary to share the load across the other servers. As we would need at least 20 of these, we could be looking at a cost of £XX,XXX plus installation. The good news is that these can be added incrementally, and we will not need to be purchasing hardware for other systems, which will at least partially offset this cost over the next 18 months.

My view, however, is that providing this acceptable web performance has always been a key requirement of the solution, including rapid access to our remote professional staff, partners across the industry, and the wider customer base. Consequently, my advice is that the failure to provide acceptable web performance should be raised as a category A fault and the prime contractor given the option to correct it. The implication of this is that it has to be corrected at the supplier's expense before the target acceptance date or the incentive payments due will not be made. We should consider that if it is not corrected, the prime contractor will then be in fundamental breach of the performance requirements in the contract and we will have the right to terminate.

Sadly, the software supplier's UK manager has now privately informed me that the prime contractor's staff were warned about the possibility of this performance bottleneck last January – i.e., nine months ago – and they have chosen to take no action until now, when we are having to force the issue with them. We could also have cancelled the order for the production hardware.

Web Publishing of Documents

It was reported to me by the contractor's project manager that no allowance has been made in the solution being delivered for the rendering of documents into an unalterable form. This was certainly a requirement made explicit to all suppliers throughout the procurement, including answering a query before contract which was circulated to all tenderers, but not, as far as I can see, clearly spelled out in the statement of requirements, which now forms part of the contract along with the prime contractor's response.

I understand that we can render to a TIF image, which may be acceptable without buying more software (otherwise the software licence required for rendering into an acceptable non-alterable format would cost around £XX,XXX, with a 20% ongoing annual licence fee). My other concern is that this means they have not included this requirement in the calculation of storage space required. This may mean that we will need to acquire extra disk storage or archive at six months after closure of each workflow case. Again, my view is that we should raise this as a category A fault, but we may need to take legal advice on this.

Company Web Portal Integration

We have asked the corporate web portal team manager to keep a watching brief on this, and she will be visiting on Tuesday. One problem to emerge is that the prime contractor is planning to use a prewritten piece of code of their own rather than the software supplier's standard functionality. The software

supplier has advised that this is not an appropriate solution and has instructed their staff not to work on it. I will be asking the web portal team manager to review this issue and to make a recommendation.
J. Smith
Project Manager
25 September

What should we advise the business sponsor to do and why? The apparent options available are to:

- *Continue with the contract, accept the changes, and fund the extra costs that the prime contractor requires.*
- *Notify the prime contractor of the category A faults with a threat of termination if they are not resolved expeditiously at the contractor's cost.*
- *Go straight to termination on the grounds that the system as delivered does not meet the specification of requirements.*

Disputes – Balancing Contractual Rights and Commercial Reality

The terms of a contract represent the rights that the different parties have agreed to in relation to each other. Notwithstanding all of our processes to help manage the contract, we may reach a point at which we have a dispute with our supplier, as in our last two site visits. The fact that we have rights under our contract does not mean that we will always chose to exercise those rights, and there may be many different reasons for this.

Taking the last site visit – the failing insurance system replacement project – the business sponsor actually decided to seek the opinion of a very senior commercial lawyer specialising in information systems contracts. This required a paper file nearly five inches thick to be prepared, which the lawyer very impressively mastered in under twenty-four hours. The lawyer supported the view that a fundamental breach of contract by the supplier could be demonstrated and that, under the contract terms, liquidated damages would be available. He also questioned whether the business sponsor was prepared for the time that would be taken, the effort that would be required by the client in preparing the case, and the potential costs that would be involved if the case was lost. He advised that as the project was mostly complete and the client would have to make other arrangements to complete the project anyway, it might be easier to pay the supplier as if the solution was finished and fully met the specification.

The business sponsor and his organisation accepted this pragmatic suggestion and brought forwards the planned recruitment and training of an internal systems development and maintenance team. With additional assistance from the core software supplier, this new team then successfully completed the project.

Another example of a client choosing not to enforce its contractual rights came about when an existing IT service supplier failed to win a contract for:

- A further period of operation of a system that they had previously developed.
- Major design, software development, and implementation work to redevelop the system to meet changing business aspirations.

During the mobilisation and handover period for the new contract, the continuing lack of co-operation of the original service provider with the new software developer was causing major issues. This lack of co-operation increased to such an extent that it was considered to be plainly in breach of the terms of the service provider's contract. The client manager responsible and the project manager were keen to resolve the issue by invoking the breach and taking court action if necessary. Whilst agreeing with their reasoning, the CEO offered the opinion that, as in the previous example, the case would be too much trouble to pursue and would still not solve the underlying issue. He expressed his decision simply as, "Just pay them to go away!" A negotiated settlement was reached, co-operation with the new supplier was forthcoming, and the contract was ended amicably instead of in litigation and delays to the start of the new system development project.

The final example of choosing not to use the rights under the contract was less amicable and a reflection of the power imbalance between a major client and a supplier. In Chapter 4, when considering poor control over requirements, there was a simple example of a dispute that had arisen from a disagreement over whether certain light fittings in a design should have been square or round. The supplier pointed out that it was entitled to payment for the rework caused by the client's design manager unreasonably rejecting the design on these grounds, even though the design provided was in line with the client's own mandatory design manual. After much sometimes acrimonious debate, the client's negotiator pointed out that "irrespective of who was right in this instance, in view of the very major amount of work that we commission from you on a regular basis, it would not be worth your jeopardising our future relationship by pursuing this issue further." The supplier reluctantly was forced to accept the situation, did not pursue its contractual rights, and wrote off the additional costs.

Keep Your Stakeholders Close and Your Suppliers Closer

Suppliers are a special kind of stakeholder in the success of the project (see Chapter 7). They are special because we depend on them to deliver – whereas most other stakeholders depend upon us. Examples in this chapter show how difficulties in project delivery can arise when we are not fully informed about how our supplier is progressing our contract. We need, therefore, to be permanently engaged with our suppliers to understand how they are managing their supply to our project. Just as we should have an open "no surprises" approach to informing our stakeholders, we must expect our suppliers to adopt the same approach with us. If they are not willing to do so, we may need to force it upon them, and we should include details of our supplier engagement requirements in the procurement and contracting processes. We often need to go beyond the step-by-step approach to contract management by positively seeking opportunities to build a joint delivery culture with our suppliers.

The way we achieve a joint delivery culture will depend upon the approach we built into our contracts. Greater separation between teams will increase the formality we have to build into the supplier management processes. For example, if we implemented partnership working with a single integrated team, our project staff and supplier staff will be sharing information on a minute-by-minute basis and our management and reporting processes will collect the information that we need on supplier performance. Equally, if we are running an agile project, we can expect information to be openly shared at the daily team meetings. At the other extreme, where we have no regular contact and are essentially waiting for the supplier to deliver, we are placing the project at risk of delays and poor quality when the supplier encounters problems that they do not share with us. Possibly the worst outcome of all is when a supplier lies to cover up internal shortcomings until the last possible moment and we then have to manage complex issues in crisis mode.

There are, however, many things we can do to keep in close contact with our suppliers, including:

- Ensure that we have regular supplier update meetings. These enable a two-way exchange of information between the supplier and the project team. If the project has multiple suppliers, then these update meetings can engage all suppliers. As ever, the timing of these meetings depends upon the type of project, the stage of development, and the scope of the suppliers' contracts.
- Foster one-to-one engagement between the project team and their opposite number in the supplier's staff. One of the advantages of partnership working is that we do not need to invest project team time in "mirroring" the suppliers' staff, but in other circumstances, it is often extremely useful as a way of exchanging information.
- As part of the one-to-one engagement approach, provide opportunities for building informal relationships between the supplier's staff and the project team. Even if it is only over coffee (and biscuits, of course) at breaks in a formal meeting, that is a start. Relationships can develop organically over time, but we can shortcut the process by giving each member of the project team key targets from the supplier to engage in conversation. Then, providing that we do not undermine change controls, this again can be a useful way of exchanging information and avoiding the build-up of tension between the project and the supplier.
- Invite key members of the supplier staff to attend the project board as observers.

Reflections

Choosing your previous project, or using the fictional bio-energy from effluent case study:

- Pick one required supplier contract and list the major critical success factors that you need to include in the procurement process.
- Prepare a bullet point briefing note for the procurement team about the project and your requirements from a successful procurement and contracting process.
- Consider the implications of the contract that you are proposing on the supplier's cash-flow. How would you balance their need to minimise borrowing and your need to ensure that payment is made on the basis of the quality and completion of the required deliveries?
- What scope is there for introducing a combined pain/gain mechanism which rewards the supplier for overperforming and reduces payments for not meeting critical success factors?
- List the top five advantages of early supplier engagement for your project. What are the inherent dangers?
- How would you create a continuing supplier engagement process?

Notes

1 Or other comparable agreement for internal suppliers.
2 See Appendix 2.
3 This may be an example of the client's legal team acting as a "business prevention" department and deterring potential suppliers from tendering.
4 See Appendix 2.

Bibliography

UK Government, 2015. *The Public Contracts Regulations 2015 Article 67, Directive 2014/24/EU Article 67* [Online]. Available at: www.legislation.gov.uk/uksi/2015/102/contents/made [Accessed 20 December 2022].

11 Monitoring, Controlling, and Reporting

Keeping on Top of Progress

Aim of this chapter: To examine the various strategies the project manager can use to control and report on the delivery of the project to meet time, cost, and quality objectives.

Learning outcomes: To be able to identify, select, and operate appropriate monitoring and control strategies and reporting arrangements for a project.

In truth, we start monitoring, controlling, and reporting progress from the very first time we meet with our business sponsor after our initial briefing, and we formally develop our control and reporting mechanisms over the early stages of the project as it grows in scale and complexity. In managing our project, we will often have to report on our progress to a variety of different audiences, and the deadlines for submitting reports can come round with alarming frequency. It is therefore not surprising that reporting can be regarded as a separate activity which takes us away from managing the project's delivery. Still, we need to try to approach reporting with a positive perspective. Instead of seeing reporting as being an onerous extra duty, we should think of it as sharing with our stakeholders selected output from our monitoring and control of the progress of the project.

This chapter is concerned with the entire monitoring, controlling, and reporting process. A working, but not snappy, definition of this process is: "The controlled way in which we share information with stakeholders on the progress of the project, resulting from both the systematic gathering of data from a variety of sources and our analysis of the impacts of that data on the future of the project and on the management decisions that are required."

Before we go any further, an urgent problem has arisen in regard to our bio-energy from effluent project.

We are at a point on the timeline about a month or six weeks since that first fateful meeting with the boss. You recall how the boss had found a room on the 52nd floor near the lift (and the toilets) to ensure that the project could be kept under close scrutiny. Well, fortunately that only lasted for a few days before the boss's attention was grabbed by something else.

Things have become more complex over the last couple of weeks. This is because the project is now engaging with a wider range of people across many levels of the company, including some of the other directors. Now the issue has been making sure that the boss is kept in touch with what's happening on a sufficiently regular basis so there are no surprises when you need to meet to consult or obtain a decision.

DOI: 10.4324/9781003405344-12

You've been staring at the wall for a while (remember, there are no windows in this former janitor's cupboard), thinking about how you can set up a more formal process to keep everyone informed about what's going on, when the door bursts open and the boss comes in.

Boss: "Listen, here's the thing. I've been asked several times by other directors and heads of department about how your project is going, and I don't have a coherent answer. I know you've been emailing me about different aspects of the whole thing, and that's great, but we haven't spoken for over a week now, and I'm just not seeing the overall picture of where we've been and how well we're progressing."

You: "I'd come to the same conclusion myself. We need a formal way of letting you know in general how things are going while at the same time highlighting anything that's coming up which I feel needs your input."

Boss: "It's early days yet, but I can foresee being questioned not only about where we've been and our current position but about what the project is likely to spend and when we are likely to complete it. What I'd like is for you to brief me regularly on all this, and to start with, I think we need to meet weekly. So, I'm blocking out an hour in my schedule from 9:30 to 10:30 every Monday morning from now on. If we don't need to meet on any particular week, we can always cancel. For next Monday, please could you come up with an overview of the project at the moment? I'll also need some ideas on a process for reporting progress regularly to me, to the project board when we have that set up, and to anyone else who needs keeping up to date."

You don't know how much caffeine the boss drinks, but you suspect it's a lot.

At least it's only Tuesday, so you've got a week to think through not only what's good practice in project reporting but also how it might need to be tailored to this high-profile project. Which will, you remind yourself with some trepidation, eventually be operating sites in multiple countries and even on different continents.

As highlighted in Chapter 1, as the project manager, we normally take a personal lead on most of the activities related to creating and gaining authorisation for the project. We co-ordinate, control, and may directly undertake preparing the required outputs of plans, strategies, business cases, and so on for approval. As time progresses and the project starts to take shape and we move progressively through the project stages into implementation, our role will change. In each stage we will involve more people, and we will delegate the completion of tasks to them. Consequently, our role inevitably grows into monitoring and controlling work undertaken by others, and this becomes a major focus of our daily activities. (In one edition of a well-respected methodology, out of nearly three hundred pages, less than one hundred words are used to describe the day-to-day work of the project manager).

As with line management, it is here in our daily, often unplanned interactions with our team and stakeholders where our management approach and style make the real difference to the successful delivery of our project. However, unlike the line manager, we also have formal and regular detailed project monitoring and progress reporting processes which we use to give a structure both to the way we direct activity at the lowest level of the project and the way we communicate upwards and outwards to managers and stakeholders. The core of this formal monitoring and reporting is our regular review of progress at a task level with the person responsible for delivering each task.

The timing of our task reviews is everything, especially in fast-moving projects. A question often asked about reporting is, "How often should we report the project's progress?" The question we actually need to ask, however, and which will give us a much clearer answer, is, "How often do we need to monitor the progress of this particular person in delivering this particular task?"

In our bio-energy from effluent case study, the boss wants a weekly progress report. The process of information discovery, analysis, and reporting will give us a cyclical structure for each week, but, especially where many tasks are running simultaneously, this could generate a large workload and become a treadmill. In traditional sequentially managed projects, weekly is normally sufficient for some tasks but may be too frequent for others, and in some cases, where a task may be at a critical point, we may need to review events on a daily basis. In agile projects, the daily team meetings can deliver this level of interaction. In the case study, the boss selected a fixed first day of the week for the report to the business sponsor. This may seem logical, but it can present scheduling problems if, as in the UK, many public holidays are fixed on a Monday. Weekly reporting, however, also needs someone able to stand in for the project manager to complete the reviews and reports during periods of leave (yes, we are allowed a break) or other absence.

In addition to business sponsor reports, we will have to provide updates for project board meetings. Often the frequency and timing of these are determined by the project board and other decision makers in the client organisation. Ideally, we will want to align at least some of these meetings and reports with key milestones in the project.

Figure 11.1 shows an example of how the different monitoring and reporting cycles could work for a traditionally managed project. Only the first two project stages are shown, and there can be as many iterations of the project board reporting cycle within each stage as required. Similarly, you can include as many rounds of the progress review and business sponsor reporting cycle as needed between project board reports.

This may look daunting, but monitoring and control are *the* major aspects of the project manager's role during the delivery stages of the project. The good news is that with adequate preparation, we need only gather information once before selecting and representing it as appropriate for our different audiences.

Our control and reporting processes depend upon monitoring current progress and matching it against our base data describing the project, as well as our set of project, stage, and financial plans. Figure 11.2 sets out the sources of and processes for collecting, comparing, and evaluating data in a traditionally managed project.

The top part of Figure 11.2 shows the data we defined in the project's early stages, including all our plans; registers for risks, issues, and changes; and our detailed work package and task descriptions. This data forms the basis against which we measure progress using the data sources shown on the right-hand side of the diagram.

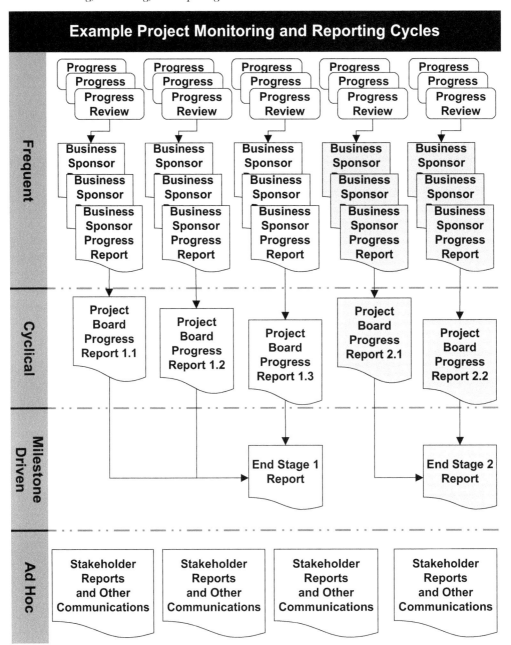

Figure 11.1 Example Project Monitoring and Reporting Cycles

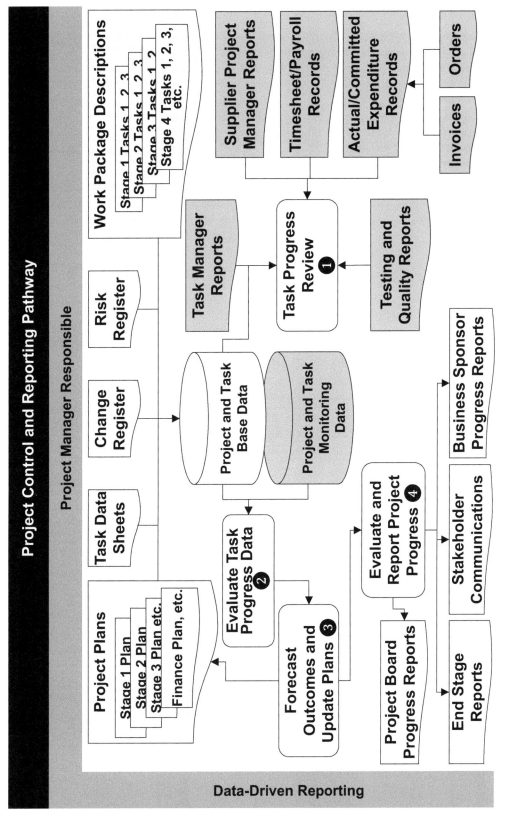

Figure 11.2 Project Control and Reporting Pathway

Our principal control activities throughout the life of the project are:

- **Task progress reviews**. Collecting progress data and controlling task delivery, usually directly with the task manager or the task's delivery team.
- **Data evaluation**. Evaluating the data obtained in the task progress review and collected by other information processes in the project and the wider client organisation.
- **Forecasting**. Taking the evaluated data and analysing expected future progress and outcomes, including the update of our project and stage plans as necessary.
- **Reporting**. Reports are prepared for different audiences and may be related to individual tasks and overall project progress.

In agile methodologies such as SCRUM, a very different approach is taken. Task progress reviews are replaced by daily short stand-up meetings in which progress the previous day, activities for today, and any issues blocking their work are discussed by each team member. In these methodologies, the project manager is in close and detailed contact with progress on each activity within the sprint. Instead of end of stage reports, an iteration review is held at the end of the sprint in which team members showcase their progress and receive feedback from the project stakeholders. Both of these activities will allow the project manager to gather and assess information on task progress, but, as in any other project management approach, task progress is only one part of the data we need in order to be able to report.

The remainder of this chapter is built around the traditional project management approach, and there will be variations from this if agile methodologies are followed. The overall objectives of highlighting and reporting current progress, and identifying where action needs to be taken to keep the project on course, do not, however, change.

Step 1: Gather the Data – The Task Progress Review

Whatever project management approach we use, whilst we are establishing the ways the project is to be operated, we need to consider how we will obtain the data we need for both controlling and reporting on the project. In Chapter 5 we looked at bottom-up, data-driven planning, which resulted in the creation of a task data sheet (Table 5.1) for every task identified in our work breakdown structure, and an overall store of data about every task in the project, including budgets, duration, and dates. In Chapter 10 we also identified that this task information should be carried forwards into the statement of work, setting out the brief for a supplier engaged to carry out a task or tasks for us. The work package description and the statement of work should have included all the necessary details of the task's outputs, timings, and budgets and the quality standards that should be met. These are the baseline parameters against which performance of those delegated with completing the task will be assessed. Consequently, our monitoring and control processes for the project are based on updating, consolidating, and evaluating data on progress against those parameters that were set for every task.

Normally, this data is collected as part of a regular formal review of the progress of every task by the project manager with the task manager – i.e., the person to whom delivery responsibility was delegated. This person may be a single individual undertaking the work, the leader of a team delivering the task, or a supplier's project manager. Always bear in mind that as project manager, there are also some tasks we will be carrying out or managing ourselves, but we still have to review, record, and report progress on these as honestly and objectively as we can.

The task progress review is an important formal opportunity for the project and task managers to discuss the progress on the task, planned future actions, and any problems which may be slowing or blocking progress. It is also the opportunity for the project manager to review the quality of both the processes being followed and the output being delivered. Within the meeting the project manager should give and record any necessary direction to the task manager to ensure not only that the task is delivered according to the work package description but importantly that it continues to mesh with other activities on the project. It is at this task level and often at this review meeting that we actually steer the delivery of the project. A task progress review with a supplier should be seen as the base level of the formal contract progress review process. On a complex contract where the supplier is supporting multiple tasks, there will necessarily be multiple task reviews which build towards the overall review of progress on the contract.

At this stage it is essential to:

- Record the data needed to assess the status of the task for the project manager's upward reporting of progress.
- Record any quality issues.
- Prepare a formal narrative note of the discussion, including any decisions about the delivery of the task.

An example format for a task progress review record that captures all this information in a summary format is shown as Table 11.1.

Table 11.1 Task Status Review

Task status review	**Task name**					
	WBS reference					
	Review period	From		To		
	Review completed by					
	Completion dates	Original		Forecast		
	Reason for variance in completion date					
	Current task status (delete as appropriate)		Red	Amber		Green
	Previous task status (delete as appropriate)		Red	Amber		Green

Escalated decisions	

Deliverables completed	Delivery status	Red	Amber	Green
	Deliverables		Due date	Done date

Financial overview	Financial status	Red	Amber	Green
	Original task budget			
	Approved changes			
	Total task budget at completion (BAC)			
	Actual cost to date (AC)			
	Funds required to complete task (ECC)			
	Estimated total task cost (ETC) (AC + ECC)			

Risks	Risk and issue status	Red	Amber	Green
	Changes in risk status			
	Ref:	Commentary		
	New risks in period			
	Ref:	Commentary		
	Risks closed in period			
	Ref:	Commentary		

Issues	New issues in period		
	Ref:	Commentary	
	Issues closed in period		
	Ref:	Description	Outcome

Quality status	Red	Amber	Green

(left label: Quality observations)

Discussion notes

Each task progress review also contains an update to the task's completion forecast, which would normally be discussed and agreed on with the task manager, including, as appropriate:

• Effort required to complete the task.
• Remaining duration (elapsed days/weeks/months as appropriate).
• Projected task end date.
• Projected spend required to complete the task.

There are two broad approaches to gathering progress information from the task manager. The first is for the project manager to collect the information during the review meeting, which may be in person or virtual. The second is for the task manager to submit the required information in advance and for the project manager to review it during the meeting. In addition to completing the task progress review record with the task manager, after the meeting, the project manager should also update the task data sheet which we created as part of our planning process (see Chapter 5). The task data sheet now becomes our formal summary record of the progress of the task. Table 11.2 shows the task data sheet now extended to include information collected and verified during each review, including, as appropriate:

• Review date.
• Effort to date (days).
• Duration since start date (elapsed days).
• Spend to date.
• Percentage of task completed.

The final part of the task data sheet is an end of task review by the project manager. This summarises the performance of the task and the amount of variance from its original parameters, including:

• Effort used to deliver the task (in days).
• Duration of the task (elapsed days).
• End date.
• Total spend in delivering the task.

Table 11.2 Task Data Sheet

Task Data Sheet			
Task/WBS reference			
Task name			
Inputs			
Activities			
Outputs			
Quality standards			
Resources needed			
Predecessor tasks			
Successor tasks			
Planned			
Effort (days)		**Start date**	
Daily rate		**Duration (elapsed days)**	
Budget		**End date**	
Actuals			
First Review		**Second Review**	
To (review date)		To (review date)	
Effort to date (days)		Effort to date (days)	
Duration (elapsed days)		Duration (elapsed days)	
Spend to date (£)		Spend to date (£)	
Percentage complete		Percentage complete	
Completion forecast		**Completion forecast**	
Effort (days)		Effort (days)	
Duration (elapsed days)		Duration (elapsed days)	
Spend to date (£)		Spend to date (£)	
End date		End date	
Task status		Task status	
Third Review		**Fourth Review**	
To (review date)		To (review date)	
Effort to date (days)		Effort to date (days)	
Duration (elapsed days)		Duration (elapsed days)	
Spend		Spend	
Percentage complete		Percentage complete	

Completion forecast		Completion forecast	
Effort (days)		Effort (days)	
Duration (elapsed days)		Duration (elapsed days)	
Spend		Spend	
End date		End date	
Task status		Task status	
Task Completion			
All outputs met	Y/N	All quality standards met	Y/N
Effort used (days)		Variance from plan +/−	
Duration (elapsed days)		Variance from plan +/−	
Actual end date		Variance from plan +/−	
Actual spend (£)		Variance from plan +/−	

The reviews as recorded in the task data sheet and the task progress review record, together with any supporting notes, form the key records which will show how the project overall is currently being delivered against time, budget, and quality plans. They will also provide critical input to any reviews of the project either during or after delivery.

Where a task review reveals risks, issues, quality concerns, or lessons learned, we need to initiate the relevant project processes so that these can be properly tracked and reported.

Are We Meeting Quality Standards?

Quality control and testing could be the subject of a series of several books, but from the project manager's point of view, we first need to establish whether the quality checks that we designed over the task been correctly undertaken. Second, we are concerned that the output from each task is in accordance with our specification and ideally that the specification remains correct. Consequently, the final questions that we have to address in our monitoring of each task is about what the task is actually delivering:

- Are the task outputs complete and correct – i.e., as set out in the (updated) project business case, contracts (where applicable), and other documents?
- Are the task outputs still fit for purpose?

These are not the same question. Often, projects can start out with ideas about the scope or the aesthetics – the look and feel – of the solution, but the practicalities of achieving our "would like" requirements can emerge during the project, as can questions of affordability when we start to understand costs more fully. Consequently, in many projects, compromises are made, and we have to settle for meeting a lower "must have" requirement instead. In a building project, it could be that we use ceramic tiles in the bathrooms instead of natural stone, or for an IT rollout, we have to buy lower-specification laptops than was originally planned. In both examples, we have reduced the project's original quality standards and therefore have to answer no to our first question. Equally, as both projects' outputs will still be operable – i.e., still "fit for purpose" – we answer yes to the second question.

Inevitably, being able to answer these questions involves drawing on the quality control and assurance processes that we set up for the project and also considering the reports from quality assessors where we have put them in place. In addition, we need to have robust methods for testing each of the outputs specified in our work packages.

Every project type will need different quality and testing regimes, frequently with reviews and testing at different stages of delivery. The last of these, usually just before handing over the project for live operation, will be some kind of user acceptance testing. The results of each of the quality reviews and tests need to be recorded and any difficulties noted during the task progress reviews.

There is a general rule in project delivery that when you are late in delivering a project and under time pressure, the things that are going to get squeezed are testing and quality control. The result of squeezing our testing period, and also having inadequate time to resolve any issues that arise, frequently means that the project will deliver poorer results. However, if we have been testing the quality of both our delivery processes and outputs at task level throughout the project, this will reduce the impact of squeezing down our testing periods.

The golden rule on quality is that if we have any doubts at any time during the task review process, we should require additional evidence and, if necessary, an independent quality assessment. Any costs and delays this would cause are more than offset by the impact of the issues that ignoring your doubts may cause, which brings us neatly to our next virtual field trip.

This visit is going to highlight the problems that poorly thought-out reviews of the quality of task deliverables can cause, even in a very simple, mundane project.[1]

You'll need your coat, a scarf, gloves, and probably a hat this time. It's midmorning in late February, and you're standing in the centre of a broad thoroughfare. The trees that line each side of the road are bare against a grey, overcast sky. The wind is blowing straight down the road from the north, it is snowing very slightly, and you're feeling glad I warned you about the coat and the hat.

It's a bizarre location for starting a field trip – by the traffic lights on a small island in the centre of the road. Immediately in front of you, and surrounded by iron railings, are two staircases. Yes, these are old-fashioned public toilets. One set of stairs descends to the gents and the other to the ladies. Wait, what's that noise? Crossing to the top of one of the stairwells, you can hear banging and shouting. Is someone in trouble? Being a concerned citizen, you head down towards the commotion. It seems that someone is trapped in one of the cubicles and the door lock has jammed. You hurry down the steps and force the door from the outside, releasing the grateful occupant.

It is now some weeks later, and we have arranged for you to meet the manager of the project in the local council offices. He meets you at the reception desk. He is so young.

You: "WHAT was that about?"

Him: "Look, I'm a trainee, I started here straight from school about nine months ago. After I'd been here about eight weeks, I was called into to see the head of finance and operations. I was told that as part of an adjustment to our charging regime, the council had decided that all the public toilets in the city were to charge a much higher rate using a new, smaller coin. All the toilet cubicle doors had mechanical coin-operated locks, and the project brief was to update the locks to take the new coins while spending as little money as possible. I had no real idea what a project even was! And yes, here I was, a project manager. Anyway, I did the research and found that the original lock suppliers were still in business and were producing a conversion kit.

"*I ordered a sample of the modified parts and got one of our mechanical technicians to update a spare lock, which worked perfectly. He demonstrated the modified lock to senior management, and we got the go ahead to proceed as soon as possible. That was it as far as reviewing the quality and reliability of what we were doing. The council got some flak from the local press for more than doubling the cost of using a public toilet, but senior management were determined to carry on with the new charging policy. So, I ordered all the parts required, I got some signs made up explaining the new charges, and the fitter put the new parts in on the agreed-on date. We were up and running, on time and to budget. Job done.*"

You: "Well, that seems a reasonable thing to do, but HOW do you explain what I saw on my field trip up the road? That poor person was stuck in the cubicle!"

Him: "The thing you have to understand is that some of the original locks were ninety years old. Ninety years old! The problem that emerged over the following weeks was that in live operation, very occasionally the coin would not drop. The lock would then jam and not allow the door bolt to be retracted. The effect of this was trapping the unfortunate occupant inside. This is what you saw. There's even a song about it. Maybe you know it? The one about a catastrophe with two old ladies being locked in a lavatory? A great example of life imitating art, and it was, at least partially, my fault."

You, thinking the poor man needs a biscuit: "Looking back, WHY do you think this happened?"

Him: "Well, I think the root cause of the problem wasn't the physical one of jamming locks or the lack of testing over a period to make sure that the locks would continue to work − although obviously it was the jamming of the doors which brought the problem to everyone's notice. The real problem was caused by the project sponsors trying to save money by not buying new locks for the new coins. Trying to fit brand-new parts into mechanisms that had been gently wearing out for nearly a century was the wrong solution, and we never saw the risk of failure coming. And, of course, them giving the job to a rookie with no training or induction into how to ensure that we had a quality solution was no way to make a project a success or even to avoid an abject failure."

You: "WHAT lesson did you take away from this failure?"

Him: "You mean apart from thoroughly assessing all alternatives and making sure that the solution selected is appropriate and fit for purpose − especially in this case where some of the locks being updated were already museum pieces? Seriously, there's also a question of the adequacy testing. A demonstration is not a test. We should have set up a task to test a representative selection of locks over many multiple cycles and included as an activity in the task of fitting locks to test each one over several cycles before they were accepted as fit for purpose. We needed one hundred percent reassurance that the solution was up to the required quality standards. In this case, this means that the locks would never malfunction, or if they did, they should have failed safe with the doors unlocked and not failed with people locked inside. I've been on a short course on project management now, and if I'd known, I would have flagged up that the doors jamming was a risk. I would also have formally recorded a quality requirement, such as each the lock should be tested for correct operation one hundred times before being accepted, and I would have put in place a quality recording system so that I could check it had been done. Obviously, the use of the coin locks had to be suspended pending engagement by the members of the council to decide what should be done. In the end the resulting bad publicity for the council led to the coin locks being removed completely. The project was a failure."

"I should add that about the same time, the council also experienced problems with a much bigger project, where the new exit gates from council car parks closed prematurely, damaging vehicles. My lock problem was comparatively insignificant for them, but it was a major lesson for me."

You: *"So HOW would you sum up that lesson?"*

Him: *"Every time someone tells you that something works OK, test it. Then test, test, test, and test again. And if there is any doubt, test some more."*

In truth, most projects do not end in people banging on the toilet door to be let out or cars being dented by car park barriers. However, failure to design and test solutions robustly at appropriate points in the project can often give rise to significant problems downstream, sometimes, as in our site visit, causing the failure of the project. Monitoring compliance with appropriate quality standards during our task progress reviews gives us the opportunity to make corrections and avoid subsequent failures of work undertaken or the very public collapse of the project.

Step 2: Understand and Evaluate Progress Data

In Chapter 5 we set out a model for data-driven planning in which the data from the task data sheets was collected in some kind of data store, defining every task in the project. This data store could be maintained within whichever software we use to control the project but could equally reside in a set of custom spreadsheets or on paper.

When evaluating the performance of individual tasks, and indeed of the whole project, there are some commonly used evaluation techniques which will help us evaluate and report overall progress against the project plan and expenditure against the approved budget.

Evaluating Progress against the Project Plan

A common question that we are asked as Project Managers is, "Are we on schedule?" The least technical assessment of the progress of a task, and hence of the project's overall progress, is whether it is meeting its planned timelines for delivery. During the review meeting for each task, we gather data on its percentage completion and anticipated completion time. We need to evaluate this data against the project plan for the task. In Chapter 5, when preparing our detailed project plans, we encountered the concept of float time. We defined this as how much later a task can finish than planned before impacting a subsequent task, and we identified two rules about float time:

- If a task already on the critical path has any delay in its completion, it will cause a delay in the project – i.e., it can have no free or total float time.
- If any other task (i.e., one not on the critical path) experiences delays and will now exceed its total float time, the critical path will change and this task will now be now on the changed critical path.

A really useful way of assessing any status is against a traffic light system of red, amber, and green. In relation to delivery against the project plan, if any task has completed, or we believe it will complete, on or before the scheduled completion date, it gets a green status. However, any task on the critical path that is due to complete late automatically switches to a red status, and we put in place strategies to mitigate the impact this is likely to have on the progress of the project overall. Of course, we then have to monitor progress against these strategies as well.

The third category are those tasks currently running late but still within available float time. These might call for more judgement. Their default status should be amber, but we should also consider whether there is likely to be any impact on the overall project delivery dates or costs. If adverse impacts are unlikely, we could consider reducing the task status to green, but equally, if we judge that further adverse impacts are likely, the task's status could be elevated to red, mitigation strategies put in place, and their outcomes monitored. Even if the task status is downgraded to green or remains at amber, we should still keep the task's progress under close scrutiny and, depending on the length of our agreed-upon reporting cycle, may need to have intermediate reviews of progress.

We should record our judgement on the status of each current task in the task's status review record and carry it over into the body of the project progress report to the business sponsor. After reviewing the progress on all current and competed tasks, we can record our evaluation of overall project progress in the headline summary box at the start of the report. This area also allows us to note any slippage in the overall project timetable, record milestones met (and missed), and highlight those areas where management decisions or actions are required.

Evaluating Progress against the Project Budget

"You mean we've spent that much already?" is a not-uncommon comment by the business sponsor. The next section of the project manager's report concerns the assessment of the project's financial performance. Before we start this section, a warning – it will get mathematical. Those of a nervous disposition will be warned when to look away, but it's safe for now.

Understanding how much we have spent against the budget and what we need to spend to complete the project are areas that are full of more pitfalls and traps for the unwary. As a line manager, controlling budgets is generally relatively simple. We know how much we expect to spend on staff each month over the year. We may have some known seasonal factors, especially if we are running ice cream kiosks on the beach or meeting increased fuel bills for winter heating, but in general terms, each year we can profile our expenditure and our income, which, with a few adjustments, will normally be broadly similar to last year. When line managers match their current position against the projected budget profile, they can make a judgement about how well they are doing as well as a prediction of their position at the end of the year. Expenditure on projects is not like this for a number of reasons:

- **Uneven payment profile.** Project expenditure is decidedly lumpy. We may have some continuing costs for running the project management and support team, but many of our costs depend upon whom we have to pay and when. That, in turn, is decided by the project plan and how well we are hitting its target delivery dates.
- **Uncertainty of task completion**. We have greater variability between the timing of actual expenditure on a project and our financial plans. For example, if the delivery of the bespoke windows in a building project is delayed, then not only will our expenditure on these not be met but we won't be engaging contractors to do the internal fixing, and we may be finishing later than planned. If we are developing a bespoke IT system, we ought not be to buying and paying for the main hardware when planned if the software development is delayed. Both

examples underline that our expenditure schedule is task dependent, and tasks are themselves subject to variation in delivery timings.

- **Volatility**. Our unit costs themselves are volatile and were based on estimates. Unlike in line management, we usually have no direct history upon which to base our estimates, and as time passes, we could easily find that we may need to increase expenditure to meet unexpected issues or decrease it if our estimates were overly pessimistic.
- **Matching delivery and expenditure**. A major problem is always the issue of matching what we have spent during the period with the work that we have actually done. The timeline of our spending might match the project's budget to date each month, but if we have not completed all the work planned, then we are overspending (much more on this later).

Consequently, we often need a method of taking all these different types of variables into account to answer the question, "Are we on budget?" Then we must answer the related, but not identical, questions, "Have we spent our money doing what we planned to do?" and "Will we have enough money to finish the project?"

Let's start with the question of whether we are on budget. The first thing we need to think about is how complete and how useful our information is when looking at what we have spent against the progress we have achieved. It can be challenging to meet the information-sharing needs of a fast-moving project using financial processes and systems designed to support a more predictable line management style of operations. Often the project manager finds it necessary to stitch together information from different sources into a spreadsheet system, and it is not only small businesses that have financial systems which need supplementing to meet the requirements of project management. For example, a particular top 100 UK company relied on spreadsheets (at least partially) because their accounting system could not cope with the complexity of forecasting volatile project-based work. This meant that managers had to spend time reviewing and consolidating multiple spreadsheets from across every operating division during each month. Another example is that of a large government organisation's processes and systems, which were based around an annual government accounting cycle and simply could not cope with projects which crossed multiple financial year boundaries for either budgeting or expenditure reporting. These examples are just snapshots of the sorts of problems we may encounter with corporate accounting systems not geared to our needs and for which we may have to compensate. Other issues with which we have to contend in establishing effective reporting of project expenditure can include:

- **Granularity of data**. As we are managing our project on a task basis, we usually need to be able to identify expenditure and time spent by every task. This is an area which may be a challenge for some accounting systems, especially if our project has many hundreds of tasks in our work breakdown structure. It is also a challenge for staff where we wish to record their time and expenses against each task. In an organisation without a timesheet culture, time recording at this level is a challenge, and in an organisation with a timesheet culture, we also have to beware of the danger of senior managers dumping their very expensive time onto our project codes at the last minute!
- **Timeliness**. Many company and other organisational accounting systems are geared towards preparing timely information for line managers at some point after a month end, and we already know that this is much more predictable than project-related expenditure. It is not unusual for the accounting system to take a week or two before the finance team have made all the adjustments that they need and then passed the information out to the line managers.

Being two weeks late with information when monitoring a task can be the difference between steering delivery to avoid a problem and having to deal with an overspending crisis.

- **Completeness**. Not all the expenditure that we have incurred will even hit the accounting system just yet. Consider the not infrequently encountered example of a supplier which delivers to us in the first week of month 1 and invoices us at the end of that month. Assuming that we pay promptly – i.e., during month 2 – the accounting system will not report the expenditure until sometime early in month 3, meaning that our information on expenditure could be up to two months behind the activity to which it relates. The situation only gets worse in those organisations that have a policy of paying 60 or 90 days after receiving the invoice.
- **Future payment commitments for completed activities**. This is another timing issue that exists even where the host organisation's accounting systems can handle commitment accounting – i.e., they have the ability to account for orders placed but not invoiced and goods and services received but awaiting payment. We also need to be able to account for other costs incurred to date, such as staff time, overtime, and expenses which the accounting system does not handle until month end – sometimes even the end of the month after that for overtime and expenses.

In summary, in order to be able to report accurately, we need our expenditure information to be as up to date as our activity information – which we will obtain from our task manager. This means that if we are reporting weekly to the business sponsor, the finance information also needs to be available weekly. It may well be that we have to settle for information from the accounting system less frequently than we would like and then have to estimate the intermediate positions and for it to be less complete than we would like and have to supplement it with records held by the project.

The second factor of which we need to be aware is that our project costs will fall into two clear categories:

- Costs based upon the content of the task – this may be materials, for example.
- Costs based upon the length of time taken by the task these costs, including:

  - All the costs associated with managing the delivery of the project (usually including the project manager's salary costs).
  - Any time-based costs for delivery staff charged to the project, including contractors – for example, the staff and other costs associated with an information systems project being delivered by an internal IT development team.
  - Hired plant and equipment.

Understanding the basis of the costs of each task (and some will have both content and time-related costs) will assist us in our evaluation process and in deciding on the red, green, or amber status for each task.

Evaluating Finance Data against Progress Data

For some business sponsors and project boards, a simple red, green, or amber status on the project's finances is enough, but we can be more precise in our evaluation. Consider a simple project lasting 24 weeks of 12 equal tasks each taking two weeks and costing £2,000 – i.e., the project budget is £24,000. We are now at the end of week 12, we have collected all the expenditure information, and we have spent £12,000. Therefore, we are exactly halfway through the project and we have spent half of our budget. So, do we get a green status for finance?

The answer can only be maybe because we also need to know how far the project has progressed. If we have delivered six tasks, exactly half the work, then we get green status. Delivering five tasks or less, then it's red, and if we have delivered more than six, it's green, plus in both the latter two cases we should provide an explanation to the business sponsor about the difference.

So, in being more precise in our examination of project finances, once we have brought together the best possible information on the actual costs incurred to date, we should examine the next question: "Have we spent our money doing what we planned to do?" The analysis tool that is commonly used to make sense of our financial position is called earned value analysis (sorry for yet more initials, but also known as EVA). If we use EVA progressively, it provides us with an early warning of potential problems and therefore an opportunity for earlier ameliorative action. Please note, such a proactive approach also makes the project manager look better in the eyes of the business sponsor!

Ignore the name for a moment and think back to our planning stage. When we put our budget together, it had three elements:

- How much we were budgeting to spend.
- What activities/tasks we were planning to spend it on.
- When we were planning to spend it.

Now consider the three types of financial information that we now have at the review stage:

- Budgeted costs of tasks that we planned to do.
- Budgeted costs of tasks that we have done (with any adjustments for partially completed tasks).
- Actual costs of tasks that we have done (with any adjustments for partially completed tasks).

Just to add to the confusion, sometimes different terminologies are used to describe the three main data items. The mostly commonly used terms are set out in the top part of Table 11.3. The lower part of the table shows the indicators that we can calculate from these three variables.

Table 11.3 Earned Value Analysis Terminology

Earned Value Analysis Terminology			
Data Items	**Initials**	**Alternative Name**	**Initials**
Planned value	PV	Budget cost of work scheduled	BCWS
Earned value	EV	Budget cost of work performed	BCWP
Actual cost	AC	Actual cost of work performed	ACWP
Calculated Indicators	**Initials**	**Formula**	
Cost variance	CV	Earned value − actual cost	
Schedule variance	SV	Earned value − planned value	
Cost performance index	CPI	Earned value/actual cost	
Schedule performance index	SPI	Earned value/planned value	

We only answer accurately that we are on budget without using an analysis tool if our project has met three tests:

- We have delivered all our planned activities.
- The project is on time.
- We have spent what we budgeted for each task.

However well we might be managing our project, usually events will have impacted our plans. We might be in front of our plans with some tasks or behind schedule in others. Earned value analysis gives us a way of taking these variations into account when tracking our performance against the budget. The simplest analysis is to plot the values of these three variables. The example in Table 11.4 is of six months' data for a project which, as we shall see, is struggling to deliver.

Table 11.4 EVA Example Monthly Data

EVA Example Monthly Data £000								
Variables		**Month 1**	**Month 2**	**Month 3**	**Month 4**	**Month 5**	**Month 6**	**Total**
Planned value	PV	125	175	200	300	350	350	1,500
Earned value	EV	75	100	175	250	325	350	1,275
Actual cost	AC	150	225	225	300	300	400	1,600

Figure 11.3 plots this monthly financial data on our project as a simple histogram. Whilst interesting, this is not particularly helpful in showing our overall financial position.

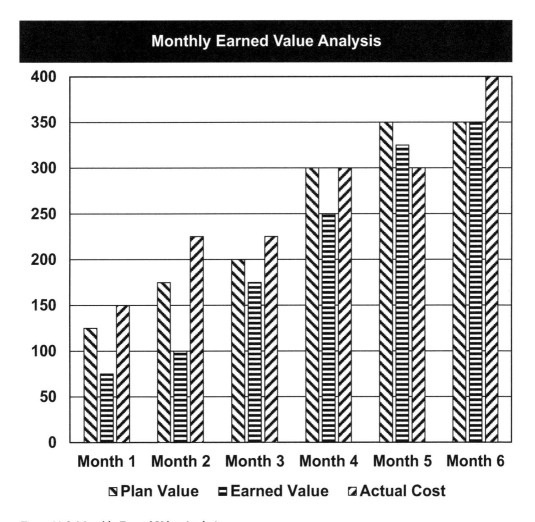

Figure 11.3 Monthly Earned Value Analysis

A much clearer picture of the progress of the project and the problems it is facing emerges as we review our budgets and expenditure on a cumulative basis, as in Table 11.5.

Table 11.5 EVA Example Cumulative Financial Data (1)

EVA Example Cumulative Financial Data £000							
Variables		**Month 1**	**Month 2**	**Month 3**	**Month 4**	**Month 5**	**Month 6**
Cumulative planned value	PV^c	125	300	500	800	1,150	1,500
Cumulative earned value	EV^c	75	175	350	600	925	1,275
Cumulative actual cost	AC^c	150	375	600	900	1,200	1,600

(Note the superscript ^{c} is used to denote that the factors we are looking at are cumulative figures to date in the project; otherwise, there is no difference in the formulas.)

Plotting this on a graph gives us the picture as shown in Figure 11.4.

Using this graph, we can show very clearly that every month:

- Actual cost (AC^c) to date is running ahead of the cumulative earned value (EV^c) – i.e., our budget for the work that has been undertaken. This means that the project is overspending against its budget.
- Actual cost (AC^c) is also always ahead of the cumulative planned value (PV^c) – i.e., the budgeted cost of the work that we had scheduled. This means that the project delivery is also running behind schedule.

As an example, assume these were the costs of an IT systems replacement project. This project contains both content-related costs (for the hardware and software) and time-related costs (for the development team). The detailed examination of the costs which we need to perform could show underlying causes such as:

- More powerful, and hence more expensive, hardware was required than originally planned. Essentially, this represents an underestimate of what was required, and so the actual cost (AC^c) of work performed is running ahead of the cumulative earned value (budgeted cost of work performed, EV^c). This is actually a finite problem. Predicted costs were too low, and now that we have been hit by the increased cost, we can be certain of the new cost base going forwards.
- The development team have encountered difficult issues in preparing the new software and have had allocate staff to the task for a longer period. This is resulting in increased costs for each extra day required for the development task. Consequently, actual costs (AC) are running ahead of the planned value (budgeted cost of work scheduled, PV). This is a measure of how far behind schedule our project currently is. Until the software development problems are resolved, the increased staff costs are not a finite cost and are likely to continue, but as we do not know for how long, we cannot be certain of the new cost base going forwards.

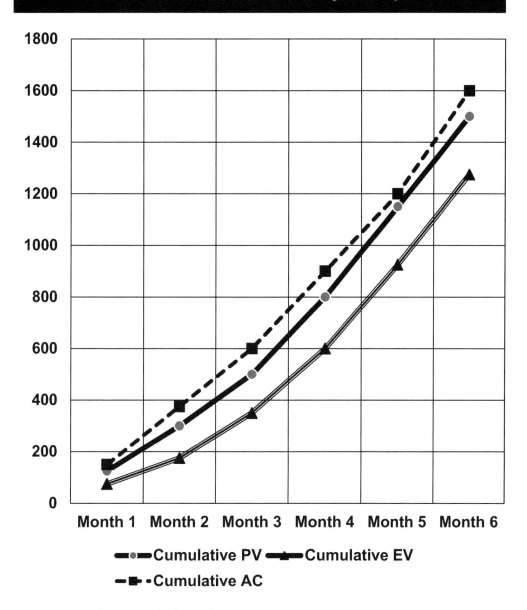

Figure 11.4 Cumulative Earned Value Analysis

THE VERY MATHEMATICALLY NERVOUS CAN LOOK AWAY NOW, BUT STICK WITH IT IF YOU CAN.
WE ARE GOING AHEAD IN VERY SMALL STEPS, AND IT WILL MAKE THE NEXT SECTION EASIER.

Whilst viewing this information in a graph such as Figure 11.4 is useful, we can also use some simple analyses to quantify the differences, as shown in Table 11.5. Analysing the cumulative position to date is probably more useful on many projects than a single-month snapshot.

Table 11.6 takes our cumulative financial information and calculates two further indicators:

- Cumulative cost variance (CV^c) – the difference between our budgeted cost for the work performed to date, i.e., its cumulative earned value (EV^c), and the actual cost incurred (AC^c) to date.
- Cumulative schedule variance ($SV^{c)}$ – the difference between out budgeted cost for work performed to date (EV^c) and the planned value (PV^c) to date.

Table 11.6 EVA Example Cumulative Variance Analysis (2)

EVA Example Cumulative Variance Analysis (2) £000		Month 1	Month 2	Month 3	Month 4	Month 5	Month 6
Variables		Month 1	Month 2	Month 3	Month 4	Month 5	Month 6
Cumulative planned value	PV^c	125	300	500	800	1,150	1,500
Cumulative earned value	EV^c	75	175	350	600	925	1,275
Cumulative actual cost	AC^c	150	375	600	900	1,200	1,600
Calculated Indicators							
Cumulative cost variance	$EV^c - AC^c$	−75	−200	−250	−300	−275	−325
Cumulative schedule variance	$EV^c - PV^c$	−50	−125	−150	−200	−225	−225

We can now restate our view of assessments of spending using the calculated indicators shown in the lower part of Table 11.6 as follows:

- If cumulative cost variance (CV) is greater than 0, we are within budget, but if it is less than 0, we are overspending.
- If schedule variance (SV) is greater than 0, we are within our timescale, but if it is less than 0, we are running late.

As an example, consider the position at month 3 in our data. We already know from looking at our graph that we are overspent and behind schedule:

The month 3 cumulative cost variance is calculated as follows:

$$EV^c - AC^c = CV$$
$$£350,000 - £600,000 = -£250,000$$

This is the amount of our overspending on work which has been delivered.

Similarly, the month 3 schedule variance is calculated as follows:

$$EV^c - PV^c = CV$$
$$£350,000 - £500,000 = -£150,000$$

This represents the amount that the project is falling behind the original plan.

Table 11.7 takes the analysis one stage further and calculates two further indicators which are both extremely useful as they allow us to better track how our performance changes over time. They are:

- Cumulative cost performance index (CPI^c) – denoting our performance against our project budget.
- Cumulative schedule performance index (SPI^c) – denoting our performance against our project plan.

Table 11.7 Example Cumulative Variance Analysis (3)

EVA Example Cumulative Variance Analysis (3) £000							
Variables		Month 1	Month 2	Month 3	Month 4	Month 5	Month 6
Cumulative planned value	PV^c	125	300	500	800	1,150	1,500
Cumulative earned value	EV^c	75	175	350	600	925	1,275
Cumulative actual cost	AC^c	150	375	600	900	1,200	1,600
Calculated Indicators							
Cumulative cost Performance Index	EV^c/AC^c	0.50	0.47	0.58	0.67	0.77	0.80
Cumulative schedule performance index	EV^c/PV^c	0.60	0.58	0.70	0.75	0.80	0.85

(Calculations rounded to two decimal places.)

In each case a calculated result of 1 or more will show that we are on track or better, but a result of less than 1 shows we are underperforming or worse.

The month 3 cumulative cost performance index is calculated as follows:

$$EV^c/AC^c = CPI^c$$
$$£350,000/£600,000 = 0.58$$

Remember, we are looking for a value of 1 or better, so a score of 0.58 shows how poorly we are keeping within budget.

The month 3 cumulative schedule performance index is calculated as follows:

$$EV^c/PV^c = SPI^c$$
$$£350,000/£500,000 = 0.7$$

Again, we are looking for a score of over 1, and although the result of 0.7 is a slight improvement over the CPI^C figure, it still shows how poorly we are keeping within the delivery plan. One way to indicate simply to a business sponsor how far the project is running behind is to divide the planned duration of the project by the SPI^c. In our example, 24 weeks / 0.7 = 34.3 weeks, an increase of 10.3 weeks.

The sorry truth behind the financial management of this project is revealed when we tabulate the results of our calculated indicators and construct a time series graph of the CPI^c and the SPI^c.

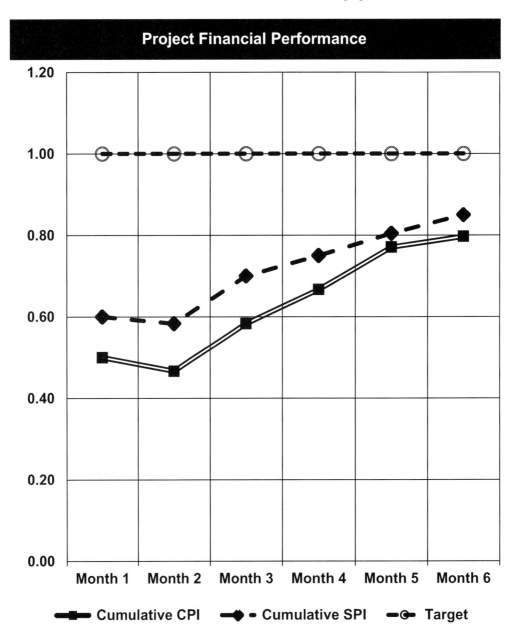

Figure 11.5 Project Financial Performance

Figure 11.5 shows how badly this project has performed to date, though admittedly it is improving marginally. However, our target is to obtain scores of both indices at 1 or better.

Remember our question at the start of this section: "Have we spent our money doing what we planned to do?" In the example, the answer to the question is, "No, we have not – for every month both individually and cumulatively. Things are getting slowly if not better at least less bad."

THE NERVOUS CAN REJOIN US NOW FOR A FEW MOMENTS. WE ARE STILL TAKING VERY SMALL STEPS.

Step 3: Forecast Outcomes and Update Plans

Estimating the Cost to Complete

When looking at project finances, we also had another question to consider: "Will we have enough money to finish the project?" There are three new variables at this stage. Again, sometimes the names may change, but the definitions are:

- The currently approved total budget to deliver the project is referred to as our budget at completion (BAC).
- The money that we will need to fund the remaining work is called our estimated costs to complete (ECC).
- The revised forecast for our total budget is known as our estimated total cost at completion (ETC). This is calculated by adding our estimated costs to complete (ECC) to our actual costs to date (AC^c).

Therefore, we now have a very simple sum written as:

$$ETC = AC^c + ECC$$

The issue is that there are several ways of forecasting our estimated costs to complete (ECC), and they may all give us different results, so we have to choose the most appropriate one.

To illustrate our options, suppose we are at month 3 in our example project and have been asked how much it will cost to complete the rest of the outstanding work. When we were preparing our business case, we estimated, and obtained approval, for a project budget of £1,500,000. This is our budget at completion (BAC).

Now, how should we calculate the estimated costs to complete? We have a number of options:

Method 1: bottom–up reappraisal. We re-estimate all the uncompleted work using a bottom-up approach, as we did previously for the business case. We now may know far more than we did when preparing the original figures, and the theory is that we should now have a much more accurate answer. We can now restate our formula as:

$$ETC = AC^c + \text{bottom-up ECC}$$

How useful this approach is can depend on the stage of the project, the quality of our original estimation process, and the nature of the uncompleted tasks. For example, if the uncompleted tasks include fixed price contracts or use industry standard indices, the estimation may be straightforward. Equally, if we are nearing the end of the project, we may have more visibility and understanding of other costs, making this a suitable method to use.

Method 2: reuse and repeat. We reuse the original estimates for all the uncompleted tasks without updating them. Effectively, we carry forwards any cost variations to date and update our estimate of total cost. In order to work out the cost of the uncompleted work estimated costs to complete (ECC), we simply subtract the budget cost of work already performed (EV) from our previously agreed-upon budget at completion for the project (BAC). The formula for calculating the amount required to finish the project (ECC) is therefore:

$$ECC = BAC - EV^c$$

The total revised cost for the project, the estimated total cost, is calculated by adding this value to our actual costs to date:

$$ETC = AC^c + (BAC - EV^c)$$

In our month 3 example, the calculation becomes:

$$ETC = £600,000 - (£1,500,000 - £350,000) = £1,750,000$$

This approach is straightforward, taking the performance of the project to date, whether that is good or bad, and assuming that in future, our financial performance will be exactly in line with our original budget (for those who stuck it out through the previous section, it means that despite previous performance, we assume that CPI and SPI will both equal 1 from now on). This approach means that any underspending or overspending to date is carried forwards into the estimated total cost (ETC).

This is a very useful approach where we have experienced a finite cost change, such as in our example where we needed to purchase more powerful hardware or perhaps where we have agreed on a contract that has a higher cost than we had budgeted. However, using this approach comes with a health warning. It assumes that any deficiencies that have occurred to date have been rectified and will not affect future project progress. In truth, if a project has encountered difficulties to date, invariably new difficulties seem continue to occur, though past ones may have been ameliorated. For example, in our site visit in Chapter 10, the supplier repeatedly sought increases in the contract price for various reasons. As we saw, only some of these were acceptable to the client organisation.

TIME FOR THE NERVOUS TO LOOK AWAY AGAIN – WE WILL BE BACK SOON.

Method 3: basic adjustment for experience. We use the original budget at completion for the project expenditure (BAC) and adjust it for our overall cost performance on the project to date. If, for example, our costs to date are double what we expected, we need to double the BAC. We do this by taking the BAC and modifying it by our cost performance to date shown by the cumulative cost performance index (CPI^c), which was calculated using our earned value and our actual costs. The formula is:

$$ETC = BAC / CPI^c$$

In our month 3 example, the calculation becomes:

ETC = £1,500,000 / 0.58 = £2,586,207

This approach is straightforward, taking the budgetary performance of the project to date, good or bad, and assuming that in future, our future costs will perform in the same way.

Method 4: sophisticated adjustment for project experience to date. This approach brings together the actual costs and moderates the estimated costs to complete by both the cumulative cost performance index (CPI^c) and the cumulative schedule performance index (SPI^c). The formula is:

ETC= AC^c + ((BAC − EV^c) / (SPI^c x CPI^c))

In our month 3 example, the calculation becomes:

ETC = £600,000 + ((£1,500,000 − £350,000) / (0.58 x 0.70))
ETC = £600,000 + (£1,150,000 / 0.406)
ETC = £600,000 + £2,832,512
ETC = £3,432,512

This approach takes the full performance of the project to date, including meeting both planned schedules and costs, as a basis and assumes that our future costs will perform in the same way. The drawback is that where different types of costs apply at different stages of the project, past performance might not be the best basis for future forecasts. It also ignores the impact that our more effective project management might have in containing future costs once issues have been highlighted by our earned value analysis!

Method 5: the hybrid. It is also possible, with more experience and caution, to take more sophisticated approaches, breaking costs down to different categories and analysing the performance of each category in detail and forecasting ETC using one of the prior methods.

AND WE ARE BACK.

The previous results are summarised in Table 11.8, which underlines the difficulties with preparing an estimate at completion as the formula-driven results can give very different answers.

Table 11.8 Example Summary of Month 3 Estimates at Completion

Example Summary of Estimates at Completion (Month 3)		
	Formula	**Estimated Total Cost (ETC)**
Method 1	ETC = AC^c + bottom-up ECC	Dependent on reestimation (but almost inevitably more than originally planned)
Method 2	ETC = AC^c + (BAC − EV^c)	£1,750,000
Method 3	ETC = BAC / CPI^c	£2,586,200
Method 4	ETC = AC^c + ((BAC − EV^c) / (SPI^c x CPI^c))	£3,432,512

Choosing the method to use to calculate our estimated total cost is a matter for discussion with the business sponsor and careful consideration of the balance between the accuracy of our original forecasts and the continuation of the financial schedule performance to date during the rest of the life of the project. Given sufficient time and a well-documented project with clear original estimates and assumptions, method 1, bottom-up reappraisal, should be the prime candidate for our consideration as it should enable more accurate estimation. The formula-driven approaches will produce estimates more quickly from data that we already have but with a lower confidence level.

Updating the Project Plan

Whilst project planning software was useful during our planning phase, during delivery its flexibility allows us to quickly evaluate the impact of changes in the delivery timing of tasks on stage plans and the overall project plan. It is essential, however, to make any changes in a disciplined way to ensure that our plans remain under control. By adopting the same controls that we use for other project documentation, we can work from the last authorised version of the plan, document the changes required to reflect current delivery status, and update the planning software with the new information.

At this point the plan should be saved, because our next step is to recalculate the plan, and this recalculation may produce unacceptable or sub-optimal forecasts. In order to overcome this, we may need to revert to the saved version and rerun the plan as model several times to examine the results of making changes. For example, these changes could be allocating additional resources to a lagging task or reordering tasks to minimise the impact of delays in material deliveries. Only when our modelling is complete and we have selected the best available option should an updated version of the project and stage plans be finalised and put forwards for authorisation. It may be that the project manager or the business sponsor can authorise minor updates, but they should still be included in the next reporting cycle to the project board.

Step 4: Project Manager's Progress Reporting

Naturally, the range of information and the level of detail that we should include in any project progress report are dependent upon the audience for it. We need to cover the audience's area of interest and give them sufficient detail to allow them to perform their role satisfactorily. The most comprehensive and detailed report (and usually the most frequent as well) should be the one we provide to the business sponsor. This report is our focus for evaluating the current status of the project, and we should consider including the following as a minimum:

- Our assessment of the progress of the whole project and information on where we may be drifting away from the project plan on the delivery of key tasks.
- Information on areas where we need a specific decision or additional support from the business sponsor.
- An appraisal of the financial health of the project. Depending on the frequency of reporting to the business sponsor and the financial information cycles in the client organisation, we may need to prepare intermediate updates based on data that we collect within the project and periodically reconcile this with the main financial systems.
- Details of how tasks currently underway are progressing and what outputs have been completed. We should also note any concerns with the quality of outputs.
- Information on changes in risks and issues.
- A forwards look into the next reporting period.

Some of the more common areas of reporting, including some of the analysis techniques discussed earlier in this chapter, are included in the example template for a project manager's report to the business sponsor shown at Table 11.9. This report format also ties closely to the information gathered in the task progress review formats shown at Tables 11.1 and 11.2. The essence of the report is a very condensed summary of progress in key areas and a recording of our judgements about the impact of that progress on the overall delivery of the project.

Table 11.9 Project Status Report to Business Sponsor

Project Status Report to Business Sponsor					
Report period	**From**		**To**		
Original completion date			**Forecast Date**		
Reason for variance in completion date					
Current project status (delete as appropriate)			**Red**	**Amber**	**Green**
Prior project status (delete as appropriate)			**Red**	**Amber**	**Green**

Status labels the section above.

Escalated Decisions

Ref:	Progress	Status		
		Red	**Amber**	**Green**
		Red	**Amber**	**Green**
		Red	**Amber**	**Green**
		Red	**Amber**	**Green**
		Red	**Amber**	**Green**

Milestones in Period labels the section above.

Task delivery status		**Red**	**Amber**	**Green**
Ref:	Progress	Status		
		Red	**Amber**	**Green**
		Red	**Amber**	**Green**
		Red	**Amber**	**Green**
		Red	**Amber**	**Green**
		Red	**Amber**	**Green**

Current Tasks Update labels the section above.

Project Status Report to Business Sponsor				
Deliverables	**Deliverables status**	**Red**	**Amber**	**Green**
	Completed in period		**Date done**	
	Due in next period		**Due date**	

Our project board progress reports will include the same headings and information but generally will cover a longer period and be in a narrative form.

End stage reports in traditionally managed projects focus on demonstrating the stage may safely be closed, as:

- Planned work has been completed (and documented) satisfactorily – i.e., that the stage outputs are in line with scope and quality requirements.
- The project is in line with the project plans and projections, including financial plans using the techniques discussed earlier.
- All risks and issues have been reviewed and actioned as agreed.

Ad hoc reports to stakeholders – including, for example, staff, supporters, and the wider public – are often designed around the needs of the particular stakeholder group, will generally not contain as much detailed data, and may rely more heavily on narrative text and graphics.

End Stage Report

Although we have been reporting progress to both our business sponsor and the project board throughout the delivery period, assessments the end of each stage in traditionally managed projects are seen as major control points. End of stage assessments (and, we hope, approvals) by the project board based on our reports are the formal points at which the client organisation decides on one of three options:

- Continue the project as planned.
- Continue the project with amendments and alterations.
- Stop the project.

It is the end of stage report, prepared by the project manager, that brings together the information needed to allow the project board to make this decision. To be fit for purpose, the end of stage report should provide:

- **Performance summary**. This should be compared against the project's objectives. This overview needs to include financial performance and performance against time and quality targets.[2] Ideally there should be two sections, the first on the project to date and the second for the stage which is completing. This performance summary should also detail:

  - Project outputs delivered during the stage.
  - Variations, including budgetary variations and the cost of approved changes. These need to be carefully examined and any future remedial actions detailed.
  - In relation to benefits, if any "quick wins" were planned for the stage, these should be identified and performance against plans reported.
  - Any planned tasks which will not be completed within the stage and are being replanned for inclusion in the next stage should be highlighted and the reasons for this explained.
  - Where independent quality reports have been received, these can be summarised as corroboration that quality targets are being met (or not).

- **Completion forecast**. The project manager's assessment of how the project will meet its targets for time, costs, and quality. Like the performance summary, the completion forecast can be split into two sections. These are overall project completion and the completion of the next stage.
- **Project strategies and processes**. An overview of how well the project is delivering the different strategies we set for the project, such as risk and issue management or communications and the effectiveness of the processes that we created to govern delivery, together with any suggested changes and improvements to them.
- **Risks and issues summary**. If risks are being quantified by the project, this should include:

  - An agreed-upon number of top risks and issues, together with any changes in their risks scores and mitigation actions either taken or still outstanding.
  - The total level of risk currently faced by the project and trends in risk scores.

- **Lessons learned**. Lessons learned during the stage can be both positive – i.e., what went well that we will repeat in future – and negative – i.e., things we will do better next time. Sometimes lessons learned should lead to us flexing our planned behaviour or activities in the current project, and sometimes, probably because we have passed the point at which they are applicable, they are only useful for future projects. In either case we will need to bring these lessons forwards from each end of stage report into our project closure report.
- **Business case reassessment**. An evaluation of the current state of the business case and the project plan, including any known changes in information and the progress of the project to date. At the same meeting, the project board could be asked to approve any required updates to the business case and the project plan.
- **Recommended actions schedule.** This should cover all the changes or updates recommended in the previous sections of the report and should detail timelines and responsibilities for taking, reporting, and monitoring actions.

The project board's acceptance of the end of stage is the prerequisite for authorising its closure and the initiation of the next stage.

Much of the data required for the end of stage report has been already gathered and shared with the project board in our regular project board progress reports, but some parts of the report, such as lessons learned during the stage, can only be prepared as the stage closes. The end stage report also provides an important building block when we come to write our project closure report. This is because, as time passes, events that need to be reported become less well remembered and are often overlaid by subsequent events, so we need to research what happened often months and sometimes years ago. Consequently, in preparing the project closure report, we always need to trawl backwards through our project reports and papers. This trawling through previous papers is made far easier if we have already prepared a well-written set of near contemporaneous summaries – i.e., our end of stage reports.

One difference between traditional sequentially managed projects and those using agile methods is that agile projects tend to prepare more frequent end of sprint reports, perhaps with lighter details, and their main focus is on the deliverables produced during the sprint. In all types of projects, however, the project manager has to meet the needs of the audience for the report. At this point, the project manager must prepare a report for the project board in the format they need and containing sufficient information to allow them to make the required decisions about the future of the project. There is often discussion about how much information should be provided in an end stage report and sometimes doubts about whether the board actually read it. It is usually better to err on the side of caution and provide more relevant information rather than less and perhaps to provide a short summary of key matters for consideration.

Reflections

As before, if possible, using the same project as in your previous reflections, spend some time considering the following exercises:

- Put together an overview of the project monitoring, control, and reporting processes, including:

  - Reporting periods.
  - List of task progress reviews.
  - Agenda for the task progress reviews.
  - Agenda for the project review meeting with the business sponsor.

- Using a task data sheet that you created previously, extend the data entered showing four interim reviews and a task completion review.
- Prepare some sample figures for the project over a six-month period and produce a graph showing planned value, earned value, and actual costs.
- Extend your analysis to prepare estimated costs to complete using methods 2, 3, and 4 (or method 2 only for the mathematically nervous).
- Pick a stage from your example project and prepare a bullet point presentation covering the main areas of an end stage review.
- Consider how you would explain to the project board that although the project is currently underspending against the monthly expenditure profile that you prepared, you are expecting the project overall to overspend by 10%.

Notes

1 You're on your own for this one. I know what's going to happen. It was my fault (partly), and I'm not going to go through it again. I still have the emotional scars.
2 Explaining the calculation of earned value and estimates to complete to a project board for the first time will be fun!

12 Managing and Realising the Benefits

Just When You Thought It Was All Over . . .
Part 1

Aim of this chapter: To explain the mechanisms the project manager can use to ensure the benefits anticipated from the project are achieved and that these benefits can be monitored and managed during project delivery as well as once the project is complete.

Learning outcomes: To be able to identify, select, and operate the most appropriate mechanisms for delivering the benefits of a project.

So, our project has completed. The business sponsor is happy, the project board agrees that we have done a good job, and now the whole thing can be handed over to the business to operate. We can close the door and move on to our next challenge with all the amazing experience from delivering this mission-critical project added to our CV. Except we have not finished yet. We have two more jobs to do. In this chapter we look at one of the most neglected areas of project management: ensuring that the delivery of the strategic and operational benefits which were the rationale for creating our project in the first place is properly managed.

Sorry to have to tell you this now, but our objective should never be to complete our project plan. The project plan, you will recall, was only ever the route map that we set out for the journey we have been on. Our true objective is the continuing delivery of the benefits which we set out in the project frame and the business case – and, in particular, meeting those critical success factors which formed the "golden thread" and which we built into the project. These benefits should have formed the "prize" which we quantified in our business case and upon which we should have been focusing our team and stakeholders since the very beginning. To be crystal clear, this means that the desired output of our project is never really the jet fighter, or the office move, or the hydroelectric dam, or the new business process, or the industrial plant. It is, in fact, improved national air defence, or a more prestigious office, or cheaper power and reduced flooding, or reduced costs and increased profits. It is ensuring effective benefits delivery that is the ultimate demand on the project manager.

Arguably, any project and any book on project management should start by considering benefits. However, unlike every other aspect of project management we have considered so far, benefits management is unlikely to be our responsibility. Our role, instead, has to be centred on preparing an effective benefits delivery and management framework for the client organisation to operate after the project is closed and we have moved on. The exception to this is those early benefits delivered before the project closes, which will usually fall at least within our area of direct influence, even if the client organisation decides that their delivery will be the responsibility of line management.

DOI: 10.4324/9781003405344-13

Why Do We Need Benefits Management?

We should regard benefits management (or benefits realisation management, as it is sometimes called) as closing the loop on the project delivery process, our initial critical success factors, and ongoing line management. Some organisations have historically taken the view that money and time have been spent on the project, the results are what they are, and now it is time to move on. This means they cannot identify whether the project actually succeeded, they miss the opportunity to learn how the new solution actually performs during live operations, and they therefore cannot easily adjust those operations to maximise benefits or even ensure that future projects perform better.

In the real world, however, not all projects are concerned with delivering continuing benefits to the client organisation. An example of this would be in residential property construction, where once each property is completed, it is sold and handed over to home buyers. In most other projects, the project may be responsible only for delivering its predicted outcomes, and continuing benefits will be delivered by the client organisation. In the rest, especially those concerned with early benefit delivery and organisational change, effective benefits management can become an integral part of the project itself. Therefore, we need to adopt a critical approach to benefits management and appraise the needs of each project and benefit individually. Fortunately, as the project manager, we are probably the best placed to lead this appraisal. We will have discussed, analysed, quantified, verified, and documented these benefits in the business case, and we will almost certainly have updated and discussed this and other documents multiple times during the life of the project.

Consider two project scenarios and the differences required in managing the benefits that they deliver:

Project 1 is to build a development of 150 homes for sale. As part of our project, we need to deliver the major benefit to the client organisation – i.e., the profit from the initial sale of the houses and flats. Therefore, we engage estate agents or employ our own sales office staff and aim to sell the houses "off plan" – that is, before they are actually built, or at the very least as soon after completion as possible.

Project 2 is to improve the processing of staff travel expense claims in a large organisation by replacing the current diverse claims made on paper with an integrated system accessible from smartphones and company laptops but still giving staff the option of paper claims. The rollout for this change requires a mix of scanning of paper claims documents and direct data entry by claimants, supported by self-scanned receipts. It is anticipated that the costs of the new system will be much lower, as there will be no staff directly employed in inputting data and checking and filing claims. Benefits will include not only staff time but the release of accommodation for the staff travel claims team.

In the first case, once every house is sold, we have achieved our main benefits, and there is no requirement for ongoing monitoring. The second case is more complex and typical of a business change project as the benefits from adopting the system will occur over its entire lifetime. They will also increase as the uptake of direct data entry and receipt scanning by staff members increases. Benefits management in this case is essential to demonstrate that benefits are being achieved and may include the promotion of the use of the online system as well as monitoring and reporting progress against the business case.

Before we go any further, we need to check back with the bio-energy from effluent project. We now know that you should have been focused on benefits delivery from the beginning. So, we are going to step back in time to a point when you've prepared the business case and you're just finalising the details and "socialising" it with various members of the project board before submitting it for final approval.

A ping comes from your laptop. An email has just arrived in your inbox from Jan, the executive assistant to the vice president of operations.

Re: Benefiting from our Investment in Bio-Energy from Effluent

The VP-O is concerned that the bio-energy from effluent project lacks details on what will happen as you transfer the operational sites from the responsibility of the project to live operations. She's asked me to share her concerns with you and explain that, in the past, we've had some major projects which looked good on paper and which were delivered on time and to budget but then got swallowed up in departmental and national reporting and we never got any view of how successful they really were. We don't really know if there was any return on our investment in these projects, other than some high-level figures about general improvements in financial performance which could easily have been due to other factors. She's looking to you to help develop a way of demonstrating the value of our investment both now and in the future.

She has time in her diary next Monday morning at 9:30 and would like to discuss how you propose that we should avoid repeating the mistakes of the past.

Kind regards,

Jan

So, first of all, you have a diary conflict. Remember that your regular progress review meetings with the boss are on Monday mornings. Second, you need to be able to present a coherent approach to tracking all the costs of your project (which will largely be under your control) and then how benefits should be identified and maximised once the project has been handed over. You'll have moved back down to managing the insurance team on the third floor from your other windowless office by then, but you're determined to leave things in excellent shape for live operations.

What you need now is a benefits management strategy and someone to make sure it happens when you move on after the project closes. You grit your teeth. No one is going to accuse your strategy of "lacking detail."

Figure 12.1 shows how our benefits management process actually begins at the inception stage of the project. After all, the core of our whole project management procedure has been understanding and then delivering the benefits from our solution.

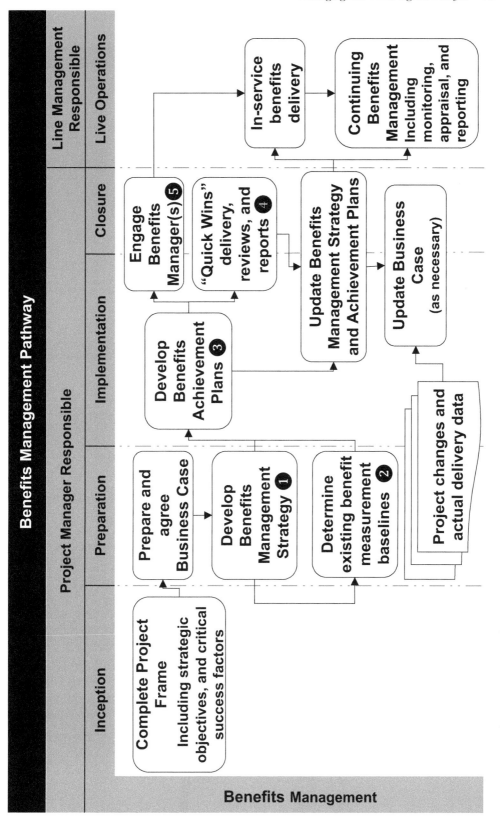

Figure 12.1 Benefits Management Pathway

When we started out on our project journey, we defined the things the project had to achieve, our strategic objectives, and our critical success factors. When preparing our business case, we quantified the main benefits we expected to achieve from the project and used the results to justify both the project and the selection of our recommended solution. During the delivery stages, when we identified changes, managed risks, or dealt with issues, we considered their impacts on our projected benefits and updated the business case. Therefore, only at the end of the project, when the final information on time, cost, and quality becomes available, can we produce the final version of the business case.

Even though we treated the business case as a living document and kept it updated, it still represented only a snapshot of the views of the benefits that may be achieved by the project. As Table 12.1 shows, this identification and quantification of benefits in the business case forms only one step in the dynamic process of benefits management.

Step 1: Benefits Management Strategy

Benefits management not only aims to make sure that our desired positive outcomes have been defined in the business case but also that they are:

- Measurable directly or by inference.
- Are actually achieved in live operation.
- Are monitored, updated, and reported throughout the operational life of the solution that our project delivered.

The objectives of our benefits management strategy are therefore to:

- Describe the benefits that the project will deliver.
- Detail the processes and procedures that will be put in place to ensure that benefits are delivered.
- Identify how benefits will be measured, recorded, and reported.

One important issue to bear in mind is that although the contents of the benefits management strategy will draw upon other project documentation, it should be designed to be read by a wider audience than many project documents and will be in use for the life of the solution, rather than the life of the project. It is essentially a high-level narrative document containing summarised information. By now you should be able to chant the answer to which seven areas the benefits management strategy document needs to cover:[1]

Why are we doing this project? Restating our agreed-upon critical success factors helps focus the readers' attention and shape their response to the proposals contained in the strategy.

What benefits will the project deliver? Again, this section is not new and is a restatement of previous work for a new or changing audience. These benefits should be as set out in the latest agreed-upon version of the business case and should include how they fit with the broader objectives of the organisation. Just to underline the point again, we should be keeping the business case updated during the life of the project, and any updates to benefits and timings also need to flow down into revisions to the benefits management strategy. Where we have quantified figures, these now become our benefit targets. It is also important to bring forwards from the business case:

- Assumptions we have made about benefits.
- Details of any risks to benefit delivery.
- Existing performance measurement baselines.

When will the benefits be delivered? There may be different time frames and trajectories for the delivery of different benefits. These should have been shown in the business case and now need to be restated in terms of set reporting periods. Additionally, we need to identify when and how often each benefit will be measured and to whom it will be reported on.

Where will benefits occur, and where will the information on benefits received be stored for subsequent reporting? In public service projects, or those delivered for the benefit of separate areas of a company, these are often difficult questions. As an example of how tricky or contentious this can be, consider the issues around collecting information on the projected financial benefits of the UK's second high-speed rail link. These projected financial benefits controversially included the value of increased productivity of business people working on the train. More simply, perhaps, for a commercial project, consider how information can be reliably collected on the benefits achieved by a new customer and sales system supporting many different company divisions, each one of which may have multiple profit and loss accounts.

Who will be responsible for ensuring that the benefits are delivered, both during the project ("quick wins") and afterwards?

In this section we should address:

- Where we have identified "quick wins," if benefits management will remain a responsibility of the project manager until project closure and, if not, how this will be handled?
- How benefits will be managed after project closure?
- If the benefits management strategy can be updated to reflect actual experience after the project is closed, who will be allowed to update it, and who will be responsible for approving any updates after the project board has been disbanded?
- Who will be responsible for monitoring and reporting on benefits achieved, and who will be responsible for receiving and actioning benefits monitoring reports?
- How any future conflicts between benefits management and operational service line management will be handled, including escalation processes?

How will we measure each benefit? Measurement and reporting should be against the target projected benefits in the business case for each period. It is also useful to set a range with upper and lower limits of expected benefits. Some benefits will be directly measurable, while some, especially in public infrastructure projects, can only be inferred. In all cases we need to set out what information will be used, how it will be collected and validated, and by whom. These will form key elements of our detailed benefits achievement planning.

How much? What methods will be used for evaluating the costs of the project, the ongoing costs of operating the solution, and the cashable benefits being obtained? We also need to identify the sources from which information on operating costs and cashable benefits will be collected and analysed. For example, we may require amendments to the corporate accounting system to allow the costs or benefits of the solution to be tracked, especially where the finances of different departments or areas are impacted.

As suggested in Chapter 4, the benefits management strategy is part of the set of project strategies that need to be approved by the business sponsor and the project board. During the research and preparation process, however, it is extremely desirable for us to engage with the senior stakeholders and line managers who will have responsibility for delivering the identified benefits. In those projects where benefits are to be achieved over a protracted future period, successfully transitioning to live operation at the close of the project will require ensuring that we have put in place continuing benefits management arrangements. The baseline for this, which it is our responsibility to provide, is a business case that has been fully updated with final details of actual costs incurred and future operating costs and that these have been fed into the final updates of the benefits management strategy (see step 2) and detailed benefits achievement plans (see step 3).

After the project, when our solution is in live operation, the continuing achievement of the anticipated benefits will require our benefits management strategy to have been embedded in operational and line management processes. This embedding needs to include monitoring to detect and evaluate those areas where:

- Operational barriers arise that detract from the achievement of our planned benefits, either on transition or through a gradual degradation over a period of time.
- Live operations reveal opportunities for delivering further benefits from the system not previously envisaged.

It is for these reasons that some organisations appoint a benefits manager for major projects.

Step 2: Baselining

We have already set out what benefits our project will achieve and how we will measure each one. There is, however, a simple question that we need to answer first: "Where are we now?" For each benefit, we need to establish the current levels of performance, perhaps supported by information from previous years. This is not always easy to achieve, as the necessary data may not have been collected or analysed in the ways we now require in order to be able to show the improvements made by the project. We may therefore have to look for other measures from which we can infer the baseline we need. This is a necessary step, but it is not sufficient on its own as there are further areas for us to consider:

- How will monitoring data be collected to demonstrate the ongoing benefits and disbenefits of the solution that we are putting in place? This can be challenging to overcome, and problems of data deficiency will be highlighted by the difficulties we have encountered in the baselining process. Non-financial benefits are often extremely difficult to monitor over time.
- What impact will other factors have on achieving our benefits, and can we separate them out? For example, a project designed to increase public transport passenger numbers could have its results enhanced by an increase in general economic activity or decreased by a work from home order during a public health crisis.

Step 3: Benefits Achievement Plans

Much of the information contained in the benefits management strategy is rightly at a high level. However, as with everything we do as project managers, we have to bring this down to a finer level of detail, and so we need to prepare a comprehensive achievement plan for each benefit. A carefully drafted benefits achievement plan should provide the basis for evaluation and reporting of benefits achieved, including both any "quick wins" and long-term benefits delivery. This development of the benefits achievement plan is another area which requires close consultation with our wider network of stakeholders, including those line managers who will be responsible for managing the operational outcomes of the project and, if one has been appointed, the benefits manager.

Our benefits achievement plans should reference every benefit identified in the business case and can be tabular rather than narrative. It is usually helpful if:

- Where necessary a benefit is sub-divided into individual achievement plans on the basis of the nominated person accountable for its achievement. This means that on many occasions, a benefit identified in the overall benefits management strategy has to be split into several achievement plans – for example, for each department head involved in operational delivery of the solution.
- Each benefit has two sets of measures. Firstly, the "as is" situation – i.e., the performance baseline for several periods prior to the implementation of the project solution – and secondly, the "to be" target performance improvements against which the project was justified. Again, these will be identified against a number of time periods.
- Disbenefits are also included for monitoring purposes – for example, increased operational costs, such as maintenance charges and energy costs, or in software-related projects, licence fees.

Table 12.1 shows what a benefits achievement plan could look like. This example format also includes the ability to record benefits achievement reviews.

Table 12.1 Project Benefits Achievement Plan

Project Benefits Achievement Plan			
Benefit reference			
Benefit name			
Accountable manager/benefit owner			
Department			
Benefit description			
Alignment to project critical success factors			
Risk factors impacting the benefit			
Baseline data (previous years)	**Year −1**	**Year −2**	**Year −3**
Measure 1			
Measure 2			
Measure 3			
Measure 4			
Performance targets	**Year 1**	**Year 2**	**Year 3**
Measure 1			
Measure 2			
Measure 3			
Measure 4			
Performance data sources			
Measure 1			
Measure 2			
Measure 3			
Measure 4			
Benefits manager (monitoring and reporting)			

First achievement review				
Date		Target performance	Actual performance	Percentage variance (+/−)
Measure 1				
Measure 2				
Measure 3				
Measure 4				
Agreed actions				
1				
2				
3				
4				
Signed accountable manager		Signed benefits manager		
Second achievement review				
Date		Target performance	Actual performance	Percentage variance (+/−)
Measure 1				
Measure 2				
Measure 3				
Measure 4				
Agreed actions				
1				
2				
3				
4				
Signed accountable manager		Signed benefits manager		

Benefits Achievements Reviews

Our benefits achievement plans should set out in detail the future targets for the benefits and the dates on which progress should be reviewed. The example format in Table 12.1 includes two reviews, but in practice there can be as many or as few reviews as necessary to demonstrate that the planned benefit has been achieved and the business case projections met.

As project managers we may be responsible for any reviews of "quick wins" up to project closure, but we can still schedule suitable review times for other benefits post-solution handover. If a benefits manager has been appointed, they may carry out the actual reviews prior to project closure, but it may also be useful for the project manager to be engaged as well, helping to smooth the transition to post-project closure benefits management. Tying the first review date for "handed over" benefits into the timetable for the project's lessons learned review can also be useful.

The report from each benefits achievement review should also contain any agreed-upon actions to be delivered in the next period to ensure that our agreed-upon benefits targets are met and sustained.

Step 4: Delivering – Quick Wins

As project managers our main focus is on what can be managed by us, including benefits delivered before the project closes. Consequently, we can define a "quick win" for our project as one falling into this special category. Often "quick wins" can be extremely valuable as we can use them to build support and momentum for the project as it moves through the delivery stages before reaching full live operation. It should be noted that the first impact of our searching our potential benefits to identify those that can at least start to be delivered as "quick wins" is that it should force us to reexamine our handover and launch strategies – especially if we had been planning a high-risk "big bang" launch.

Let's call to mind our site visit in Chapter 5. This visit was to that first joint planning meeting for a major project aiming to revolutionise the way a national government-to-citizen service was delivered. The team had planned to launch a series of web transactions over three years of the project to replace the forms currently used, gradually building to a fully digital system for all its users. Benefits delivery, including reduced staff costs, would have started when the first web-based transactions went live because data would now be entered by the clients directly instead of the current situation of completing and posting manual forms that then required input by the client organisation's staff. This productivity gain would have increased as other web forms were launched until the project achieved its final objective of a fully internet-based service. We can now see that this was actually a plan for a series of "quick wins," which the deputy CEO effectively killed by introducing a totally artificial deadline. So instead of building support for the project as the solution moved to live operation, a "big bang" approach was forced onto the team. It was this "big bang" that ultimately failed.

Step 5: The Benefits Manager

The line manager responsible for operational services after project closure will normally be accountable for delivering the planned benefits. In practice, however, there may be multiple managers involved if the project delivers a solution that runs across several areas. If this is the case, we should argue strongly for the allocation of the responsibility for co-ordinating monitoring and reporting on benefits to a benefits manager on behalf of all the line managers involved. This may not be a full-time role, but it is one that we should consider including when we prepare our business case, up to and including possibly reducing the total cash benefits of the system to fund the role.

Ideally a benefits manager is appointed as early as possible, as they can have a major role in supporting the project manager in defining and quantifying benefits and can be delegated responsibility for leading the realisation of any "quick wins."

In our site visit to Daniel in Chapter 6, he actually highlighted the issue and assigned the role of benefits manager to the project leader role that he outlined. He said:

> I also thought that to give them ownership, I might appoint them not to the project but to a role of service manager with long-term responsibilities for managing the two solutions and ensuring that we get the projected cash and non-cash benefits from them.

In that case the project leader included the role of benefits manager and operated jointly with the project manager. Where a separate benefits manager is identified before the project closes, they should be expected to be part of the project team and report to the project manager, and after the solution is transferred "business as usual" status, they should report on benefits directly to the business sponsor.

Reflections

As before, if possible, using the same project as in your previous reflections, spend time considering the following questions:

- How would benefits best be managed for your project, and to whom should the achievement of benefits be reported?
- What would be the mechanism for resolving any conflict been line management and benefits management?
- Consider the timing of the delivery of a sample of five benefits identified in your business case, including intermediate achievement targets.
- For each benefit selected:

  - What are the risks to benefit delivery?
  - How will you measure the benefit?
  - Are there possible "tripwires" associated with the intermediate achievement targets?
  - Who should be responsible for delivering the benefit?
  - Populate a sample benefits achievement plan for each of the five benefits.

Note

1 OK, if you still can't remember the answer, please feel free to check your phone's lock screen or the tattoo on your inner arm. Somewhere it should read What, Why, When, How, Where, and Who – and How Much.

13 Project Closure

Just When You Thought It Was All Over . . .
Part 2

Aim of this chapter: To explain the various tasks the project manager must perform in bringing the project to a controlled close.

Learning outcomes: To be able to identify when a project should be closed and to operate the most appropriate mechanisms for closing a project.

Project closure is the orderly shutdown of a project. Closure is the final stage of the project, not a single point in time, and is an essential element of our control of the project, even if it has to close early.

Before we consider when to close a project, and the detailed steps in the closure phase, it's time for you to don the (by now slightly battered) hard hat of the bio-energy from effluent project manager for the last time.

The project, under your expert guidance, has moved on. You're well into the implementation and rollout of new plants in various locations around the world, and one by one, you've been transferring them across to the operations division. It's late in the afternoon, and just as you were wondering whether you could finish on time for a change, there's a knock on the door. It's the boss. Since when does the boss knock before barging in? This is a first.

"I was wondering if you could spare me a few moments. I've just come out of a very long board meeting, and there are a couple of issues arising from the project which we need to discuss as soon as possible."

This is not a good sign. You brace yourself. The boss continues, "I'll get some coffee and biscuits organised, and you can come to my office and we'll have a chat. Let's say in about ten minutes?"

You're relieved that you're not immediately on the spot.

"OK, see you in ten minutes."

The boss leaves, and you have a bit of a think.

"Coffee and biscuits, is it? Well, that's unusual. Perhaps it isn't bad news. Or perhaps it is bad news, perhaps they're closing the project early and I'm being softened up before being told my services are no longer required. Perhaps security are already on their way to escort me from the building."

You check you've got your security pass with you, in case you have to hand it over.

There's no way to tell how this conversation is going to go, so you start to prepare for the worst-case scenario. You pass ten anxious minutes in tidying your desk and finding a cardboard box to load your few personal items in to bring home. As zero-hour approaches, you take a very deep breath, smarten yourself up, and, possibly for the last time, leave the janitor's former cupboard and make your way along the carpeted executive corridor. Exactly at the appointed time, you arrive at the boss's office. The boss's PA waves you straight through. Was that a congratulatory smile or one of sympathy?

DOI: 10.4324/9781003405344-14

As you go in, you take another even deeper breath and try to slow your heart rate.

"Here, help yourself to coffee and biscuits." The boss indicates a tray laid out on a side table. You notice these are the good coffee cups and not the usual slightly chipped staff mugs. Well, that doesn't help. Determined to go out in style, you take both chocolate biscuits.

"So, as I said, I just came from a board meeting. I'll cover our decision on the future of the project first. As I understand it, the implementation is going well. You now have ten sites up and running, and you've got a routine process running right from the identification of new sites all the way through to commissioning full operations. I also understand there have been no more unfortunate incidents of spraying of leading local politicians with . . . let's say, 'product,' however much they might deserve it."

Having just bitten into the second chocolate biscuit, you can only nod. The boss continues, "The board felt that this whole bio-energy from effluent thing is now mature enough to transfer to operational management and that we can stand the project down. This is a measure of the great success that you personally have achieved in making this strategic project happen. The board has asked me to congratulate you on a job well done and to convey their thanks to you."

Swallowing quickly, you choke out, "Thank you. That was very kind of them."

You find yourself waiting for the boss to add something like, "But as your department is now managing very well without you, and the project doesn't need you any more, I'm afraid that you are now surplus to requirements and security are on their way up. . ."

But he doesn't. He says, "I'd be grateful if you could draft an updated plan to close the project and hand over to the divisions by, say, early next month."

Getting your thoughts together a little more quickly this time, you manage to respond, "I can't say I'm not disappointed to not be carrying this on through to the final achievement of our objectives, but yes, I can draft a plan for you by early next month."[1]

The boss replies, "Well, we'll come back to timing later.[2] I did say that there were a couple of things coming out of the board meeting, and this next part is highly confidential and market sensitive. Please don't repeat any of this outside of this office for the next couple of days, at which point it will all be made public. The chair is standing down with almost immediate effect. Apparently, something rather murky from his past is about to surface in the press, and to limit the fallout on us, the decision has been taken that he should leave before it becomes public."

Your eyes grow wide. You don't dare interrupt.

"This means that we have to bring forwards our long-standing plans for the CEO to become chair. In turn, I'll be moving down the corridor and becoming the new CEO from the start of next week. I'm telling you this so that you're prepared, because the board were so impressed by the way that you handled this project that they'll be inviting you to take over from me as director and to become a full board member. Congratulations. You will, however, also get a new area of responsibility – strategic projects. You'll have a brief to develop the way you have delivered bio-energy from effluent into the standard way we run projects. You'll be tasked to build a team of like-minded project managers as an internal Centre of Excellence for Project Management and Delivery. Welcome to a permanent place on the executive floor. You'll move in to this very office – don't you like the view? – and the janitor will finally get his cupboard back. Now, how soon did you say that can you get me a closure plan?"

Ambushed again, you need to think very quickly through what a closure plan for this project needs to contain, how closing it early will impact on its contents, and how soon can you get it done and move out of the janitor's cupboard.

Figure 13.1 sets out the key steps in the closure process. As ever, not all steps will be required by every project.

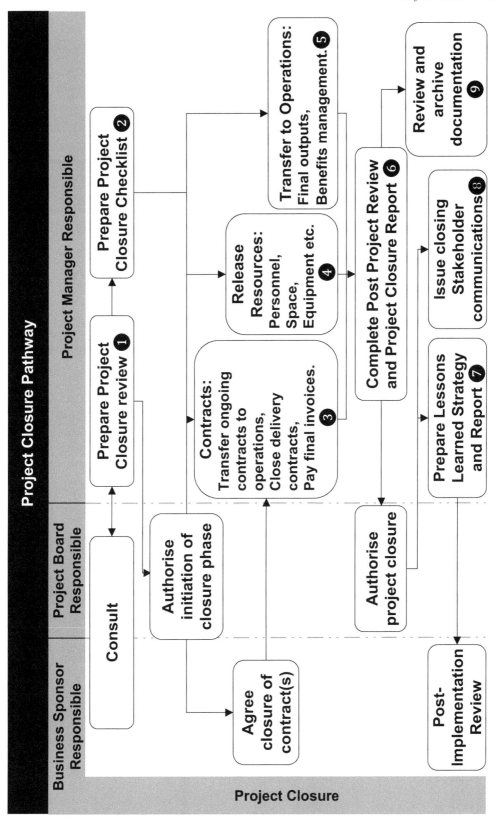

Figure 13.1 Project Closure Pathway

Step 1: Project Closure Review and Report – Are We Ready to Close Up?

Closure is often described as a handover for operational management, which may be true in many cases. But at the end of a project, we may instead be handing over directly to a client or to multiple clients or even to another project as part of a wider programme. Our orderly shutdown always involves a number of similar key tasks, although the details of each task are highly project specific. Our first step is to review the current state of the project and arrive at a decision as to whether the project should be closed. There are a number of basic situations which should trigger project closure:

- We have successfully delivered the project's objectives as contained in our project frame and our business case, or, as in the bio-energy from effluent case study, the project is in a strong state and its implementation can be completed by line management.
- Our reforecast of the project's costs and benefits contained in an update to the business case indicates that, even considering all possible changes to the project, there is no justification for continuation.
- An increase in the total risk levels of the project to an unacceptable level.
- A change in the host organisation's financial position means that it no longer wishes, or perhaps is unable, to fund the remaining work on the project.
- A change in host organisation's corporate strategy means that the project is no longer required.
- A change in external circumstances means that the host organisation no longer wishes to achieve the project's objectives. This could include, for example:

  - A company takeover or liquidation.
  - Technology changes rendering the chosen solution obsolete.
  - Market changes reducing the potential return from the project.

When the project has progressed successfully through its different stages, the role of the project closure review can be amalgamated with the last end of stage report. Here, we will have provided the evidence required for the project board to give permission to move to the next stage, which in this case is to initiate closure. Where the project is closing early for any reason, a separate project closure review is a good practice. The objective of the report is to highlight to the project board the reasons for closing the project, its impact, and the closure actions required.

Step 2: Prepare Project Closure Checklist

The project closure checklist is highly individual and reflects the project's objectives, history, and current status. It details those tasks that will need additional actions either to be completed as part of the closure process or to enable them to be transferred to the client organisation as part of business as usual. Although we would usually discuss this checklist with the business sponsor, it is essentially an assessment for ourselves as the project manager of the tasks we need to complete in order to deliver closure. As a minimum, the areas shown in the following three steps should be reviewed:

- Contract management.
- Releasing resources.
- Transfer to operations.

Tasks and actions should be identified for inclusion in the project closure checklist, and there may be other tasks necessary depending on the specific project.

Step 3: Contract Management

The closure of our involvement in contract management as project manager has two distinct aspects:

- Transfer continuing contracts, such as maintenance, continuing support, materials, and any software licences.
- Closure or transfer of solution delivery contracts.

Where the continuing operation of our solution requires continuing contracts for items such as maintenance, leases, or licences, we should ensure that these are fully documented and passed to operations. Additionally, we should inform the contracted suppliers that responsibility for the contract has passed to line management and ensure we provide the suppliers with full contact details. Where necessary we may even arrange a modified form of kick-off meeting between the supplier and the line managers to whom we are handing over.

Closing delivery contracts, including arranging final payments to suppliers, is more complex than transferring continuing contracts to operations. As there are at least two parties to a contract, the contract closure requires that we check that both parties have completed all tasks and met all the conditions that were included in the contract:

- The supplier should have completed contracted obligations, including those in any agreed-upon contract variations. These obligations include:
  - Completeness of delivery.
  - Timeliness of delivery, especially where penalties for later delivery are specified.
  - Providing all goods and services to the contracted quality standards.
- On the client side, the key requirement is to have paid, or agreed to pay, for the goods and services they have received. This normally means that the business sponsor, or sometimes the project board, should have accepted all the deliverables provided and agreed that they fully discharge the supplier's contractual obligations. At this point any final payments can be made to the supplier and the contract formally closed. Sometimes, however, there may be final payment retentions until the end of an agreed-upon period of operation which will delay formal closure of the contract. These payments will be handled by the host organisation as the project will have already closed. It is not unusual for there to be discussions, sometimes heated and possibly occurring in court, about whether all contractual obligations have been fulfilled. Effective contract management can reduce the likelihood of differences of opinion at this stage but cannot always eliminate them. Of course, these can happen at any time. We do not have to wait until project closure to enter into a dispute with our suppliers!

Where some deliveries are still expected under the contract, the client organisation needs to identify the line manager responsible and ensure that the processes to control deliveries and payments set up by the project are properly transferred. As with continuing contracts, it may be necessary to set up another kick-off meeting, this time between the supplier and the line managers to whom we are handing over.

Step 4: Release Resources

All resources cost money, often based on the time or period involved. Consequently, the sooner resources are released, the lower the overall cost to the project. By the time we reach project closure, we should have been releasing some of the resources used by the project as their special-ist contributions were completed. Some resources will be required to complete final testing and handover tasks, and still more will be needed for staffing and workspace for ourselves and the remainder of our project support team.

We may have drawn on staff, accommodation, and equipment from inside the organisation on temporary transfer, or we may have engaged these externally for the duration of the project, or they may have been provided as part of a contract with a supplier. In each case, these resources need to be subject to a controlled release. Our first step is to plan the dates on which each indi-vidual resource can be released. Staff and suppliers should be informed directly of the date on which the release will happen. For project team members, in particular, this is best done on an individual basis. Where the project has been progressing to plan, this step is merely confirm-ing the expected dates. In practice, there can be issues in retaining staff members, particularly temporary ones, during the closure phase as they will necessarily seek to find other opportuni-ties and avoid personal employment gaps. The release of other resources, including space and equipment, needs to be scheduled and may require agreeing on contractual dates with suppliers.

Step 5: Transfer to Operations

If our project delivers an ongoing solution to the business, we need to work with the line man-agers who will be responsible for its operation to ensure they are ready to accept it. We cannot metaphorically just "throw our solution over the wall" to operations and sidle off. We need to work with them so they can prepare their resources and have them ready for the transfer day. We may also already have been gradually transitioning elements of the solution to them and avoiding a risky "big bang" approach. We need to ensure that at least:

- Appropriate staffing is in place to operate the solution.
- Handover and operating instructions and manuals have been completed and trialled prior to handover.
- Any supplier warranties and guarantees have been passed to operations and, where necessary, suppliers informed of the contact details of the responsible managers.
- Staff have been trained prior to handover, and the training has included any new or changed safety requirements, including any necessary staff certifications.
- Where the new solution requires the handover of design or safety documentation, that this has been prepared and reviewed for completeness and accuracy.
- Other resources required to operate the solution are in place.
- Operations management have signed to accept the transfer of the solution from the project. Ideally it should be the manager who will be responsible for continuing operation of the solution who signs a handover form accepting the solution and releasing the project manager from responsibility.

In addition to handing over the solution, we should also ensure that we have implemented the final stages of our aspects of the benefits management process (see Chapter 12) and that arrangements are in place (where appropriate) for continued monitoring and management of benefits. If this is not in place, it should be highlighted in our post-project closure report.

Step 6: Post-Project Closure Report

Our objectives in preparing a post-project closure report are firstly to demonstrate that we have completed all the activities set out in the project closure checklist and secondly to provide the project board with a complete overview of how the project has met its objectives so that they can agree to its formal closure. This review is not the same as a post-implementation review, which is carried out some time later by the client organisation, but our review should be a feed into it.

Not surprisingly, our post-project closure review can be formulated in terms of Kipling's six honest serving men, plus How Much. However, there remain two big questions which this review must answer and which may be uppermost in the minds of the project board and our business sponsor and stakeholders:

- How well do the project's outcomes meet the objectives that were agreed to for it?
- How much has it cost?

The most appropriate document against which to review and judge the project is the most recent, and indeed final, authorised version of the business case. This should contain all the information against which to compare our outcomes in terms of quality, scope, timeliness, and cost. For the first time in the project, we are now dealing completely with actuals. At last, we no longer have to go through gymnastic contortions to predict our financial outcomes or estimate when the project will complete. The only unknowns that should be left are those benefits arising during live operation that the organisation has still to deliver.

The post-project closure report is the final major document which we prepare for the project. After we have moved on, this post-project closure report is often a key starting document for stakeholders, business managers, and even auditors when they seek to understand:

- How we organised the delivery of the project.
- How our outcomes were delivered.
- What decisions were taken that influenced the outcomes and why.

Consequently, is important to include in our report a brief history of how the project achieved its outcomes. This section may include both details from previously published documents and a narrative history, for which the project manager's log is a very helpful source of information.[3]

Other records, including the logs of risks, issues, and changes, are essential inputs to the report. Our post-project closure report will likely, therefore, include the following elements:

- **Performance summary**. This should summarise the final position against the project's objectives. As with the end of stage review, this overview needs to include financial performance, performance against time, and performance against quality targets of the project to date. This performance summary should also detail project outputs delivered and variations, including quality standards met, budgetary variations, and the cost of approved change.
- **Completion statement**. The project manager's assessment of how the project met/missed its targets for time, costs, and quality on closure.
- **Risks and issues summary**. An examination of any outstanding risks and issues which need to be transferred elsewhere in the host organisation.
- **Narrative history.** A brief overview of the history of the project's delivery and the challenges and obstacles that were overcome.
- **Recommendation to close the project**.

The post-project closure report also sets the context for our lessons learned summary, which is the second essential part of our assessment of the project.

Step 7: Lessons Learned Strategy and Report

Lessons Learned Strategy

Our lessons learned strategy should have recognised the importance to the client organisation of learning from the way our project was delivered. As ever, it should address our standard seven key questions outlined in Chapter 4:

- **Why** is a separate lessons learned strategy required?
- **What** are the main objectives of this strategy, and what processes will we use to deliver them?
- **When** will activities take place? In this case it is useful to highlight that lessons learned fall into two categories: those that can be adopted during the life of the project and those that will benefit future projects.
- **Who** will be responsible for ensuring that activities designed to meet the strategy are undertaken in a timely manner and for monitoring delivery? Usually this will be the project manager but will include the role of any separate project assurance team or independent gateway review team.
- **Where** in the project development and delivery process should lessons learned be logged, reviewed, and reported? In preparing our review, we should draw on the lessons learned that we have logged and reported as part of the end stage processes throughout the project. Although it can also be useful to hold a final workshop to review lessons learned with those directly involved, in reality, many of the project team may have already been released, and bringing them back together may not be achievable.
- **How** will lessons learned be shared across the wider client organisation? This is an important area and should be in a form which will enable other projects to learn from our experiences, positive and negative, as there is a natural tendency not to want to share lessons learned that arise from situations which we could have managed better.
- **How much** time should be invested in the lessons learned process?

Lessons Learned Report

Sometimes the review of lessons learned is included as part of the post-project closure report, but as it can have a wider circulation than the closure report, it could also be prepared as a separate document.

Our lessons learned report should be a detailed examination of the lessons arising from the project, and our review process provides the opportunity to take a wider retrospective view of the lessons learned; in our report, we should consider grouping them by category rather than merely including a sequential summary. One additional area to address specifically is feedback on any lessons that we adopted from other projects and to assess how well, or not, they worked for us.

Our lessons learned strategy will also have to overcome the inherent barriers preventing the use of lessons learned. Individuals, managers, and teams may not wish to face their own shortcomings or share their success. There are many reasons for this, including:

- Negative events – As both individuals and teams, we would all prefer to cover up where we have fallen short.
- Positive events – If we share our successes, it looks like self-promotion or even boasting.

Also, many organisations are reticent about opening issues to public scrutiny, some things are genuinely confidential, some might be market or politically sensitive, but most are none of these – they are just plain embarrassing. There is also a procedural problem, as if organisations don't capture lessons learned well, they are not in a position to share them, or if they do capture them, there may be no adequate mechanism for feeding them into the training of the managers of future projects.

We may have been recording our thoughts on lessons that we are learning from delivering the project since day one. Our strategy, however, should set out and obtain agreement on how we will formally identify, document, analyse, store, report, and retrieve lessons learned. Recording and reporting lessons learned is not sufficient, because although we may improve our own practices the next time we manage a project, the client organisation also needs to be able to feed our lessons into the wider improvement of its management of projects. Consequently, our strategy should also address the integration of our lessons learned processes with the client's wider culture to ensure that the lessons we learn can be shared with and accessed by other projects as early as possible. We need to address, in particular:

Lessons identification. Whilst some lessons may seem obvious to us as project managers, others may not be, and for still others we may only have a partial view. Appendix 1 sets out different approaches to running discovery exercises, and these approaches can be applied equally to lessons learned and, as we shall see later, to risks. Failings take up far more of our time and attention than areas which ran without problems. More publicly, project failings are scrutinised by auditors, reported in the press, and, when project partners fall out, are examined exhaustively in courts in jurisdictions around the world. Successes are largely ignored because the end result was what was expected. When running lessons discovery exercises, which we could do at the end of each stage or, in the case of agile projects, at the end of each sprint, we need to focus on three key questions, ideally always finishing on the positive lessons:

- Where did we fail in a way that impacted the running or outcomes of the project?
- How can we improve areas where we performed only satisfactorily?
- What things did we do right that we should share widely for other projects to adopt?

Lessons analysis. Each lesson needs to be reviewed in the context of other lessons from the project and elsewhere in the organisation to check if the lesson is a true "one-off" or if a pattern of related lessons is emerging which may require action at a higher level than the project. In order to support this analysis, further information may be required to be stored on each lesson. The ability to do this will depend on the maturity of project management within the client organisation.

Lessons storage requirements. At the basic level, lessons learned should be stored as a clear component of the project's documentation architecture. Ideally, however, the client organisation will be able to provide a joint repository for lessons from all projects.

Lessons retrieval and sharing. Paradoxically, we need to start with a retrieval exercise, because we need to start our management of lessons learned by reviewing those from previous projects. These lessons may be from within the host organisation or from more widely published case studies and academic papers – some of which may be obtained from open access journals. Where the organisation has a joint repository for lessons learned, a method of searching and grouping similar lessons is useful either by a structured system for recording data about each lesson or through a free text search.

Lessons documentation. There are many ways that lessons can be captured and documented from unstructured narrative text through to highly structured data, and our options may be limited by the host organisation's own policies – if they have any. Also, some organisations may require that lessons learned are logged as they occur and reported on throughout the life of the project. We have to strike a balance between retaining confidentiality and including sufficient information for the reader to be able to profit from our experience. Each lesson needs to be prepared in a shareable form – i.e., written in an easily digestible format – with any commercially confidential information, which may limit the circulation of the lessons learned, avoided, or redacted. Quantifications, however, will always make it easier for readers to relate their situation to our lessons, and where possible we should try to include them.

The following three examples of lessons learned from real projects include one positive one, one can do better, and one negative. They all use a very simple structure of: context, action, result, lesson.

Example Lesson 1: User Testing – Done Well

Context. *A public body was making major changes to its internal processing to ensure that it was able to leverage the latest technologies to improve services. As part of this, it had run a major consultation campaign with all its partners, including their engagement in design and testing of the solution. The new service included internet access for members of the public to interrogate systems and review documents online.*

Action. *Whilst the final user interface for members of the public was felt to be excellent by the developers, the business sponsor decided that it should be independently tested to ensure that it was usable by individuals with no knowledge of the service provided or the systems and data being used. The independent assessment of the proposed interface was commissioned from a specialist testing company, which provided facilities to recruit and observe users operating new systems. In terms of the whole development costs, the additional costs of the testing were negligible, and the additional elapsed time was only two to three days in a three-year project.*

Result. *The observational testing vindicated most of the design but highlighted some areas which caused the testers some confusion, requiring improved explanation, and some areas which needed minor systems adjustments to improve clarity. The resulting system became widely acclaimed as a leader in its field and went on to win multiple awards.*

Lesson. *Even though the project team and those engaged directly with it may feel that the solution being provided is adequate, an external objective view may lead to otherwise unrecognised improvements being made, thereby increasing the project's level of success.*

Example Lesson 2: Cost Reduction – It Was OK, but We Could Do Better Next Time

Context. *A local authority had contracted out its highway maintenance and improvement services. Professional highways engineering services were provided by a major engineering consultancy and routine highways maintenance by a specialist contractor. Larger projects, such as resurfacing works, were let on the basis of either annual or one-off contracts, mostly depending on the size and complexity of the contract. The local authority client was concerned about reducing costs of resurfacing and other major highways repair projects and asked the engineering consultancy to identify where the project processes could be made more efficient.*

Action. *All parties were asked to comment on their needs from the project's main design and delivery processes, and very quickly it emerged that a significant change was possible. The time-honoured practice had always been for the consultant engineers to survey the roads concerned and mark where repairs were required on the carriageway. On returning to the office, these engineers then prepared drawings of the repairs and schedules detailing each required repair. In turn, the drawings and schedules were passed to the chosen contractors to estimate costs. Discussions revealed that only the schedules were used by the contractor for the estimation process and that the detailed drawings, which were time consuming to produce, were not used at all.*

Result. *The preparation of plans was removed from the process and instead a "walk, talk, and build" approach was used. The contractor continued to use the schedules provided for estimating the work, but before each project commenced, the contractor, the consultant engineer, and occasionally the client would now undertake a joint site visit and review where the repairs were required. This change allowed the project processes to be speeded up by removing an unnecessary time-consuming step, produced an estimated saving of between 7% and 10% of the total cost of repairs, and ensured that there was no misunderstanding of requirements.*

Lesson. *Understanding the information requirements of each participant in the project and adopting new approaches, rather than slavishly following "the way that things have always been done," can produce major cost savings and shorten timescales without adversely impacting the controls over the project.*

Example Lesson 3: End-to-End Communications – It Was Bad, but We Overcame the Challenges

Context. *A supplier was contracted to develop a bespoke software system containing unique functionality to support the client's role. This involved consolidating and analysing complex data under a development and operating contract worth several million pounds. The requirements were contained in a document supplied to the developers by the client. These requirements were patiently explained by the client's expert staff in a series of discovery exercises and were recorded by the supplier's senior business analyst in both a narrative and a structured format. As part of the approval of a wider systems design document, the client manager was asked to agree with the business analyst's output, which the manager did. On this basis the supplier's developers programmed the required functionality to create a single database from multiple sources and build a set of custom interrogation tools.*

When the system was delivered, the client was disappointed that the developers had not understood the complexities of the calculations required in different parts of the system and that part of the system failed unit testing. The client was considering terminating the contract and seeking to invoke penalties. Internal investigations by the supplier and an in-depth discussion with the client's staff showed that the problem had arisen because the client staff had different understandings of some of the terms used than the business analyst and the developers, which, in turn, had a major impact on the system and its calculations.

Action. *The project manager called a halt to all development, and in a series of joint workshops, all the client's lead professionals, an external adviser to the client, and the supplier's development manager, analysts, and developers worked through every requirement and agreed on precise definitions and meanings of the words used. Elements of the system were then revised to meet the new definitions, and the client professionals and the developers were in daily contact to first check on understanding and issues arising and subsequently to unit test and final system test the development.*

Result. *The client was finally able to accept the updated system, but the project delivery was delayed and the contractor incurred considerable extra costs in revising the system to meet the actual user requirements.*

Lesson. *Failure in the development process to close a communications loop and check on understanding resulted in a near collapse of the project. It is necessary to ensure that communications between all parties are clear, direct, open, and comprehensive. This includes providing the opportunity for every requirement to be questioned, and hopefully clarified, by the project staff responsible for delivering the requirement with its originators in the client – in this case through joint meetings which were followed by continuing contact between the developers, the analysts, and the client lead professionals.*

Step 8: Final Stakeholder Communications

The final communications from the project with each group of stakeholders should follow the project board's acceptance that the project can be closed. The communication should inform them that the responsibility for the solution is being transferred to line management, with any required contact details, and thank them for their co-operation in the project.

Step 9: Documentation Completion

Our penultimate act is to ensure that the final versions, and any intermediate approved versions, of project documents are stored in our project filing system, however that was organised. This should include reviewing and closing any registers still open and ensuring that we have captured and recorded the final documents covering the closure of contracts and the making of final payments. The lessons learned report and registers should also be shared through whatever mechanism the client organisation has in place.

Post-Implementation Review

The very final thing we need to do is to agree with the business sponsor on the timing for a post-implementation review of how well the solution provided is meeting its objectives and to judge how well the project was managed. Essentially this is not our responsibility, it is a management post-mortem review with which we, as project managers, may be asked to co-operate. However, we should be able to discuss and agree on an appropriate time for this review after the system has operated for a suitable period after solution go live. This period of operation needs to be a time sufficient for the solution to "bed in" and reach a stable enough operation to allow a judgement to be made about how it will perform in the medium and long term.

The benefits from undertaking such a review to the organisation are that any necessary changes to the solution to improve its operability or increase the benefits obtained can be identified and lessons learned can be reassessed.

Now It Really *Is* Over

We can hand back the keys and turn off the lights. Our final task is to close the door on our way out.

Final Reflections

As before, if possible, using the same project as in your previous reflections, consider the following questions:

- What would be the main areas to be included in your project closure checklist, and would these vary if you were requested to close the project early?
- How would you ensure that operations management were aware of the tasks and complexities in operating the solution that you plan to transfer to them?
- What steps would you take to ensure that all supply contracts are closed and responsibilities for continuing contracts and warranties are passed to the correct managers?
- Draft two lessons learned, one negative and one positive, in a structured format. What would be the principal barriers to getting an organisation to act on each of the two lessons?
- How would you frame a final communication with your stakeholders, and would you need different communications for different groups of stakeholders?

As before, if you record the results of your reflections with previous reflections, this will help you build an example project case file.

Notes

1 Congratulations on trying to exploit the boss's ambiguity to buy yourself some time here. Well done.
2 Drat, you have been rumbled.
3 I promise that after several months of project management experiences following rapidly after each other, you will not recall any except the most memorable events in sufficient detail to write a solid narrative history. I certainly never can.

14 Project First Aid

How to Stop a Project Bleeding to Death

Aim of this chapter: To outline an approach to rescuing a failing project.
Learning outcomes: To be able to identify, select, and operate the most appropriate mechanisms for bringing a failing a project back on track.

So far, we have dealt with project management as starting at the project inception and finishing with project closure and handover. Sadly, life is not always that simple for the project manager. Often, we can be called upon when the delivery of a project has started to concern the business sponsor, and we get asked a question along the lines of, "Hey, you're an expert project manager.[1] This project is failing. What can you do to fix it?"

We're going on a special, bonus site visit. Now in your role of director of project excellence, you've been called on to assess the problems with a failing IT development project in a subsidiary company. We're invited to meet the local project manager in a non-descript modern office for a discussion before they meet with the supplier. You are served stale coffee from a flask, and the biscuits come in those tricky little packets.[2] Obviously your new director status does not mean that the quality of the refreshments offered improves.

The project manager is clear and direct. "Thanks for coming. I'd like you to confirm agreement with the actions we're proposing to take. We let a contract for an integrated development followed by three years of operation of this system, extendable to five years. We went through a very extensive selection exercise, and to be completely honest, we got the wrong result. The supplier is incompetent – they are very late in delivering, and the solution is not what we wanted. There were "off the shelf" solutions available, but we'd chosen a bespoke development."

Documents are then shared revealing that both of these views are factually correct. It seems clear that the supplier is failing.

The project manager continues, "I can't put up with them anymore. Obviously, our only choice is to go to contract termination and seek liquidated damages. I've already had initial talks with the procurement and legal teams, who say that we could have a case."

Before giving an opinion, you decide to delve a little deeper.

"I think you need to be cautious," you tell the project manager. "Our experience shows that termination is not always an easy route. It can be time consuming, expensive, and the courts don't always agree with us. And it doesn't deliver the project either. Putting the problem of supplier non-performance to one side for a moment, can I just ask – what issues have they raised with you?"

DOI: 10.4324/9781003405344-15

'Well, they claimed that our system requirements document was not fully fit for purpose because it was based on assumptions about the availability and standard format of input data from multiple 'feeder' systems from partners which were not valid. We agreed with them that they should undertake a large amount of extra work to research data formats and availability and to set up mechanisms to gather and translate data from multiple sources each day before it could be brought into the new IT system. We paid for these as contract variations.'

You think for a moment. *"So, you accepted a major change to the project deliverables and timeline. Was there anything else that they raised with you?"*

"They said that many of our stated requirements were not precise enough to avoid ambiguities, and whilst we might have meant them in one way, they had interpreted some of them differently. It was only when they asked us to test parts of the solution that we realised they weren't giving us the results we expected."

You think a bit more. *"It seems that rather than being a simple case of supplier failure, a range of more complex issues may be involved. If so, it puts a different light on litigation and probably points to us needing an alternative strategy. I'll come to the meeting with you, and let's see if we can find a way to solve this. If we do have to sue them later, at least we'll have shown that we tried to get a resolution in good faith."*

So, what are your first steps going to be before that meeting with the supplier?

This real project was saved because, after negotiations between the client and the supplier, instead of pursuing contract breach, a recovery solution was agreed on. A project pause allowed a detailed reexamination of the differences between the client's needs and the current solution. This was followed by the redesign of certain elements and new project control processes requiring much closer direct liaison between the client's experts and the supplier's developers.

Like virtually every other area of project management that we have considered so far, rescuing a failing project can be approached as a simple process. Adopting a stepwise approach allows us to both break down the project's problems into manageable chunks and also get appropriate support and agreement as we go along. There are, however, pitfalls if we are not careful.

A potential template for a project rescue process is laid out in Figure 14.1. Like everything else in the management of projects, the rescue process adopted must be determined by the organisational environment and the constraints of the project itself. The templated process involves a new role, which for the sake of the process is called the project investigator. The role could equally easily be called consultant, project expert, or adviser or even could be an incoming project manager. It is possible that the role could be undertaken by the existing project manager, although as this is the person who has been unable to prevent the project arriving at its current situation, the project investigator is more likely to be a fresh mind and a fresh pair of eyes.

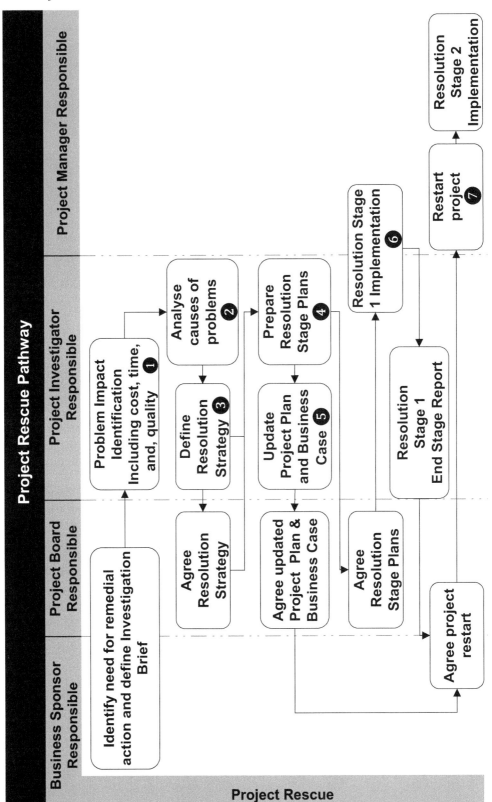

Figure 14.1 Project Rescue Pathway

Usually, our investigation into a project will reveal that it falls into one three types: predominantly fine, not salvageable, or requiring major changes to delivery and possibly expectations:

- **The good**. The project is not perfect but is delivering as well as possible. The current issues are to do with expectations and, whilst we may find that some improvements are desirable to help improve outcomes, essentially the client could continue the project "as is" with no significant changes.
- **The bad**. There is no chance of turning the current project around to get it to deliver any significant results, and no number of major changes will make it viable. It needs to be cancelled.
- **The ugly**. We now understand the issues that the project is facing and can recommend ways the project's delivery processes and organisation, outcomes, or expectations should be adjusted to deliver acceptable results.

At any time during an investigation and rescue process, we may conclude that the project will fall into "the bad" category, and we should initiate serious discussions with the business sponsor about its future.

A word of warning before examining each of the steps in the rescue process. Maintaining an open mind for as long as possible is important. It is both easy and dangerous to make early assumptions in an investigation into a project failure and even more dangerous to reach conclusions and recommendations without first gathering and sifting the evidence necessary.

Step 1: Problem Impact Assessment – What's the Real Problem?

Many of the solutions that we have available to us are contained in the earlier chapters. But first, we need to form a diagnosis of the problem or problems faced by the project. That is, we need to clearly establish what is actually going wrong. So, like any good expert, we always answer a question with at least one other, and our response to that opening question from the client of "What can you do to fix it?" should always be to start to clarify the issues along the lines of, "What do you mean by failing?" We do this by asking

> There are three, and only three, types of project failure, and you need to ask which one, or ones, of the following apply to your project: Is the solution that you were expecting not working out for you? Are you not delivering on time? Are you over budget?

Now, before we go any further, consider the difference in the language used between the client's initial question and our response. It's no good being impersonal. If the client is close enough to a project to be worrying about it failing and asking for help, then they are probably close enough to own at least part of the problem and be part of its resolution. They have already taken the first steps – recognising a need for action by asking for outside help. So, each question is deliberately reflected directly back at the client organisation and tries to understand their perspectives on the nature of the project's failure.

In most cases, the worst thing that can happen to a project that is perceived to be sinking is for it to carry on pushing forwards in what may be the wrong direction, and that direction is almost certainly going to be downwards, backwards, or both. If, therefore, the business sponsor has not already considered one, our next step should be to discuss instituting a pause in the project to allow time for our review. Achieving a complete pause is not always easy to achieve, partly because projects have their own momentum and partly because there may be delivery dates contracted with suppliers and delivery expectations from stakeholders. The former we may have to renegotiate with the suppliers and maybe even accept that additional costs will be involved. The latter we have to manage, and this will require careful stakeholder communications.

The investigation into a project's problems is the first part of a process to change the way the project has been delivering. Once again, we are adopting satisficing behaviours in order to minimise the elapsed period required and to maximise the return on our time, which is inevitably limited. We can often achieve this by undertaking a rapid overall review and highlighting the impacts of the different problems being faced by the project. We can then concentrate our investigations, effort, and time on the critical areas for improvement. A blow-by-blow forensic audit of the project in all its gory detail can come later, if necessary, but our focus for now is understanding if we can change the dynamics of the project and get it moving forwards.

First, we use the review pause to investigate and take stock of the answers to our three questions and to see how they can lead us on to identifying the steps required to bring the project back on track.

Question 1: Is the Solution That You Were Expecting Not Working Out?

It all seemed so hopeful at the start, when the client organisation and the project manager mapped out exactly what the project was going to deliver and how they were going to deliver it. Sadly, somewhere along the line this has stopped happening, and their solution is now not meeting expectations. The implications of a solution not working out properly are that the timeline is going to drift off and the additional resources required to resolve the issue are going to push the client organisation way over budget.

We need to dig a little deeper in this area as "not working out" – i.e., not meeting the needs of the client organisation – can have several meanings:

- The solution just does not work, meaning, "We had expectations for technical performance which are not met."
- The needs of the business have changed, meaning, "The solution proposed no longer fits our business."
- The solution was not right in the first place. This may be caused equally by the client being over ambitious, by being too timid when the solution could not answer the whole problem, or, as in our example of our last site visit, by the client just not fully understanding and communicating their requirements.

If the answer is that the solution just does not work – i.e., does not meet technical performance specifications – we also need to ask the client the further questions:

- Is there a technical problem that could be solvable within the scope of the current project?
- Are the technical specifications both correct and necessary?

We may need to take independent technical advice on the nature of the issue and on potential options. We also need to be aware that it may not be one problem causing the technical issues but an interlocking matrix of problems, not all of which are immediately apparent.

We can reassure the business sponsor that failure on the grounds of not being fit for purpose is fairly common and affects even major projects. Projects failing in one of these ways may often be turned around. Let's look at some examples:

- **The US C-17a Globemaster III aircraft project**. Here, the design of aircraft had to be modified after the wings failed structural tests. This design issue was resolved, and the aircraft went on to be the major strategic airlift workhorse for the United States Air Force and many other air forces for decades. At one point during the 2021 Kabul airlift, a flight was reported as carrying 823 passengers. This was more than twice the number it was usually configured for on humanitarian work – obviously, they did a good job on those wing redesigns.
- **London's famous pedestrian Millennium Bridge**. Nicknamed the "Wobbly Bridge" after it opened, it was "discovered" that people crossing the bridge tended to unconsciously walk in step, causing the bridge to sway alarmingly. After being open for two months, it was closed for two years for modifications.[3]
- **The travel ticketing system** whose departing project manager we visited in Chapter 3. Here, the country had decided to introduce a national card system and banned the project's own planned card. The project manager felt that the project had also been far too ambitious a development because the city leading the project wanted to be seen as a world-leading pioneer of accounts-based ticketing systems and the new systems development was not working in time. It was a "fitness for purpose" failure on two grounds: not meeting the (now revised) needs of the organisation and not being technically complete.

Question 2: Are You Not Delivering on Time?

At some point in the past, someone sat down, mapped out the necessary tasks, possibly did some research, and then drew up a timeline for the project which, usually, someone else approved. That timeline was drawn up on the basis of the information available at the time, and it may also not have been a thoroughly prepared project plan corresponding to our view of what a project plan should be (see Chapter 5). Even if the timeline was an absolutely perfect project plan, we know that the realities of project delivery can cause any plan to need to change. Remember Prussian general Helmuth von Moltke's view was that no battle plan ever survives first contact with the enemy (see Chapter 5).

A project plan doesn't need to come under enemy fire, because it can be thrown off by something as simple a failing to get the request for the project's approval on the board's agenda for the planned month. It is, after all, not unusual for a project manager to receive a call from the board's secretary saying something on the lines of, "Sorry, your paper has had to be dropped this month. It's now going to be a short meeting as that day the chair now has to fly out to Davos for the World Economic Forum. I can probably get you a slot on April's agenda? Well, May for sure. Or possibly June?"

Or maybe a supplier has let the project down: "Yes, I know we said that your custom windows would be ready by September. I'm sorry about that, but I can genuinely promise you that they will be on site by the end of November. One hundred percent certain." All the planned interior work, of course, had to wait until the building was weathertight, and if the project was in the Northern Hemisphere, the site would probably already be in the grip of winds, rain, ice, or snow long before the windows arrive. That is if they actually came in November and the next call from the supplier was not something like, "I know I said November, but it will be later now, and oh, we shut for two weeks at Christmas. I should think around mid-January is the earliest that I can do. One hundred percent certain this time."

We have to tell the client to take heart on this one, as however far a project is falling behind its planned dates, it is in good company. Many projects that we now consider a success were delivered very late. For example, rebuilding the Houses of Parliament in London was planned to take 6 or 7 years but took 32 to complete. No one remembers those extra 25 years now when they are taking their selfies outside or listening to pontificating politicians inside.

Question 3: Are You Over Budget?

Like timing, the project budget was a matter of opinion at the time, not fact. As the project progresses, the client will have learned more about the reality of the costs of delivery and the impact of timing of costs on the budget. The client perception of the project failing to be within budget may or may not be incorrect. We can use earned value analysis (see Chapter 11) of the current budget and expenditure to understand what the financial position actually is. Again, we should reassure the client that if they are over budget, they are in good company:

- **The UK Houses of Parliament (again)**. When being rebuilt following a fire in the 1830s, it cost over 2.5 times the initial budget.
- **The Anglo–French Concorde supersonic airliner**. This project had an initial estimated development cost of £70 million and a final bill of up to £2.1 billion.
- **The Channel Tunnel between the UK and France**. When completed this project was 80% over budget, a cool £4.5 billion extra. Rest assured, on a stormy day, no passenger cares about a cost overrun on building the Chunnel. In 45 minutes, it still gets you between Felixstowe in the UK and Calais in France without risking being seasick on a car ferry for hours.[4]

Step 2: Analyse the Cause of the Problems – Finding the Why

Restating the Problem

Of course, the project may be suffering from not one but several problems, and we can borrow the concept of a formal problem statement from the Six Sigma methodology. Surprisingly, it seems that a good problem statement will reflect most of our seven questions. When restating the problem, it is useful to draw on the client organisation's own wording from the brief so that the project investigator can describe the problem in the client organisation's own terms. The following is an example of a very simple form of such a statement:

> The AA project has been subject to increasing costs and time delays. There is now doubt by both the project board and the main company board of directors over whether the project will be able to generate the return on investment estimated in the approved business case.

We can then expand this with our observations, based on the answers to our questions, and add our considered conclusions on the current state of the project. We should always identify and quantify those areas where there are issues and, if there are any, acknowledge those where there are no problems:

> The AA project's predicted costs have increased from X in the approved business case to Y now. The project is currently three months behind its project plan, and the deliverables to date from the supplier have not fully met specifications, giving rise to a contractual dispute. The project management regime had not detected the extent of the quality and other problems that the

project was facing, and those problems of which the project manager was aware had not been detected early enough to enable corrective actions to be taken. There is now doubt by both the project board and the main company board of directors over whether it will be able to generate the return on investment estimated in the approved business case.

This problem statement may become the basis of how we will focus the project rescue.

Having agreed on the situation that the project has actually reached across a number of areas, we now have to investigate why it arrived at its current difficulties. A good starting point is a documentation review, searching for evidence of how the project has been managed. A failure to provide adequate documentation is automatically an indictment of the quality of the project management and will naturally inform our recommended resolution strategy. Consequently, a great first request to the project team or project manager (if one is still in place) is *always*, "Please, can I see the latest version of the project plan?"

Cause Analysis

Our first objective is to find the earliest point at which the project started to drift off track and ask, "Why did this happen?" We can again borrow ideas to help our analysis, and a fishbone diagram, as shown in Figure 14.2 (also called an Ishikawa diagram after its inventor), is a useful way of recording and communicating our analysis of the problems faced by the project and their underlying causes.

It is the causes of problems that are our prime interest, but our analysis of these causes has to be based on demonstrable facts and not hypotheses of what may or may not potentially have happened. This is not only because our hypotheses may be totally incorrect but also because there is a danger that if we start with a hypothesis, we may then subconsciously look for evidence to support it. Starting from a hypothesis of what happened results in our narrowing our investigation, and thus we may miss or not adequately weigh key facts. The simplified example here shows how a valid hypothesis may mislead us and suggest a false cause:

Hypothesis – The project manager selected for the project was not competent:

- **Fact 1 discovered**. The project manager had not undertaken a review of progress on key tasks during the last three months.
- **Fact 2 discovered**. The business case missed a number of significant costs.
- **Fact 3 discovered**. There were no risk and issues registers or mitigation processes.
- **Fact 4 discovered.** There was no process in place to assess deliverables from the supplier as meeting quality requirements.

We have proved our hypothesis and the analysis may be halted. However, there were other factors that we failed to reveal:

- **Undiscovered fact 1**. The project manager was only released from their full-time role to the project for three hours per week.
- **Undiscovered fact 2**. The project manager had never received any training or guidance in the management of complex projects, and their request for attendance at a training course was refused because there was not one suitable at a time when the new project manager could attend.

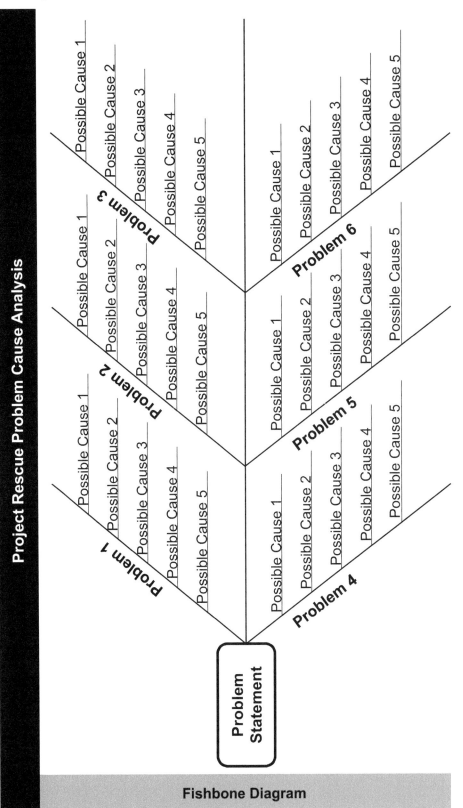

Figure 14.2 Project Rescue Problem Cause Analysis

- **Undiscovered fact 3**. The only experienced project manager who could have been a suitable mentor left the company to work for a competitor just as the project manager was drafted in.

Starting from the hypothesis of incompetence, we managed to prove it, but by having a narrowed field of analysis, our investigations did not reveal the true cause of the lack of project management skills.

Each project will present specific areas to consider, but generally our investigation will need to examine at least time, cost, and quality:

Time – Planning. To properly understand and be able to demonstrate the impact of delays in specific tasks on the whole project (if the existing project management has not done so already), we may have to prepare an updated Gantt chart showing how the critical path has changed. Comparison of a current Gantt chart and one drawn up based on the project's business case should highlight where and when the issues with project delivery times have arisen.

Cost – Funding. Our investigation into project funding issues may reveal a variety of situations, some of which can reinforce each other. Perhaps:

- The projected costs were wrong in the first place – maybe the supplier quotes were much higher than expected or labour rates or raw material costs have risen beyond expectations.
- Something unexpected has happened. If a problem that required additional funding to resolve is now over and the expenditure has been incurred, we should treat it as a one-off cost and update the forecast total project costs. Examples of this might be a construction project where expert ground investigations had not revealed problems which were later uncovered by the digger operator, or in a project involving IT systems (such as on one of our site visits), advances in technology might require increased hardware capacity.
- The client organisation has increased the quality or scope of the outcome (remember our old friend scope creep).
- Just possibly, the project is not actually overspending at all.

In the first three cases, our investigations should clearly show what has happened to cause the extra spending. In the fourth case, where the project is not actually overspending, a common cause could be a mismatch between when it was expected to spend money on the project and when the client actually has had to pay out. For example, did the client have to agree to deposits, or mobilisation fees, or stage payments to a supplier when the business case had expected payment on delivery? So, we need take a closer look at where the money has been spent so far by applying earned value analysis.

Quality. If there are no suitable answers as to whether the quality problems are technically solvable, we need to look back and try to analyse where in the project the problems started to occur. We have to help the client answer the question, "Was a decision or decisions taken *then* that led the project down one path that can be changed *now* to resolve your problem?"

Our cause analysis has to be like our approach to a work breakdown structure in that the analysis needs to proceed to the level at which we can determine possible remedies and potential actions to be taken. Our high-level fishbone diagram (Figure 14.2) included only one level of analysis of the causes of a problem, but Figure 14.3 shows how we can break these down to further levels if we feel it to be necessary. It takes our simplified example, where our hypothesis was that the project manager was not competent, to further levels of cause analysis. Each of the other causes shown could be similarly taken down to lower levels.

Step 3: Define a Resolution Strategy

Now that we have defined the problems facing the project and identified their likely causes, our next task is to work out a strategy for resolving each of the issues impacting the client. Our objective is to bring all the resolution actions together into a resolution strategy. This strategy should highlight the actions required to bring the project back to deliver within acceptable bounds of time, cost, quality, and scope.

It may be, of course, that we cannot demonstrate the project will be able to deliver effectively. In that case, we will report to the business sponsor that we consider that the project should be closed. The most positive outcome that we should seek from step 3 is an agreement to prepare an updated business case and project plan (step 4) to justify project completion; we should not, at this stage, seek an agreement to restart the project. Essentially our resolution strategy is reframing the project (see Chapter 2).

In order to justify each action that we consider as required to resolve the project's problems, our resolution strategy has to assess the impact of each one and demonstrate the differences that each would make to the project both individually and collectively. Whilst our fishbone diagram helped us visually identify and communicate the causes of the project's problems when identifying resolution actions, it has two clear shortcomings:

- It does not make clear how we can move from analysis to action.
- Being a linear, top-down analysis, it is not easy to show where a cause has impacts in multiple parts of the project.

Taking the information in Figure 14.3 and putting it into tabular form enables us to overcome both of these shortcomings, as we can cross-reference items and build the list of the actions required that will form the basis of our resolution strategy. In the example in Figure 14.3 now reformatted as Table 14.1, even where we have only examined the causal root of the problem in one area to any depth, it shows where some of the required actions are becoming synergistic – in this case focusing on information systems, external support, and internal resourcing.

In addition to our three questions on timing, budgets, and quality, our investigations should have also revealed if the project organisation is capable of delivering the project. Whilst all detailed conclusions and recommendations will be project specific, it is useful to summarise our findings about weaknesses, conclusions, and overall recommendations under a series of main headings, including:

- Fitness for purpose.
- Organisational capability.
- Timing and planning.
- Funding and financial viability.

Fitness for purpose. We have to consider the current situation of the client organisation and the project and conclude whether the project can ever deliver on the client's expectations. It may be that the scope of the project requires alteration or that the way the project is being delivered has to be changed – for example, replacing a non-performing supplier or team.

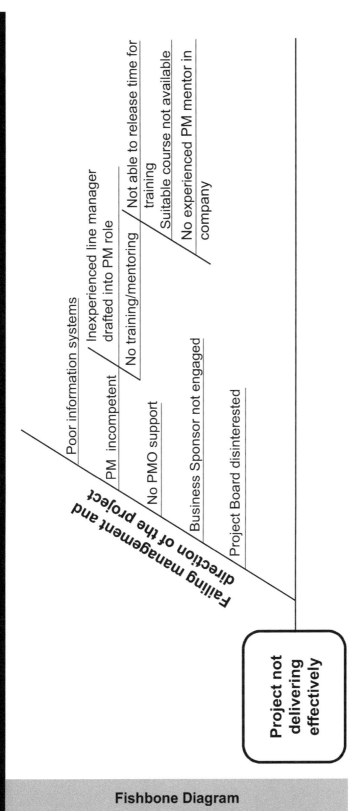

Figure 14.3 Project Rescue Problem Cause Analysis Example

Table 14.1 Cause Analysis Matrix

Perceived Problem	Cause Level 1	Cause Level 2	Cause Analysis Matrix		
			Cause Level 2	Cause Level 3	Recommended Actions
Failing management and direction of the project.	Poor information systems for supporting project-based working.				Short term – Identify suitable subscription-based project management information system to support current project.
					Short term – Ensure project manager receives appropriate training in the project management information system.
					Medium term – Undertake a needs analysis and a system selection exercise for an organisation-wide project management information system.
	Project manager incompetent.	Inexperienced line manager drafted into project manager role.			Short term – Bring in external expertise to lead the project.
					Team existing project manager with external support in a learning capacity.
		No training or mentoring.	Not able to release time from the project management role for training.	The project manager was only released from full-time role to the project for three hours per week.	Short term – Identify backfill for the project manager and increase time available to the project.
			No experienced project management mentor in company.	The only experienced project manager who could have been a suitable mentor left the company to work for a competitor just as the project manager was drafted in.	Short term – Use external support as mentor for the project manager.
					Medium term – Train selected line managers in project management.

Cause Analysis Matrix (Continued)					
Perceived Problem	**Cause Level 1**	**Cause Level 2**	**Cause Level 2**	**Cause Level 3**	**Recommended Actions**
			Suitable course not available.	Request for attendance at a training course was refused because there was not one suitable at a time when the new project manager could attend, during the three hours per week released to the project.	Short term – Identify suitable flexible online training courses as an alternative to face-to-face training.
	No project management office (PMO) support.				Short term – Release and backfill existing staff members to fulfil PMO role.
					Short term – Put in place appropriate PMO training and training in the project management information system.
					Medium term – Create permanent PMO team with organisation-wide responsibilities.
	Business sponsor disinterested.				Short term – Consider reallocating sponsorship role.

(Short-term actions: actions taken in 0–3 months in relation to rescuing this project; medium-term actions: actions taken in 0–12 months to improve ability of the organisation to deliver projects.)

If, during our investigation of causes, the answer was yes to our question of whether a specific decision had been taken which led the project into the pathway which eventually failed, we have to ask, "If we get the client to change that decision, how much work would have to be redone? Or, can they change the way the solution or the project operates to remove the problem, even if they have to reduce the project's scope?" A classic decision in this area would have been opting to go for an all-encompassing "big bang" approach to project delivery instead of a phased solution, again as in our ticketing system site visit. If the technical problems exist in later areas of project delivery, rewinding the project to a point where incremental delivery would become possible might be a potential option. Another ultimately failing decision might have been an increase in scope which caused disproportionate costs or difficulties. For example, planning a road or rail project to join two towns might be straightforward, but the decision to extend it to a third town through much more difficult terrain might be the one that is undermining the feasibility of the whole project.

If we discovered that the project was suffering from the technical failure of the solution, we also have to consider the amount of work that might be required to resolve it and whether there are actually multiple problems to be resolved. On one much smaller project to deliver real-time predicted bus arrival information, the presenting problem was that predicted arrival times were not accurate. These predicted arrival times were displayed using on-street displays (as in our site visit in Chapter 8). Consequently, this was a very public failure and was reported heavily in the press. It turned out that there were a number of disparate underlying problems which had to be resolved one after another, including needing to define bus stop locations more accurately and changing how drivers operated the locator equipment on the bus. Chipping away at each issue over time not only solved the overarching problem but got the system into the highest echelon of prediction accuracy internationally. In this case, as some issues masked other issues, the final amount of work required to deliver fitness for purpose was not immediately apparent and could not be estimated.

Organisational capability. As in the example in Table 14.1, we may find that the environment within the organisation was not properly set up to deliver the project that is now failing. This is generally referred to a problem of project management maturity. Fortunately, there are a wide variety of models available to help assess project management maturity, and many of the knowledge areas that these models assess have been covered in the preceding chapters. They generally assess different areas of project management, such as management control, finance, or benefits, against a standardised scale − for example, a five-point scale running from an initial level 1, where project management processes are undocumented and ad hoc, through to an "efficient" level 5, which includes planned optimisation and improvement of processes. Using an assessment model can help us frame our recommendations for actions in the medium term to improve organisational capability where it is clear that a lack of project management maturity is impacting on the delivery of project outcomes. Remember, though, we are also looking at the widest definition of the project organisation, not just the project team. This will also include suppliers.

Particularly with organisational capability, in order to deliver the project successfully, some of our recommendations may need to be implemented prior to a project restart. Examples of this could include where a new project manager or key project team members are required (and these roles need to be filled short term whilst recruitment takes place) or a new procurement needs to be started as quickly as possible to replace a supplier that is not delivering.

Timing and planning. If there is an issue of project slippage, we need to conclude how this has occurred. We understand that all planning is based on assumptions about the future and that events do not always align with those assumptions. Our cause analysis will have revealed those "hot" areas where events have pushed the timeline, and we need to take a view on whether any of this time can be recovered. We also have to reflect these in a revised high-level plan and add in the implications of all the recommendations we are making in the resolution strategy.

Funding. Based on the information uncovered and the recommendations being made, our conclusion on funding has to be based on reviewing the total potential costs of the project. To do this, we should take the earned value analysis which we carried out in step 2, add in any additional costs stemming from our recommendations elsewhere in the resolution strategy, and estimate a new cost to complete the project. This requires preparing an estimate to complete in one or more of the four different ways set out in Chapter 11.

The resolution strategy can then address which of the usual financial strategies are required:

- Relaxing, because there is actually not a problem with the overall financing of the project and what we uncovered were issues of variations in the timing of expenditure between the business case and live project. We can prepare a conservative estimate to complete showing that the project is still expected to come in on budget. Sadly, project budgets nearly always seem to melt away faster than an ice cream on a summer day, so this is not the most frequent course of action.
- Arguing that a new project budget will be necessary to reflect where the project actually is now. Even the most conservative of our estimates to complete shows that the client will need to find additional funds to complete the project but that it will still be financially justifiable.
- Trimming off the "nice to haves" to get back closer to the original budget. This includes challenging both any scope creep and any original requirements that seem not to add sufficiently to the overall outcome. It is always worth asking the client the equivalent of, "Do you really need gold-plated lampposts?
- A mix of both additional funding and trimming.
- Closing the project as there is no chance of it ever coming anywhere near its financial objectives.

Organisation-wide recommendations. If, during our preparation of recommended actions, we identified areas where the client organisation could implement more wide-ranging changes to improve performance outside the particular project under investigation, it is useful to pull these together into a separate section.

Formal recommendations. As in other major reports, it is useful to start the resolution strategy with a brief summary of conclusions and recommendations. It particularly helpful to the reader to understand our overall judgement before detailing findings, which will often be less than positive.

Cancelling the Project

It can sometimes be that our cause analysis shows that the project is no longer viable and we are unable to recommend sufficient suitable steps to rescue it. Then there is no point in proceeding with updating the business case and project plan. This means that the best approach is to recommend bringing the project to a controlled stop and salvaging what can salvaged, rather than letting it continue onwards and then crash spectacularly in flames later. Deciding to halt a project is not an easy decision as there may be many people, both within the client organisation and the external stakeholders, who have invested their time, effort, and reputations in the project's success.

Step 4: Resolution Planning

A high-level plan for the implementation of the resolution strategy is required to feed into the updated project plan (step 5) as well as to support detailed planning at a task level.

If the project is to continue, it is useful to divide project resolution into two clear stages, and each can be subject to end stage assessments:

- **Resolution stage 1**. Changes required in order to allow the project to restart.
- **Resolution stage 2**. Changes which are implemented after project restart.

The objective of resolution stage 1 is to improve the project management and control processes to a level of adequacy sufficient to enable the project to be safely delivered. By contrast, the objective of resolution stage 2 is to enable the adjustment of the project's planned delivery tasks to ensure a successful delivery of the required solution. As an example, bringing in a new experienced project manager is a resolution stage 1 task, but running a procurement to find a replacement supplier is a resolution stage 2 task.

Adopting this two-stage approach gives the project board and the business sponsor the opportunity to review the progress of the resolution tasks in order to improve the management of the project's delivery and take a view on its uprated ability to deliver the project immediately prior to agreeing to a project restart. We should also recognise that there may be some stage 1 tasks that have commenced but are still uncompleted as the project restarts. These may include, for example, a training programme for project team members or a programme to co-locate different elements of the project team.

If project closure is recommended, however, only a single resolution stage is required detailing required actions to initiate a safe project close. Any subsequent reporting will become part of the project closure process (see Chapter 13).

The resolution stage plan or plans should be prepared in accordance with a structured planning process (see Chapter 5). There is an issue, however, regarding how far to engage the existing project manager and project support team in preparing the resolution stage plans. The level of their engagement should depend on the recommendations in the resolution strategy about the future of the current project management structure and processes. If the current team is to remain in place, fully engaging them in the resolution planning process will foster support for its implementation.

Step 5: Update the Project Plan and Business Case

During our investigation, we will have gathered sufficient information for the business case and the project plan to be updated. There are two particular areas to consider:

- Are the current business case and project plan fully fit for purpose, or do they need revising to bring them up to a suitable standard?
- Are the current project manager and project support team sufficiently capable to prepare the updated documents that are required?

The more negative either answer is, the greater the degree of support that the client organisation will require when preparing documents of a suitable standard to support the decision on project continuation, project amendment and restart, or outright cancellation. Updating the business case and project plan should include a complete review of alterations necessary to reflect the current state of the project, as well as recasting assumptions, estimates, and predictions throughout, including:

- A new estimate to complete, including updated costs arising from the resolution stage 1 and resolution stage 2 plans.
- Revised analysis of the financial justification of the project, based on the new estimate to complete.
- Changes to the delivery timeline, including any commencing resolution stage 1 activities, prior to restarting the project.
- Changes to the delivery timeline to reflect resolution stage 2 recommended changes in the wider activities and tasks of the project, in order to complete project delivery successfully.
- Updated analysis of the capability of the project organisation to deliver the project, including the impact of the changes to be made in resolution stage 1.

In order to justify any required extensions to the overall delivery time frame, when considering the replanning of the project, we need to ensure that the impact of tasks in resolution stage 2 on the project's critical path are clearly identified and explained. This is because any time extensions will inevitably be seen as slippage in delivery dates. To partially offset the adverse impacts of such extensions, including reputational damage, it is useful, where possible, to identify opportunities for "quick wins" which were not in the original project plan and to prioritise any which had been previously identified.

Whether the current project management team prepares the documents alone or with external support, we need to argue that a controlled process development and authorisation process must be followed. We should also ask the project board and business sponsor to agree that the updated business case and project plan should be subject to an independent quality review before being submitted for approval.

Step 6: Implement Resolution Stage 1

As the adviser on how to rescue the project, our task could now be seen as complete. We have identified What the problems are, Where improvements or changes are required, shown the client How to implement them, set out When this can happen, explained Who should do it, forecast How Much it should now cost, and justified Why the project should either continue or be closed. Implementing the resolution stage 1 plan, however, represents a transition in the way the project is to be delivered. We should discuss with the business sponsor whether, and how, that transition process should be supported to ensure that the project is adequately structured, resourced, and supported by relevant processes to be delivered successfully.

Step 7: Project Restart

After the successful closure of resolution stage 1 (and approval of the end stage report if the project is shown to be still viable), the delivery of the new project plan, including the changes recommended for resolution stage 2, can start immediately. We need the client organisation to recognise that this is not the same as an initial project launch and to understand the impact that the project's previous potential failure will have had at a human level. Very often concerns about project failure over the last weeks or months will have spread throughout the project team, their confidence levels may have diminished, the atmosphere may have become unhappy, delivery partners may have started seeking to blame each other, and suppliers may be focusing on other more lucrative or less troublesome contracts. So, it can be very helpful to get the business sponsor's agreement to set up a reboot event for the project team and, with a fanfare of trumpets, full of energy and enthusiasm, we can emerge with the new, or at least the reinvigorated, project manager, waving new plans and budgets (or at least with them on a set of PowerPoint slides), and so relaunch the project. We need to have already worked with the project manager,[5] so they now understand the importance of working with the project team and show them each day that progress is being made and successful delivery is now not only possible but highly likely. One of the mechanisms for doing this is to borrow the idea of a short daily team meeting from agile methodologies (see Chapter 3).

The project's stakeholder community may have become similarly disillusioned, and it may be useful to repeat the relaunch event to reassure and reinvigorate them as well. At the very least, careful consideration will need to be given to how the relaunch and the project's future progress will be reflected in stakeholder communications.

Reflections

As before, if possible, using the same project as in your previous reflections, spend some time considering the following exercises based on a situation where some, or all, of the risks and issues that you identified in your reflections on Chapter 8 have caused the project to be halted:

*   Prepare a fishbone diagram showing the principal problems that the project has faced and identify their principal causes.
*   Tabulate your results and identify the remedies for each that you would recommend.
*   Prepare a brief relaunch presentation to the project team to show them how the project will be different going forwards.

Notes

1   Well, at least you are if you have read through the book this far. If you are starting the book here, you are going to have to loop back and read the chapters dealing with particular areas of concern.
2   There never seem to be any chocolate ones.
3   Actually, harmonic vibration of bridges caused by pedestrians was not an unknown phenomenon. In England in 1831, marching troops were reputed to have caused the collapse of a suspension bridge in Salford outside Manchester, and troops forever afterwards were instructed to "break out of step" when crossing bridges. The Albert Bridge, which like the "Wobbly Bridge" is over the river Thames in London, has long had a notice posted on it to this effect. It is believed that the problem has been known since as far back as the Romans.
4   Personally, four long, long hours in a force 10 gale.
5   Never underestimate the likelihood of the business sponsor asking if, as now you know so much about the project, you will take over and deliver it.

Appendix 1

Facilitating Discovery

One of the biggest challenges in delivering many projects is building a consensus around What it is that the project actually has to deliver and How and When. The objective of using a facilitated discovery process in a project is not making a decision; that's usually the responsibility of the project board. Instead, our objective is to develop recommendations for the project board to consider by fostering effective engagement by a range of stakeholder and to maintain a record of how those recommendations were reached.

Whilst the objective of the project – build a bridge, move a factory, implement a new business process, install a citywide public transport fares system – can often be very succinctly stated, designing the required detailed solution will often require a myriad of choices and decisions among competing priorities, methods of delivery, and timing. Building a shared view of the best approaches to be adopted amongst the individuals who will be engaged with our project can be as important as the project board's decision. We also need to remember that on many projects, the individuals and organisations with whom we will be engaging will often be the ones working with the project's outcomes in the long term.

There are many points in the project where the facilitated discovery process can be used to develop recommendations that will shape both the design and the solution and the way in which it is delivered. These include areas such as:

- Identifying the project's critical success factors.
- Prioritising and weighting detailed requirements.
- Assessing different options for the ultimate solution that will be implemented.
- Identifying and quantifying risks and their mitigation.
- Making choices between suppliers.
- Formulating lessons learned.

Discovering, quantifying, and balancing these areas, including choosing between competing options, is an issue for many projects. This is especially true when there is lack of clarity about the impacts of different choices or there are strongly held divergent views. To take just one example, let's consider choosing between costs and aesthetics in designing a large-volume public access building. If costs are agreed to be the most important factor, the final building is likely to end up as a largely rectangular box, perhaps looking like an out-of-town supermarket. If, however, aesthetics are seen as being the principal driver, the project may deliver a building such as a Sydney Opera House or a Bilbao Guggenheim Museum.

Consensus is normally best achieved through engagement of those people and organisations who will be affected both by the delivery of a project and by the solution that it will deliver. These people and organisations are called our stakeholders. At the very least, genuinely engaging with them throughout the project will help us build both acceptance of the outcome and support during implementation. At the best, their knowledge, experience, and expertise will lead us to identifying, developing, and delivering a far better solution.

Our panel of participants in a facilitated discovery process can be drawn from very different groups depending upon the area or industry in which the project will operate and the type of project itself. Consequently, our first questions to answer are about panel membership:

- Who should be involved in the panel?
- When we should engage with them?

Answering these questions may require consultation and guidance from the business sponsor and possibly members of the project board. It is important that decision makers are accepting of the membership of any panel when considering its recommendations, for obvious reasons.

The facilitated discovery process may end up involving a fairly large number of people. In one project to implement a new information system in a UK public sector service provider, the panel concept was extended. The panel had 16 members, each of whom was made a representative of their local team, which averaged a further 10 staff members. Panel members were mandated to discuss issues for panel consideration with their local team and reflect the team's view in panel discussions. The discovery process therefore directly engaged 176 staff.

Not all individual participants will be involved throughout the project, and our panel membership may change over time. In Chapter 7 there is an example of this changing composition of participants; in another public sector digital service project, representatives of staff groups and interested parties were involved throughout the process – in all, around 20 people. However, when it came to making a choice between system suppliers, it was decided that the entire staff of eight hundred were to be engaged by inviting them to view, comment, and score the solution being offered by each of the shortlisted tenderers. The five tenderers agreed to run rolling demonstrations throughout a two-day exhibition which were attended by the staff and other stakeholders.

Mostly, consensus is built in one or more meetings of our panel in which all attendees feel they have contributed to the decisions taken. There are, however, major drawbacks to using traditional open discussion and debate to determine the output from a discovery session. These drawbacks often arise from the inequalities of persons participating in an open debating process, including:

- Hierarchical inequalities can mean that some participants may either moderate or not express their views if they are at variance to more senior participants. As an example, the requirements for an accounting system to be developed in-house by a hospital was clearly directly based on that of a local authority where the hospital's finance director had previously worked as a junior accountant years before. Consequently, there were major doubts about whether the discovery process had been subverted and the hospital staff had all moderated their views so as not to disagree with their director.
- Inequalities arising from different issues including confidence, culture, and gender bias.
- Participants feeling unable to express their views equally in open debate. Often there is a view that "the person who shouts loudest wins," and the voices of others are drowned out.

Overcoming these inequalities in the engagement process can be helped by using a few structured discussion and facilitation techniques which help ensure that the widest range of views are collected and evaluated. It is possible to engage a trained facilitator, or even a facilitation team, to lead discovery meetings. Engaging external support can be necessary where the project manager wishes to stand outside any anticipated conflicts, and a team is very useful where the discovery process is to be run as an event with multiple parallel streams.

Simple Techniques to Try

The following are examples of practical techniques which can be used at most stages in the life of the project, often on a "mix and match" basis. The techniques usually work, although some panels become "stuck" and may need further help to resolve issues.

Most of these techniques seem to involve sticky notes in some way. The advantage of using sticky notes is that they both keep answers short and the answers can be stuck directly on a wall or on the ubiquitous roll of brown paper, itself stuck on the wall – much more of which later.

About the Example

The example project that we are using to illustrate the techniques is the replacement of a company accounts system. We are trying to develop a list of critical success factors which we can use in our options analysis. This project has both technical and business change aspects which will affect the internal financial working of the client organisation and the operations of its various departments and branches. Panel members should, therefore, be drawn from across the client organisation, and they must include the staff inputting or processing the information as well as the managers who will be using it. The techniques, however, can be used in building a consensus view at any point in a project where one is required.

Technique 1: Building Consensual Statements

The start of the process, at whichever point it is used – e.g., as now, discovering critical success factors – is determining the initial views of the panel members and building these into a consensus view for the panel.

One of the issues in extracting creative thoughts is allowing the participants to be released, at least partially, from their preconceived ideas and prejudices based on the current situation. Muscles need a brief warm-up before exercise, and creativity is the same. Therefore, if we are introducing something novel that is outside the experience of some of the members of the panel, like a new IT system or a new method of working, it is worth starting a discovery session with relevant training or an information-giving session on possible solutions. In our example, it could be a presentation on the generalised features of a modern accounting system – which, as many panel members will only be aware of the existing failing company accounting system, may come as a revelation.

In briefing the panel on the exercise, it is useful to ask our questions in a way which requires the participants to move away from just restating present difficulties. We want them to be using their imagination to describe the significant positive features of the future state when our project has delivered. Experience also shows that we will often obtain more engagement throughout the discovery session if we use a written exercise as an opener, rather than any form of discussion.

So instead of asking a direct question, such as, "Please write down each of the four things that the new project must deliver in order for it to be successful," we should start to harness their imagination and cast them into a positive frame of mind by asking them to visualise a future situation:

> I want you to imagine that it's three years from now. It's a very foggy morning, and when you walk into the reception area, it's never seemed so warm and welcoming. You get two of your favourite hot drinks and biscuits and take them over to your desk, where Future You is already sat working. Giving Future You one of the drinks and the first choice of biscuits, you sit alongside your future self and ask, "What are the four best things about the new process and system in helping you do this job?" Please write down each answer.

Depending on the project and the participants, we can even up the level of imagination required by getting the participant to consider the impact of the system on a different role – for example, the customer service team members talk to a future someone from invoicing and vice versa.

Within a very short time, each participant should have four things to report. Each will be worded differently, and some may come from a different perspective. The facilitator's next task is to be able to bring these disparate words together, to create energy in the meeting, and to prepare the participants for the next stage of the work.

We can use either sticky notes or action cards to record their views, and there are advantages to each method. Often putting the answers on cards and arranging them on the floor, with participants standing around looking down and discussing the results, can be more a dynamic process than using sticky notes stuck up on a wall somewhere. The problem with the cards on the floor approach is that the facilitator has to record the placing of the cards whilst everyone else breaks for coffee and a biscuit. It is also usually very difficult to get a photo or series of photos which adequately captures cards strung out across a large piece of floor. Of course, it is possible to stand on a table, but that wouldn't exactly set a good example of working in a safe way.

Enter the roll of brown paper attached to the wall. Actually, it can be anything – flip chart pads are also useful but are generally smaller and so may work better for smaller groups. Being able to roll up the group's output and take it away for further analysis can be very useful, as can the ability to mark up time frames, notes, and connections between ideas on the paper itself. A practical word of caution when using brown paper stuck on a wall: some marker pens may "bleed" through the paper and leave traces on the wall. Testing each one first can help us avoid the acute embarrassment of having to apologise for ruining the decoration in the board room! It is also useful after the session to tape sticky notes firmly to the paper so that when writing up the session later, we are not left wondering where these orphan notes which have fallen off actually were.

Whichever approach is chosen, the next steps are identical. Seemingly at random, the facilitator asks each member to come forwards, share their four answers, and then group them with other similar ones. The order of speakers needs to be carefully chosen, with some more confident members of the group going first and preferably the senior staff last. It is always advisable to avoid going round the room in order (known in some circles as "creeping death") because of the stress that it can generate in less confident participants as their turn to speak gets closer. This means that they listen less to the previous speakers as they are thinking and worrying more about what they will have to say.

There is one further adjustment that can be made when speakers are still reluctant to express their views clearly. Before inviting any speakers, the sticky notes or cards are rotated several places to the left (or right). Speakers then present the views they have received rather than their own, which may be perceived by them as being less exposing. Choosing to move the sticky notes or cards several places also means they will normally be presented by someone outside any pre-existing group of attendees which has chosen to remain together. The facilitator should be able to observe such groups and pick the number of places for the rotation based on the largest apparent group.

The last steps are for the facilitator to summarise the grouped answers into words and then seek confirmation from the group that appropriate words are being used. Any remaining conflicts can be resolved using technique 2, facilitated discussion, as described following.

The output from several facilitated discussions in our accounting system example produced a list of the main critical success factors for the system's replacement. These were summarised by the facilitator at the end of the exercise as follows:

- High availability – The system should be available at least 99.99% of the time during office hours.
- Short timescale to implement due to the age of the hardware and the end date for support by the software provider.
- UK-based team is preferred for local accounting knowledge and retaining a close relationship with the supplier.
- Limited in-house IT development – There is limited knowledge within the IT team, and the solution should not require programming skills.
- Revenue funded – There is greater availability of revenue funding, and a capital-funded project could be delayed.
- Web front end to minimise user training and to ensure that Windows PCs from all locations can be used for system access without requiring individual loading of programmes.
- Integration with other systems data must be passed freely to and from existing systems and potential future systems.
- Flexible reporting tool using a commercially available product that can be used in future to report on data held in a wide variety of systems.

Technique 2: Facilitated Discussion

This is a form of structured debate using some very simple rules. We state these up front and are clear that we will rigorously enforce them in order to make sure we can collect and give equal weight to everyone's views:

- There will be two rounds of discussion on a topic introduced by the facilitator.
- In the first round, all panel members have up to two minutes to state their views. Ideally, we should ask them to write this down first and then read it out so that we hear their views uninfluenced by those of other speakers.

- In the second round, each speaker has normally either one or two minutes, depending on the time available and the size of the group, to adjust their view, bearing in mind the points made by other speakers.
- Interruptions and cross-table discussions are not permitted.

Once again, the facilitator is careful to select the order of speakers, and at the end of the second round, the facilitator sums up the views of the panel.

The danger in using this technique is that cross-table discussions can start to break out, usually led by more senior or vociferous staff. In order to preserve the process, the facilitator has to be rigorous in policing the meeting. In order to time limit this sort of discussion, one facilitator used a musical reindeer (it was the week before Christmas, after all!). In another case a facilitation team used loud music to put a very effective stop to a speaker, though this is definitely not the best approach if you are in a room next to the CEO's office. Without resorting to such tactics, we can also gently bring the discussion back on track without being too confrontational by using phrases such as, "Thank you for that, it's a very useful point. Let's capture it and return to it later," "Thank you for that, it's a very useful point. Let's take it offline," and "Thank you for that. Before we move on, do you have a final point?" As a last resort, you can always use, "I know that it's a little earlier than planned, but let's go to our coffee break now."

Variations on the facilitated discussion technique can be used in different circumstances. For example, if after using technique 1, building consensual statements, there are still conflicting views, the facilitator can identify volunteers to speak for up to two minutes in favour of one of the conflicting issues. Every other participant then has one minute in which they must give their views in favour of one of the issues. After everyone has spoken, the facilitator sums up the mood of the discussion, which has hopefully resolved the conflicts. If conflicts still exist, it may be necessary to move to the different levels of technique 3, paired comparison analysis, which will enable scoring and ranking the views of the panel.

Technique 3: Paired Comparison Analysis

Whilst the previous two techniques are good for drawing out and summarising views from a group, they are less good for comparative evaluation. When assessing completely different things such as critical success factors, or deciding between shortlisted options, or assessing risks, or evaluating supplier's tenders, we need a simple, open, and clearly communicated method of ranking the panel's views. In our IT accounts system example, we may need to build a consensus around questions such as, "Is the time required to complete the project more important than the ease of maintenance of the delivered solution?" and "Are these more or less important than ease of access via a web front end and revenue costs?"

In this case, both judgements are subjective, and – unlike, say, financial information – there are no hard, objective facts upon which to base our decision. Instead of attempting to decide on overall absolute values to assign to the different factors/options, we should assess them relative to each other.

There are two methods that are very useful for facilitating groups views of relative desirability:

- Ranked lists of options.
- Weighted list of options.

Producing a ranked list of options. There are a number of alternative ways of scoring different options to produce a ranked list that reflects the panel's combined views. It is certainly possible to ask all participants to produce their own ranking and then to combine these into an overall ranked list. However, participants can struggle with accurately reflecting their views in a ranking of multiple options, which will require multiple decisions about their preferences.

Paired comparison analysis provides us with a way to determine each participant's views of each option and then to build a group view very quickly. The approach works by structuring the multiple decisions required of the panel member into a series of assessments of the relative values of pairs of options. In its simplest form, we list all the options to be considered and usually assign each a letter for convenience. The participant then answers the deceptively simple question, "Do I prefer item A to item B?" The participant repeats this question for every possible pair of options. It is important for the panel member to fully consider each pair but equally not to overthink the exercise. Table A1.1 shows all the possible preference pairs in our accounts system replacement.

Table A1.1 Paired Comparison

Paired Comparison							
Option	**A High availability**	**B Short timescale**	**C UK-employed team**	**D Limited development**	**E Revenue funded**	**F Web front end**	**G Integration**	**H Flexible reporting tool**
A High availability		BA	CA	DA	EA	FA	GA	HA
B Short timescale			CB	DB	EB	FB	GB	HB
C UK-employed team				DC	EC	FC	GC	HC
D Limited development					ED	FD	GD	HD
E Revenue funded						FE	GE	HE
F Web front end							GF	HF
G Integration								HG
H Flexible reporting tool								

In practice, for each possible pairing, each panel member considers the question, "Do I prefer the option in the *column* or the option in the *row*?" For each of the possible pairs, the panel member places the answer into a simple results comparison table. If the panel member has no preference, then they leave the cell blank. Table A1.2 shows one panel member's answers.

After the table is completed, the number of appearances of each option in the whole table are counted by the panel member (or sometimes their neighbour) and entered into the score column.

Table A1.2 Paired Ranking Analysis

Paired Ranking Analysis											
Option	*A High availability*	*B Short timescale*	*C UK-employed team*	*D Limited development*	*E Revenue funded*	*F Web front end*	*G Integration*	*H Flexible reporting tool*	Option score		Option rank
A High availability		B	A	A	E	A	G	A	**A**	4	2
B Short timescale			C	B	B	B	B	B	**B**	6	1
C UK-employed team				C	E	C	C	H	**C**	4	2
D Limited development					E	D	D	H	**D**	2	7
E Revenue funded						F	G	E	**E**	4	2
F Web front end							F	H	**F**	2	7
G Integration								G	**G**	3	5
H Flexible reporting tool									**H**	3	5

The facilitator now has to total the scores from the panel for each option and then rank them in order to synthesise the view of the panel. This can be done by passing all of the scored tables to the facilitator to calculate and then announcing the scores of each option and hence the priority ranking of the options. This takes some time to do, so we miss the coffee break whilst doing the arithmetic. Does anyone ever think to bring a coffee and biscuit back for the facilitator? No one, not ever. If we do manage to finish the maths before the end of the coffee break, only the rubbish biscuits are left.

The session is, however, more dynamic and engages the participants more if we run the totalling of the scores as an interactive session with each panel member calling out their own – or possibly their neighbour's – scores. This enables us to build some excitement around the emerging scoring. Think of the Eurovision Song Contest as a model – five points go to Option A, three points to Option B, and Option C null points. And, if instead of doing the maths yourself, you use this approach, as the facilitator you can get the coffee break as well. Furthermore, you can appoint one of the panel members to record the scoring on a flipchart and then position yourself strategically by the door just as you call the end of the session. That way, you might even get to the chocolate biscuits first.

Producing a weighted list of options. Quite often we need to move beyond ranking the options in order of desirability and recognise that each of the options can have markedly different levels of impact on the project beyond the simple numerical order that we obtained in a ranking exercise. The question is how we might reflect these different levels of impact in our evaluation. For example, when assessing tenders, we need to arrive at a single number that expresses how well the tenderer has met our overall list of requirements. Our answer is usually to use two factors to assess the response to every requirement:

- The importance of the requirement to the overall project.
- The degree to which the supplier has met the requirement.

Extending paired option comparison can help this process by determining the weights that the panel wish to give to each of the requirements.

Instead of just indicating which of the options is preferred, as in the ranking system, the panel member also has to answer a second question: "By how much do you prefer the option selected?" In the example in Table A1.3, only a factor of 1, 2, or 3 (maximum) are used. If there is no preference, the cell is left blank. Wherever in the table an option receives a value, this is added in to the options total score column.

The table is shown complete with weightings, calculated as each option's score as a percentage of all scores. Although the example shows the weightings, we would not normally work out the percentages on the individual panel member's score sheet but only after totalling all responses.

Table A1.3 Weighted List of Options

Option	A High availability	B Short timescale	C UK-employed team	D Limited development	E Revenue funded	F Web front end	G Integration	H Flexible reporting tool	Total score	Weight
A High availability		B1	A2	A1	E3	A2	G1	A2	7	13.5%
B Short timescale			C2	B2	B3	B1	B1	B2	10	19.2%
C UK-employed team				C3	E1	C2	C2	H1	9	17.3%
D Limited development					E1	D2	D3	H2	5	9.6%
E Revenue funded						F2	G1	E3	8	15.4%
F Web front end							F1	H2	3	5.8%
G Integration								G3	5	9.6%
H Flexible reporting tool									5	9.6%
										100.0%

Are We Being Completely Ridiculous?

Having completed the scoring, it is useful to close with a short, facilitated discussion as a sense check to confirm that the panel's views were captured correctly. For example, if the output from our discovery exercise is a ranked list from a paired comparison analysis, we could hold a single round of one-minute discussions using the question, "If you could move just one item in the list up two places, which one would it be and why?" Or if we have completed a weighted list, we could ask the question, "If you could increase the score of just one of the options by three points, which one would it be and why?"

In passing, note that these are both positive questions. Whenever asking for positive and negative comments, it is always best for the participants to finish with the positives, rather than have them leave a section of the discovery process, or even the complete meeting, on a negative note.

Recording the Output

In Chapter 2, a suggested format for recording the views expressed during any discovery exercise was set out. This is the discovery progress sheet set out as Table A1.4, but whatever format is chosen, the output from every session should be recorded.

Table A1.4 Discovery Progress Sheet

Discovery Progress Sheet	
Issues raised	
Resolved	*Unresolved*
Closed and agreed to take forwards *Issues and activities closed because agreement has been reached that they should be addressed by the project.*	Issues and activities requiring further discussion *More discussion needed because agreement was not achieved or discussion revealed new items that needed further research or consideration.*
Closed archive *Issues and activities raised and closed because discussion has revealed them as not helpful or not required.*	
Next steps	
Suggestions for future activity.	
Prepared by	**Date**

It is also useful to provide a short summary feedback report to each of the participants attending a session, which will act both as a final sense check and be a useful input for summarising the views uncovered in the whole discovery exercise.

The Next Steps

It is unusual for a single discovery session with one group to yield all the answers we need. Several facilitated sessions may be required, some of which may deal with different areas of interest and others which cover the same areas but with different groups of participants. At the end of the process, we will prepare a document which consolidates the output, including resolving any differences between outputs from different groups. This may sometimes require further consultation, remembering all the while that we are trying to build consensus amongst our various stakeholders.

Our consolidated document forms the evidence base from which we will extract information to include in our formal project documentation. This may be determining our critical success factors, prioritising and weighting detailed requirements, assessing different options, identifying and quantifying risks, or making choices between suppliers. It is these project documents which are submitted to our project board for approval, rather than the output from our discovery process.

Appendix 2

Partnership Working in UK Construction: A Journey from Conflict to Collaboration

Introduction

This appendix looks at the lessons for the project manager who needs to create a relationship with one or multiple suppliers. These lessons are drawn from a consideration of the recent history of the UK construction industry, which has been on a journey of considering how we can transition from project delivery based on traditional confrontational contracts to collaborative partnering. In the first section, lessons are drawn based on published information. In the second section, lessons are drawn from anonymised examples of contracts which are operating on a partnership basis between clients and suppliers. These are based on confidential discussions with project and other managers.

A note on definitions: in this appendix, and across the UK construction industry, the terms *partnering* and *alliancing* are used interchangeably.

Section 1: Moving Forwards from Conflict in Contracting

UK in the 1990s

The Heathrow Tunnel Collapse as an Example of Non-Integrated Working

It was decided to improve the connection between London's main airport, Heathrow, and the London transport network by constructing an express rail link to London's Paddington Station. As the main Heathrow terminals were clustered in the centre of the airport, this new Heathrow Express (HEX) rail link required the construction of tunnels beneath the airport. On the night of 21 October 1994, a tunnel collapsed, causing a large crater to appear between the airport's two main runways and also damaging buildings and car parks. The damage was so severe that the clean-up took months.

A month-long trial revealed the major problems in delivering the contract caused by the client, the contractor, and the engineering consultant not co-operating together. The working practices of the main contractor, Balfour Beatty, were heavily criticised, and the company received a fine of £1.2 million. Austrian consultant engineering firm Geoconsult was also fined £500,000 for its "less culpable role" in one of the worst civil engineering disasters in the UK in the last quarter of the 20th century.

During the trial, the prosecution's expert witness said at least six engineers qualified in the new tunnelling method (New Austrian Tunnelling Method, or NATM) being used were needed to oversee work. The judge later asked, "How is it there was only one qualified NATM engineer, and he was home at the time of the collapse?"

Reportedly, 26 August was a major turning point in the disaster. According to the court reports, the shift handover report for that day, written by Balfour's NATM engineers, read, "Central terminal area concourse: first section of invert remedial work complete. Was it agreed that invert repair need not progress further toward shaft?" Against this question was written "NO!" by Geoconsult's NATM engineer.

Balfour Beatty could not account for why this section of tunnel was not checked even though the original instructions were to check back until a sound section of tunnel was reached.

Between late August and the collapse, a series of meetings was held between Balfour Beatty and the HEX management team, during which HEX warned the contractor that several problems needed to be addressed urgently and the tunnelling method reviewed. But much of the ensuing correspondence was not copied to Geoconsult.

In one letter sent in early September 1994, the HEX team told Balfour, "We believe that the full extent of the failure in the lining has not been exposed." But this was not acted upon.

During his summing up, the judge said that Balfour Beatty were not able to explain why they had not acted on this letter from HEX, which was an opportunity to avoid the catastrophe.

A report for the UK Health and Safety Executive on Tunnelling (2000) found that poorly drafted procurement and contractual arrangements can conflict with statutory obligations; impose unfair and unreasonable conditions; create ambiguity, misunderstanding, and doubt; and cloud areas of responsibility from the very beginning.

Jubilee Line Extension – A Second Example of Non-Integrated Working

The London Underground (LU) Jubilee Line Extension is the extension of the line eastwards from close to Buckingham Palace to Greenwich and Stratford. The extension comprised 16 kilometres of mostly underground rail plus 11 new or extended stations. It was constructed during the 1990s and opened in stages from May 1999. There was a deadline to deliver the line by 31 December 1999 in time for the UK's millennium celebrations in Greenwich.

The delivery strategy for the project was for LU to manage a project consisting of over 30 independent design and delivery contracts. The project was:

- Over budget 67% (£3.5 billion vs £2.1 billion) on an initial estimate of £800 million, 25% of the final cost.
- Over time by 40% (74 months vs 53), and the last station, Westminster, only opened on 31 December 1999. There is a link to the HEX case as LU had proposed to use NATM – and had to wait until it was reassessed for health and safety. At worst this may have cost 6 months of delay out of the 20 months of overrun on a 53-month project.

The model of the client managing the project had to be changed, and the delivery was eventually brought in by appointing independent project managers.

UK in the Early 2000s

Wembley Stadium – An Example of Contractual Conflict

The construction of the iconic home of English football was beset by problems. The structure of the new Wembley Stadium was to be dominated by a 133-metre-high arch. The arch's fabrication and construction was a key works package, and the key site relationship was between the main contractor, Australian Multiplex Constructions, and the steel specialist company building it, Cleveland Bridge UK.

Cleveland Bridge's involvement in the Wembley construction was simply defined as building and erecting the iconic arch. One version of the story is that Cleveland Bridge walked off the job in 2004 because they did not believe they would be paid for materials and there were irrevocable difficulties between the two parties.

Another version is that Multiplex was unhappy with Cleveland and how much it was paying for the work and so they sacked them. In any case, Multiplex recruited Dutch contractor Hollandia to finish off the arch and roof structure, at premium rates. Multiplex and Cleveland sued each other for breach of contract – Multiplex sued for £45 million and Cleveland Bridge sued for £22.5 million. In September 2008, Multiplex won the case and received £6 million from Cleveland.

In December 2008, Multiplex also sued the stadium designer, Mott MacDonald, for £253 million, saying that it was denied access to key design information that led to increased steelwork costs – this was settled out of court two years later.

The stadium was supposed to be completed by May 2006 for the FA Cup Final, but Multiplex was unable to complete the stadium within the scheduled time and had to pay penalties. The stadium was finally completed in March 2007, ten months late.

Whilst the construction company suffered losses, the Wembley project was extremely good business for the legal industry.

Evolutionary Change in Culture – From Conflict to Collaboration and Partnership

Let's put these examples of conflicted relationships into the context of evolving thinking in the UK construction industry on how to resolve the difficulties of delivering projects involving multiple suppliers with competing objectives – and legal advisers.

- 1994 Heathrow Tunnel collapse delaying £500 million project by six months, cutting Underground links.
- 1994 Latham Report criticised the adversarial approach inherent in traditional construction contracts.
- 1998 Egan Report demonstrated progress in partnering in the public and private sectors.
- 1999 Jubilee Line Extension complete.
- 2000 ICE Partnering Addendum – bilateral contracts.

- 2001 National Audit Office report, *Modernising Construction*.
- 2001 NEC3 contract Option X12 published.
- 2007 Wembley Stadium opens.
- 2010 BS11000 for collaborative business relationships published, squarely recognising the need for behavioural change.
- 2017 ISO 44001.

1994 Latham Report

This report was titled *Constructing the Team: Final Report of the Government/Industry Joint Review of Procurement and Contractual Arrangements in the United Kingdom Construction Industry*. It was led by Sir Michael Latham, a former member of the UK Parliament with a background in housing and construction. The report saw that the client should be at the core of the construction process and that the industry should move away from its adversarial structure, adopting a more integrated approach with greater partnering and teamwork.

1998 Egan Report

Rethinking Construction, the report of the Construction Task Force to the UK's then deputy prime minister, John Prescott, on the scope for improving the quality and efficiency of UK construction, was led by industrialist Sir John Egan. When considering the nature of relationships in construction, the report concluded that what was needed were integrated processes and teams and replacing competitive tendering with long-term relationships.

2001 Modernising Construction Report

The UK National Audit Office, which has a remit to study propriety and value for money across all public expenditure, published a report entitled *Modernising Construction*. The report's conclusion included:

- Risks involved in any project needed to be placed with the parties best able to manage them.
- Outcomes could be improved through different forms of partnering with contractors committed to continuous improvement.
- Reliable performance measures were needed to ensure that benefits are being achieved.

2001 NEC Contracts – Partnering Concept

The NEC family of contracts are one of the standard forms of construction contract in the UK. In the third generation of the NEC3 contract, a partnering option was introduced which made a significant contribution to formalising relationships. This is in terms of:

- The contract form, with an optional partnering clause, X12,[1] common to all bilateral contracts with the client.
- In the governance relationships, introducing the concept of a core group to manage the work drawn from all partners.

See Figure A2.1.

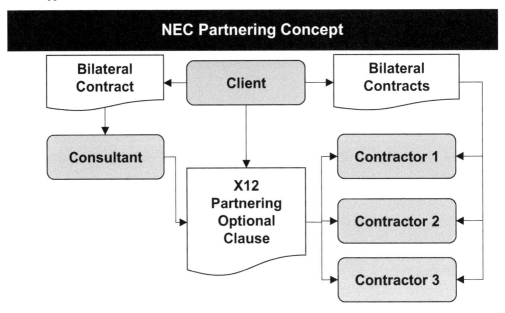

Figure A2.1 NEC Partnering Concept

The detail contained in the partnering data that supports the contract sets out how the relationships work, including features such as:

- Shared offices.
- Management meetings.
- Risk management.
- Value engineering.
- Individuals involved.

The membership of the partners is dynamic and changes over the life of the project as it moves through design and construction phases – e.g., the groundworks contractor and the roofing contractor may not be needed on the team at the same time.

In June 2018 NEC published its next generation for contracting which advanced the partnership concept. The NEC4 Alliancing Contract is multiparty, and the NEC considers that their approach has the benefit of delivering a deeper collaboration between all project participants because they will be bound by a set of defined common interests. Therefore, there will be reduced grounds for dispute.

In practice the NEC4 contract built the previous NEC3 X12 partnership option into the core contract and had some advanced features:

- The contract is governed by an alliance board with members from each partner, which is required to act unanimously. This board sets strategy, resolves disputes, and appoints an alliance manager.
- The client is a member of the alliance and still retains some additional client powers which it exercises separately.

- The alliance manager manages the contract on behalf of the alliance and acts as the project manager.
- Risk is shared between alliance members. Legal claims are made in limited circumstances – e.g., deliberate breach of contract.

The contract is based on defined costs, and all partners are incentivised to achieve alliance objectives through a performance table with an associated pain/gain mechanism.[2] Whatever form a contact takes, it is only ever the legal embodiment of what the parties have agreed to achieve together.

2010/2017 BS11000/ISO 44001 Collaborative Business Relationships

Moving beyond the contractual underpinning of the alliance or partnership approach, collaboration between organisations may take many forms, from loose tactical approaches through to longer-term alliances or joint ventures. BS11000 was the first standard of its type in the world. It formalised how organisations approach mutual relationships and focuses on the thinking about how to construct and operate partnership. The international standard that derived from it, ISO 44001, was first published in June 2017.[3] The original standard:

- Allowed organisations to collaborate successfully and outlined different approaches to collaborative working that have proven to be successful in businesses of all sizes and sectors.
- Showed how to eliminate the known pitfalls of poor communication by defining roles and responsibilities and creating partnerships that add value to business.
- Established a consistent and structured approach to facilitate a focus on integration of collaborative working within operational procedures, processes, and systems.

There are eight key stages in the BS1100 approach which have been carried forwards into the ISO standard:

Strategic

- Awareness – fit with your business objectives.
- Knowledge platform.
- Internal assessment of your own capabilities.

Engagement

- Partner selection – not only can they do the job, but will they fit in?
- Agreeing on formal foundation for ways of working together.

Management

- Value creation.
- Staying together – maximising effectiveness and continuous improvement.
- Exit strategy – how will you disengage?

Section 2: Working a Strategic Partnership in Real Life

The following section contains a number of brief examples showing how alliances have been used to deliver multiple project contracts and capital programmes in the UK. These examples are based on interviews with managers delivering alliance relationships who were asked to be candid about what they felt were the key attributes in making their partnership successful and what they found most challenging. There is no science here, no wide survey and deep analysis, just a summary of the subjective views of managers with experience of working in one or more alliance partnerships. There are sufficient commonalities in their experience to suggest that many of the same lessons came up in all the partnerships

Local Highways Partnership

This was a multi-year strategic alliance for highway maintenance and repair among a local authority, a highways contractor, and an engineering consultancy. The total annual spend on highway maintenance by the authority was around £30 million revenue plus a capital maintenance programme. The alliance ran multiple projects and services to deliver an improved service to the local residents.

Client objectives. Objectives were to:

- Reduce headcount by removing "person marking" in which members of the client staff literally shadow and check all the work done by an equivalent in the supplier.
- Make contracts more outcome driven.
- Incentivise genuine efficiencies.
- Ensure that only genuine costs incurred in carrying out the service are paid.

Solution. The client had already won awards for its work on costs reduction but accepted that it needed to work in a different way if it wanted to achieve more. The framework for this alliance was designed to be delivered through a target cost–based integrated contract. The intention was that contract would ensure that the supplier shared the benefits from efficiencies in delivery and earned the right to contract extensions.

Results. The manager's view of the results of the alliance method of working included:

- The contract had delivered against targets of improving cost efficiency and was extended several times.
- Efficiencies were delivered, but the contract definition of efficiency needed clarification.
- Governance mechanisms needed to evolve.
- There was flexibility in service delivery with transfers between client and provider.
- The alliance had delivered against a four-year plan.
- Working practices and relationships had survived changes in senior personnel on both sides.
- The norm was joint working between client and provider managers and staff to resolve issues, but like all relationships, you have to keep working at it.

There were still issues as even after four years of operation of the alliance the client organisation's legal team still requested internal contract discussion meetings prior to formal alliance meetings. This was a clear indication that they did not see themselves as part of the alliance team but as part of the local authority's contract management team. They still had not realised that the essence of integrated working is between organisations as well as individuals.

English Water Company (1)

In England the supply of water, the drainage, and the treatment of sanitation are mainly provided by 12 regional water companies. These are subject to a high level of regulation and monitoring, including over their capital investment programme. This client was perceived by the industry regulator as being particularly poor at delivering a five-year capital programme.

Client objectives. Making a step change in delivery performance for the its next five-year capital programme, which totalled £265 million. To achieve this improvement, it needed to reduce project delivery costs, targeting an overall figure of 10%. Part of the challenge of regulated capital programmes in the UK water industry is to achieve a fast start enabling the year 1 investment targets to be met – a very rare occurrence in the industry. The company's drivers were challenging, and it recognised that its previous approach of contracting engineering consultants to work on each individual project within the programme was one of the causes of its failure to deliver to the satisfaction of the industry regulator.

Solution. A competitive procurement for a partnership with an engineering consultant to work as part of an integrated team, working with in-house staff to design and deliver the next five-year capital programme. The bespoke contract included incentivisation aligned to achieving the company's efficiency targets.

Results. The service delivery manager reported that during the first capital programme cycle they had:

- Co-located an integrated team of over 70 staff.
- Cut staff costs as a proportion of the final built costs of schemes by 27%.
- Achieved the investment target for year 1 and for every subsequent year.
- Raised the acceptance by the regulator from "poor" to meeting 100% of the regulator's targets.
- Changed health and safety culture into one of zero tolerance of accidents, cutting their frequency rate by 73%.
- Improved processes to cut the rate of inadvertently striking other utility services during excavation by 70%.
- Delivered savings in management time – one consultancy, no duplication, no unnecessary procurement.
- The integrated team won the Managing Director's Award for Department of the Year.

After the completion of the initial five-year capital programme, the engineering consultancy was reengaged to deliver the next five-year programme.

UK Electricity Supply Company

The drive for moving to partnership working with an engineering consultancy came from a board director who wanted to show that this method of collaboration would be a successful way of helping the company succeed in delivering its capital investment programme, rather than relying on a traditional transactional approach of letting a separate contract for each individual project.

Client objectives. To achieve more competition than from existing suppliers and to generate more resources to meet the workload for a major investment programme, whilst delivering greater efficiency.

Solution. The electricity company decided on an approach based around a delivery partnership with a unified central team equally staffed by the client and an engineering consultant. The incentivised alliance contract had a complex 50% pain/gain sharing arrangement based on achieving key performance indicators (KPIs). The team was headed by an alliance manager and a deputy, one drawn from the client and one from the supplier.

Results. Staff from both the client and the supplier had been successfully co-located in dedicated offices. There were, however, initial teething issues with integrated working, but these were worked through and settled. He also felt that the policy on joint working was determined by the individuals from the client and their previous experiences of partnering. Overall, the electricity industry regulator, however, decided that the arrangement did not guarantee that the client continued to get best value for money over the life of the investment programme.

In addition, the delivery manager summarised the history of the contract as:

- A downturn in the engineering consultancy and construction market after the alliance contract had started meant that it started to look expensive compared to those then currently being bid.
- The client changed priorities for its investment programme to refurbishment, which required less consultancy input.
- In year 4 of the contract, there was a new client management team which wanted a clear differentiation of responsibilities between supplier and client and had different views on how to achieve value, which was by being focused on individual transactional costs.
- Client staff were pulled out of the integrated team.

English Water Company (2)

The company supplies to a large UK region. In order to support a particular major urban regeneration scheme, it was necessary to deliver a ten-kilometre-long trunk main. This installation would normally take two and a half years to complete but had to be completed in only 18 months to meet the needs of development. The client and a major contractor decided to use the project to be an exemplar partnership working and radically improved health and safety.

Solution. The creation of an integrated team including client, contractor, and an engineering consultancy. The client paid attention to the Latham and Egan Reports and moved away from lowest bidder tenders to contracted relationships. The two suppliers forming the partnership were incentivised to deliver through a pain and gain mechanism.

In selecting team members, staff from all partners were interviewed and assessed against criteria covering working attitudes and technical skills. To enhance partnering behaviours, the whole team was located in one suite of good-quality offices. As far as possible, these were open plan with staff grouped by function to facilitate easy, direct, and immediate communication not by employer.

A team-building event agreed on a set of team values to which everyone signed up and set contract KPIs, including the team's perception of its performance against these values. Team members who subsequently failed to make the culture change were removed. The team also replaced company corporate branding with a team brand.

Results. The service manager was proud of the way a major problem was resolved by the team working together. A misalignment of a major element put the scheme at risk of a major rework. Rectifying this error was dealt with under the partnering ethos with costs being minimised by using resources to fix the fault as they became available. The final costs were absorbed by one of the partners rather than pursuing contractual rights.

The project was a success in that it delivered on time and 10% under budget. Quality was reflected in the fact that there was no major snagging phase required.

Lessons Learned from the Movement towards Partnership Working

Lessons from the examples given can be grouped into two main areas:

- Needing to build a contractual framework to deliver the right organisational behaviours.
- Understanding that success in working in a new way stems from individuals building new types of relationships and behaving in different ways.

Contractual Needs

Contracts can be characterised in many ways. Looking at contracts on two axes of payment mechanisms and delivery mechanisms in Figure A2.2 shows the difference between the traditional combative approach to delivering individual projects and the alliance model of delivering through longer-term partnerships (see Chapter 10). This analysis also underlines the lessons from the real-life examples for all actors in delivering an alliance working arrangement to behave as if they form part of an integrated team.

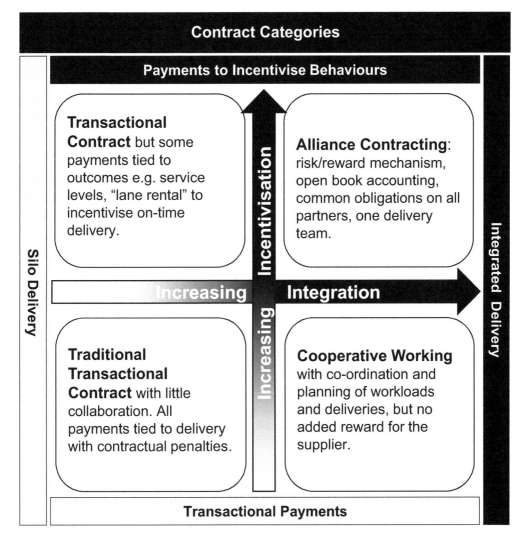

Figure A2.2 Contract Categories

In discussions, the managers agreed the following key elements must be addressed in the partnership contractual framework:

- Using suitable contract conditions to align objectives among the partners, including pain/gain sharing.
- Selecting suitable KPIs helps reduce the tensions between the corporate goals of the partners and those of the delivery team.
- The contractual framework needs to address governance issues, as these are not totally replaced by partnership working and trust.
- Open-book accounting is essential.
- Traditional contracting includes quality assurance controls, and some of these can get lost in partnership working and need to be replaced.

Not Everyone Pulls in the Same Way

The examples show that there are issues with integrated working, and continuing strong leadership is needed from the client director in charge who understands the drivers for the alliance or partnership. Over time personnel and views change. In the highways example, the alliance has survived personnel changes, and methods of operation have evolved and been refreshed over time. By contrast:

- In the electricity example, new objectives for capital investment and management changes in year 4 of the contract brought in new managers who motivated by different client objectives.
- In the second water example, the sale of the client brought in new owners who reverted to a more traditional approach to contracts.
- Team building and identification with the team rather than the employer are important in successful delivery. For most alliance staff, their role in the alliance is one part of their career with their employer. Client staff, when the alliance contract comes to an end, will revert to the client organisation or will be transferred to work with new alliance partners. Supplier's staff may be moved into roles in other teams working with other clients. Building a strong team ethic and close working relationships can overcome this if there is a zero tolerance of staff not performing in accordance with the partnership ethos.
- Even in the more successful alliances in building the team there are opportunities for developing a tension between team members and their employing organisation. Although each organisation contributes full-time staff to the project, it will still have departments and managers engaged with the contract that remain outside the integrated project team. For these individuals, their focus is still on their corporate objectives, thereby creating the opportunity for tension between the alliance team members and the managers remaining in their employing organisation.

In summary, the lessons about changed working approaches are virtually all about helping people used to operating in a combative environment to deliver better in a collaborative one, rather than about effecting the contractual mechanisms to permit collaboration to happen. This means that, unlike creating an effective contracting arrangement, delivering changed working approaches is a continuing effort during the life of the partnership, or, as one manager summed it up, "Like all relationships, you have to keep working at it."

Notes

1 In NECX contracts, X clauses are optional clauses that can be added in to the main contract terms.
2 For further information on NEC contracts, see www.neccontract.com/.
3 For further information about ISO 44001, see www.iso.org/standard/72798.html.

Index

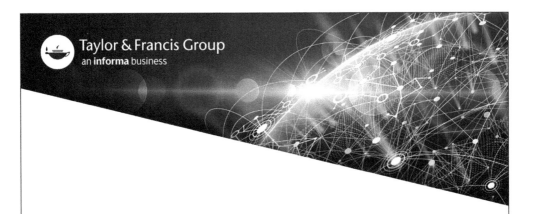